Applied Groundwater Hydrology

Applied Groundwater Hydrology

A British Perspective

Edited by

R. A. DOWNING

and

W. B. WILKINSON

CLARENDON PRESS · OXFORD
1991

Oxford University Press, Walton Street, Oxford OX2 6DP
Oxford New York Toronto
Delhi Bombay Calcutta Madras Karachi
Petaling Jaya Singapore Hong Kong Tokyo
Nairobi Dar es Salaam Cape Town
Melbourne Auckland
and associated companies in
Berlin Ibadan

Oxford is a trade mark of Oxford University Press

Published in the United States
by Oxford University Press, New York

A catalogue record for this book is available from the British Library

Library of Congress Cataloging in Publication Data
Applied groundwater hydrology : a British perspective / edited by R. A.
Downing and W. B. Wilkinson
Includes bibliographical references and index.
1. Hydrogeology.   I. Downing, Richard A. (Richard Allen).
II. Wilkinson, W. B.
GB1004.A67   1991   551.49—dc20   91–10993
ISBN 0-19-852139-1

Set by CentraCet, Cambridge
Printed in Great Britain by
Bookcraft (Bath) Ltd
Midsomer Norton, Avon

This book is dedicated to

NORMAN SAVAGE BOULTON

1899–1984

Professor Emeritus of Civil and Structural Engineering
University of Sheffield

For significant contributions to the theory
and practice of groundwater hydraulics

Preface

This book reviews recent advances in groundwater hydrology in Britain. It has been prepared by the British section of the International Association of Hydrogeologists and follows, after a decade, a similar volume published by the Royal Society in 1981 under the title *A Survey of British Hydrogeology 1980*. It has been organized on behalf of IAH by a small committee comprising Professor W. B. Wilkinson, Dr R. A. Downing, Mr R. W. Simpson, and Mr C. P. Young. The forty-one authors are drawn from universities, government agencies, the water industry, research institutes, and consulting engineers. Their contributions describe the work of British hydrogeologists, but the themes are international in their application.

The wide range of the subjects reflects the continued expansion of hydrogeology, embracing, as it does so, new specialized applications of other disciplines. Many of these have arisen because of the growing concern about the quality of the groundwater environment. Being 'out of sight, out of mind', groundwater has not received the attention from the public at large that its importance merits. About one-third of potable water supplies in England and Wales are provided by groundwater, but in many countries the proportion is even higher and, indeed, in some it is the only readily available source of water. Regrettably, evidence is accumulating that contamination of this vital resource is increasing, and it can no longer be accepted as the pristine source of water that was once the case.

This book will draw attention to the complexity of groundwater reservoirs, the long time that water is stored in them, and the difficulty, if not impossibility, of restoring them to their original state once they have been polluted. It is to be hoped that the 1990s will see a wider recognition of the importance of groundwater and the need to protect such a strategic resource from further deterioration, and thereby ensure its availability for future generations.

Wallingford   R.A.D.
November 1990   W.B.W.

Contents

List of contributors

J. N. Andrews
University of Reading

R. P. Ashley
University of Birmingham

J. A. Barker
British Geological Survey

J. P. Bell
Institute of Hydrology

N. C. Blakey
Water Research Centre

F. C. Brassington
Hydrotechnica Limited

N. A. Chapman
Intera Sciences

P. J. Chilton
British Geological Survey

L. Clark
Water Research Centre

J. D. Cooper
Institute of Hydrology

W. G. Darling
British Geological Survey

R. A. Downing
formerly British Geological Survey

W. M. Edmunds
British Geological Survey

S. S. D. Foster
British Geological Survey

C. M. K. Gardner
Institute of Hydrology

D. A. Gray
formerly British Geological Survey

H. G. Headworth
Southern Science Limited

D. G. Kinniburgh
British Geological Survey

R. I. Jeffery
British Coal

G. P. Jones
University College London

A. R. Lawrence
British Geological Survey

J. W. Lloyd
University of Birmingham

T. J. McEwen
Intera Sciences

R. Mackay
University of Newcastle upon Tyne

M. Morgan-Jones
National Rivers Authority

C. Neal
Institute of Hydrology

N. H. Neill
University of Birmingham

P. E. O'Connell
University of Newcastle upon Tyne

M. Owen
National Rivers Authority

J. M. Parker
British Geological Survey

R. H. Parker
Camborne School of Mines

C. E. Reeve
Hydraulics Research Limited

H. D. Robinson
Aspinwall and Company Limited

D. Savage
British Geological Survey

R. W. Simpson
Simpson and Partners

A. C. Skinner
National Rivers Authority

J. M. West
British Geological Survey

P. G. Whitehead
Institute of Hydrology

W. B. Wilkinson
Institute of Hydrology

G. M. Williams
British Geological Survey

C. P. Young
Water Research Centre

1. Groundwater—the present and the future

R. A. Downing and W. B. Wilkinson

Diversification

During the 1980s, the theme underlying progress in groundwater hydrology has been diversification. From the dominance of water resource programmes in the early 1970s, groundwater quality began to assume increasing importance towards the end of that decade with emphasis on contamination from nitrates and landfills. This trend continued in the 1980s but the scale and range of pollution problems increased, involving pesticides, organic solvents, acid rain, radioactive wastes, as well as increasing problems associated with nitrate pollution, caused by more intensive agricultural practices, and too wide an application of the 'dilute and disperse' philosophy for landfill disposal of wastes—both domestic and more hazardous industrial wastes.

In the last ten years major studies have been undertaken on the agricultural pollution of groundwater, the impact of waste disposal by landfills, the disposal of radioactive waste, rising groundwater levels in urban areas, and the potential for geothermal energy. The possible impact of climatic changes on both groundwater resources and the long term effects of groundwater movement from radioactive waste repositories has led to programmes anticipating the future. This diversification has been accentuated by the increasing and welcome involvement of other disciplines including chemists, physicists, mathematicians, engineers, and economists—a trend first seen in the 1960s. Hydrogeological programmes have become more complicated requiring the overall co-ordination of many strands by the hydrogeologist who is now very often cast in a management role.

A good example of the increasing diversification is the recent application of noble gases and natural radioactive elements to hydrogeological studies. The palaeoclimatic conditions at the time of recharge of ancient groundwaters have been assessed from the amounts of dissolved atmospheric noble gases and regional flow systems have been identified by the progressive dissolution of radiogenic helium. Nitrate reduction which has occurred within an aquifer has been determined by the measurement of the resulting increase in the nitrogen/argon ratio. Changes in the argon isotopic ratio caused by the release of radiogenic argon–40 have been used to assess the importance of alteration processes and to identify possible genetic sources for hydrocarbons. The hydrogeochemistry of uranium and its decay products have been related to changes of the redox conditions in an aquifer and have been used to estimate residence times for some radionuclides in solution. The noble gas radionuclide, radon–222, is especially useful in its hydrogeological applications. Radon concentrations in groundwaters may be used to assess semi-quantitatively the relative importance of intergranular and fracture flow in aquifers with similar lithologies. In fractured crystalline rocks, radon has been used to estimate the size of fracture apertures and hence the extent of the water–rock interactive surface. These techniques have practical applications for the assessment of groundwater recharge and resources; for estimating changes in groundwater age and quality; for the evaluation of sites for radioactive or toxic waste repositories; for earthquake prediction, and for the development of the geothermal energy stored in 'hot dry rocks'.

Advances have also been made in recognizing the nature of the complex chemical reactions that occur between rock fabrics and groundwater. Such reactions can modify the permeability of hot dry rock reservoirs and thereby influence the operating economics of a system; they are of fundamental importance in predicting the effectiveness of rocks as barriers to the migration of radioactive nuclides from waste repositories, as well as being the basis for secondary porosity in aquifers and hydrocarbon reservoirs.

The departure in recent years from the emphasis on water resource problems can be illustrated by the programmes to assess geothermal energy potential in both the UK and overseas and to assist in the design of safe repositories for the disposal of radioactive waste.

The considerable increase in oil prices in the 1970s directed attention towards non-conventional sources of energy including geothermal energy. This led to a programme to assess the geothermal potential of the UK. The programme had two aims: to study hot, although relatively low temperature, groundwaters and hot dry rocks.

The Permo-Triassic sandstones were soon recognized as the most favourable low enthalpy aquifers likely to contain groundwaters at temperatures in excess of 60°C. The resources of the sandstones in the deep sedimentary basins were estimated and their nature and potential were investigated by exploration boreholes at sites in England and Northern Ireland. The programme investigated selected sedimentary basins with a single borehole. Holes were drilled in the Wessex Basin, the East Yorkshire and Lincolnshire Basin, and at Larne in Northern Ireland, each being essentially a wildcat. Coastal sites were selected to allow disposal to the sea of the hot brines that were produced. The programme was not intended as a resource development programme but to demonstrate the validity of the concept and provide information that would hopefully encourage development. The only borehole that was drilled for production purposes was the Southampton No. 1 Well in Southampton city centre. This well has now been developed by Southampton City Council and it is providing some 26 000 GJ a^{-1} from a capacity of 1400 kW. The entire programme proved that hot groundwaters do exist in the Permo-Triassic sandstones and can be developed at the high rates necessary. However, at the present time, the resource is not competitive with conventional fuels.

The experimental investigation of the hot dry rock concept, being undertaken by the Camborne School of Mines at Rosemanowes near Truro, is probably the largest hydrogeological experiment at a single site in the UK. The object of the programme is to study problems associated with creating a hot dry rock reservoir by hydrofracturing

techniques. Three boreholes have been drilled to 2–3 km depth in the Cornish Batholith. Flow paths have been created by hydraulic stimulation of natural joints and water is currently being circulated through the fractured reservoir. The drilling depth was selected because it was within the capability of rigs readily available and yet the earth stresses were high enough to confine the reservoir. In addition to the main objective of drilling and fracturing the granite at depth, many substantial, supplementary programmes have been necessary. These include geophysical logging to identify fractures and flow paths, microseismic activity, reservoir analysis (by hydraulic analysis, hydraulic and thermal modelling, microseismic emission, and the use of tracers), and geochemical studies, including high temperature and high pressure laboratory studies of water–rock reactions that could both increase and decrease fracture permeability. Currently engineering design and economic studies are concerned with a full-scale system requiring boreholes drilled to depths of about 6 km, where temperatures exceed 200°C. The programme has demonstrated that heat can be extracted from hot dry rock reservoirs but an economic prototype has yet to be developed. Problems also remain with the stimulation of effective fractures and the subsequent manipulation of such fractured reservoirs.

The increasing concern about global warming may ultimately lead to a more encouraging economic prognosis for geothermal energy in the UK especially as it has a relatively benign impact on the environment. Certainly development can be anticipated overseas wherever high temperature resources occur and the involvement of British hydrogeologists and engineers in the Caribbean, Central and South America, East Africa, the South Pacific, and elsewhere is likely to continue. The cost of energy from relatively small schemes can be very favourable and of benefit to many small communities.

The need to dispose of radioactive wastes safely has opened up the study of the hydrogeology of low permeability rocks—such as crystalline igneous and metamorphic rocks, argillaceous sediments, volcanic tuffs, and salt deposits. Large funds have been made available to examine the nature of these rocks, their chemical composition, flow patterns, and possible water–rock reactions. Field investigations of suitable sites in the UK have been limited but there are major programmes in a number of countries including Sweden, Switzerland, the USA, and Canada and British scientists have been associated with some of these programmes.

In the face of public concern, the investigation in the UK of possible high level waste sites in crystalline rocks was abandoned in 1981 and the initial studies of possible low level sites in argillaceous rocks at Killingholme, Bradwell, Fulbeck, and Elstow were also not pursued for the same reason. Currently field programmes are focused on the preliminary investigation of sites for intermediate and low level wastes at the existing nuclear installations at Dounreay and Sellafield. At these sites a repository would be in hard basement rocks below sedimentary cover, one of the environments recognized as suitable for the storage of radioactive wastes; the rate of groundwater flow at such sites would be expected to be very low.

Despite the limited field programmes in the UK, British scientists have carried out extensive theoretical and laboratory-based programmes as well as being involved in the international studies. The low permeability of rocks suitable for repositories necessitated the development of equipment suitable for measuring the low flows that occur naturally or could be induced. Detailed studies of host-rock geochemistry and its relationship to groundwater chemistry, particularly with regard to the retardation of radionuclides, have led to sophisticated field and laboratory experiments. New types of hydraulic tests have been developed along with new mathematical tools for their analysis. Tracer tests have played an important role in studies of fractured rocks and their interpretation, in conjunction with new geophysical techniques, has helped to recognize flow patterns.

Because of the large volumes of rock and the long periods of time involved, together with the complexity of the various interrelated chemical processes, models are essential to forecast the possible migration of radionuclides from a repository. The requirement has led to significant developments of computer models for describing transport processes in rocks with a wide range of hydraulic characteristics. These take into account advection, hydrodynamic dispersion, solution–precipitation reactions, sorption, radioactive decay, rock–matrix diffusion, and flow in fractures. Three dimensional models have been used to study regional groundwater flow and determine characteristic path lengths and travel times and these parameters are used in one dimensional transport models to study more complex geochemical processes and diffusion and fracture flow. Solution of the problems associated with the disposal of radioactive wastes requires prediction modelling over long periods of time and it is difficult to make such long term predictions on the basis of short term data. Care must be taken to ensure that the interpretation of the results from models does not exceed their geological basis; the hydrogeological insight must always be sound. Consequently, an important aspect of current modelling studies is to quantify the uncertainty that exists in the predictions of the performance of repositories.

Climate change

High level wastes remain hazardous for a very long time and the aim of national programmes is for isolation for more than 500 000 years. Consequently, the effects of future geological processes are very relevant to decisions about suitable repository sites. This has led to assessments of future climatic, geomorphological, and hydrological changes that could influence the rate of groundwater flow and hence the time of travel of nuclides from a repository to the biosphere.

Future climatic changes will also have an impact on groundwater resources. In 1979, the World Climate Conference, organized by the World Meteorological Office (WMO), identified a potential problem of global climate change associated with the increasing emissions of carbon dioxide and other 'greenhouse gases'. Scientific work on global climate change progressed steadily through the 1980s but did not begin to attract public attention until the publication in 1987 of the report of the World Commission on Environment and Development chaired by the Prime Minister of Norway, Mrs Brundtland. 'The Brundtland Report', as it has become known, expressed in terms that were readily understandable to the public, the disastrous consequences that the impacts of climate change could have on almost every aspect of our daily lives.

Since then the pace of research has quickened. In 1988, the United Nations Environment Programme (UNEP) and WMO set up the Intergovernmental Panel on Climate Change (IPCC). The working groups have been assessing the evidence for climate change, particularly likely future scenarios, as well as considering the impacts and examining the appropriate responses that nations could make to such impacts. The IPCC Committee reported its findings in late 1990.

Much research has been proceeding at a national level in parallel with the intergovernmental activity. Estimates of how the global climate model will change due to increasing concentrations of carbon dioxide and other 'greenhouse gases' are made with numerical models derived from models used in weather forecasting. Such general circulation models (GCMs) estimate values of pressure, temperature, humidity, precipitation, etc., at regular grid points distributed over the surface of the globe. The models are complex and require a large computing facility. There are five centres which have developed GCMs including the UK's Meteorological Office. The predictions from the models differ but the IPCC concluded that there is likely to be an increase in global mean temperature of about 1°C by 2025 and 3°C before the end of the next century. This will have serious hydrological consequences and indeed it has been estimated that global rainfall may increase by 7–11 per cent although local values may show much greater variability.

The grid size of the GCMs is too large at present to be able to predict with any accuracy the change in rainfall or evapotranspiration patterns over the UK. However, one current view is that while the winters will be wetter, the summers will be drier and more prolonged, particularly in the south and east. Winter recharge of groundwater may increase but soil moisture deficits in the summer and autumn can also be expected to increase, and the net impact of the changes on groundwater resources remains to be resolved. However, greater demands on groundwater for water supply and irrigation would reduce the storage in aquifers, particularly at the end of drought periods, which would become more frequent. In this context, it is very relevant that storage in the top 50 m of the Chalk, that is in the main water-bearing zone, only represents two years of average annual infiltration.

Less infiltration would increase the risk of contamination of aquifers by reducing the dilution of diffuse pollutants in them. A reduction of recharge and a rise in sea level would affect freshwater/salt water relationships in coastal aquifers and would also reduce the volume of fresh groundwater available. The uncertainties emphasize the strategic value of groundwater and the need for its care and protection in a changing environment.

Although a substantial amount of research is proceeding on the impact that such scenarios would have on surface water regimes, little work has so far been undertaken on the groundwater front. The UK is not alone in having devoted so little effort to this important area. The WMO Conference on Climate and Water in Helsinki, 1989, and the European Geophysical Society's Assembly in Copenhagen, 1990, both sought contributory papers on the impact of climate change on the hydrological cycle and on water resources in particular. Although there were many papers on surface water hydrology and resources, few considered groundwater. It is indeed surprising, in view of the crucial importance of groundwater resources in so many parts of the world, that so little research is being undertaken. Perhaps it is because the research funding agencies mistakenly perceive groundwater systems as being more robust than their surface water equivalents that they are unwilling to support work in this area. However, groundwater resources are vital in many areas and as has been seen over the past decade, are much less robust than was previously considered. It is, therefore, inevitable that the questions raised above will be addressed in some considerable detail over the coming decade.

Water management

Water management in the UK is based on river basins and it involves the comprehensive management of all aspects of the hydrological cycle of a river basin. In England and Wales, the legal basis for this is the Water Resources Act of 1963, the Water Act of 1973, the Control of Pollution Act of 1974, and the Water Act of 1989. The government's policy of converting the public water authorities into private companies in 1989 led to a division of the functions that were previously the responsibility of the water authorities. The privatized water companies now provide water supplies and are responsible for sewerage systems and sewage treatment and disposal. A new regulatory body, the National Rivers Authority, assumed responsibility for water resources planning, control of abstraction through licenses, pollution monitoring and control, prevention of environmental pollution of rivers and groundwaters, land drainage and flood prevention, and certain aspects of water recreation and navigation. The separation of the two functions—regulation on the one hand and water supply and its disposal on the other—was a logical improvement to the management system, removing the role-ambiguity of both 'poacher and gamekeeper' that existed previously. It will work well, given the foundation established since 1963, provided the National Rivers Authority has adequate funds to pursue what is an environmental protection role with regard to water in the widest sense. A significant test will be whether the new dichotomy proves effective in the role of developing regional and national water resources while maintaining the water environment, in terms of such factors as water quality and providing acceptable river flows.

Overall, the most disappointing event in the 1980s has been the continuing decline in the quality of groundwater. The great benefit of groundwater has always been that it is cheap to develop and is of good quality requiring minimal treatment. There is increasing evidence that this situation is changing. Groundwater resources are being polluted from both point and diffuse sources associated with urban and industrial developments and modern agricultural practices. The pollution of groundwater is insidious for it occurs slowly, almost imperceptibly. Once a resource is polluted expensive forms of treatment are necessary, which increase the cost of development and can eventually make it uneconomic for potable supply.

In the early 1970s, it became apparent that nitrate concentrations in British groundwaters were increasing. In the ensuing 15 years hydrogeological research established that changes in arable farming, aimed at increasing productivity, were having a profound impact on groundwater quality. Serious effects are already apparent in the principal aquifers, especially where the unsaturated zone is thin. Evidence suggests that nitrate concentrations in groundwater will continue to rise slowly in numerous regions for many years. In many unconfined aquifers, they will eventually exceed 50 mg l^{-1} as NO_3 (the maximum admissable concentration given in the EC Drinking Water Directive) and, where aquifer outcrops are extensively used for arable agriculture, in areas of low infiltration, concentrations are likely to exceed 100 mg l^{-1}.

Some groundwater supplies, especially in eastern England, have also been found to contain pesticides above the guide-line limit proposed by the European Community. Incidents of groundwater pollution by organic contaminants have also increased markedly. This can be due to leaking underground storage tanks, accidental spillages, casual disposal of liquid effluents, and other activities. One of the greatest threats appears to be from the chlorinated alkanes and alkenes, which although only slightly soluble in water, are toxic, resistant to biodegradation, and once below the soil zone may not be strongly retarded by interaction with the aquifer matrix. These compounds are synthetic solvents in widespread use by industry. Ironically, they are a less serious threat to surface waters because of their volatility; such losses from groundwater are less likely and the high density of many of the compounds allows deep penetration into aquifers where they can remain continuously contaminating groundwater for a long time. Small amounts can contaminate very large volumes of groundwater.

It would be misleading to say that the decline in the quality of groundwater was not anticipated. The risk from rising nitrate concentrations was realized as long ago as the mid-1970s and the fear that pesticides and organic compounds would increase was also anticipated late in the decade. It was clear that these compounds could move through the unsaturated zone into groundwater and persist in detectable amounts. However, it is only in the last five years that analytical techniques have become widely available allowing detection of the low levels of organic solvents, specified in the EC directive. Indeed the directive only became effective in the mid-1980s prior to which there was no statutory requirement to examine water supplies for these constituents.

A major factor in the deterioration in quality has been the difficulty of taking remedial action until irrefutable evidence is available. In the case of groundwater pollution, once the evidence is irrefutable it is often too late because of the long residence time of water in aquifers. The extent of a pollution incident may be uncertain but the consequences are not in doubt.

The need for protection zones around public supply wells, particularly in aquifers at outcrop, was an obvious early necessity and it is to the credit of the water authorities that they attempted to cover the absence of well-defined legislation by introducing aquifer protection policies that acted by persuasion rather than through legal control.

Generally, the capacity of the environment, including groundwater, to assimilate and cope with wastes has been misjudged. In the case of groundwater this attitude has been compounded by the 'out of sight, out of mind' syndrome and the blithe assumption that aquifers have an infinite capacity for absorbing and neutralizing waste. The long-term gain of maintaining groundwater quality must be greater than the short-term gain of allowing practices that pollute aquifers; the latter policy just creates problems for the future.

The implementation of legislation in the Control of Pollution Act 1974 for the disposal of wastes has not been entirely successful. County and metropolitan authorities were made responsible for both the disposal of wastes from households and industry as well as regulating waste disposal by issuing licences to themselves and other tip-owners. This legal and administrative system has not been satisfactory and furthermore, some authorities have not yet completed details of their waste disposal plans.

The monitoring facilities and pollution control at many landfill sites are also unsatisfactory. The British practice of disposing of industrial and domestic wastes at the same site is not universally accepted. From a groundwater viewpoint it is encouraging that the emphasis in the management of landfills is moving towards containment of wastes followed by treatment of the collected leachate.

Under the new environmental pollution legislation, councils will have to establish companies to dispose of waste on a commercial basis. The licencing of waste disposal sites to ensure that they meet environmental standards will be run separately. The government now accepts that waste disposal will be more expensive (and more profitable). It hopes that this will encourage higher standards but the problem will be to ensure that costly disposal does not lead to greater evasion of the regulations.

In the past, the full cost of pollution has not always fallen on the polluter. This is a recipe for disaster. Polluting activities commonly produce private benefit but in the future social costs must be taken into account. It is a question of taxing antisocial activities, for example by using a pricing mechanism to discourage such activities, while at the same time anticipating the evasions that will follow such a policy. Groundwater can cope with some pollution but a high price should be applied to polluting practices to ensure that too much does not occur. Maintaining the quality of groundwater is essential—it is not a luxury. Skimping costs on waste disposal can lead to much greater costs for remedial action.

Arresting the decline in groundwater quality will depend not only on the existence of a strong regulatory body supported by adequate legislation, but on increasing public awareness about groundwater and the need to protect it because of its strategic value in droughts and during short-term pollution incidents, such as, that caused by Chernobyl. There is a need for better public relations and education.

The future

Hydrogeological programmes today require more detailed information about the nature of aquifers, the groundwaters they contain, their flow patterns, and the means by which they are replenished. To give but one example—a major research effort is necessary to identify the contaminants that occur in groundwater, where they occur, and at what concentration. In the future new tougher environmental regulations will be necessary. These will have to be supported by appropriate, enlightened research and monitoring programmes which bear in mind that although hydrogeology is a practical subject, the intellectual aspects must also be encouraged. Funds made available for research should bear a closer relationship to the capital invested in groundwater and the cost of remedial action if such a strategic resource should become widely polluted.

The success of the next decade is likely to be judged by the extent to which groundwater quality has been maintained. But a further yardstick will be the benefits provided to the developing world by technology transfer. One of the disappointing outcomes of the 1980s has been the limited rate of progress with the International Drinking Water Supply and Sanitation Decade. Some 50 per cent of the world's population does not have access to a clean water supply. Groundwater, a cheap, reliable, and generally good quality source in the developing world, has the potential in many areas to help meet this demand, providing funds are made available. The sums necessary are small in relation to the potential benefits for health and the quality of life, and in relation to expenditure on luxury items in wealthy countries. It is beholden upon such countries to ensure that a clean, reliable water supply, as well as at least basic sanitation, is provided for all, for water is a primary basic need.

The threat to the environment from global warming will increase the value of groundwater storage world-wide and this reality further emphasizes the need to maintain its quality. Hydrogeology has expanded and diversified in the 1980s but towards the end of the century the provision of water supplies and the implementation of measures that anticipate and prevent the degradation of groundwater quality are likely to be the dominant issues.

2. Groundwater—legal controls and organizational aspects

A. C. Skinner

Introduction

This chapter is neither an authoritative legal treatise, nor a guide to the institutions responsible for the management of groundwater in the UK. Reference works are available on legal aspects (Ackroyd 1986; Howarth 1990) and the relevant legislation is identified in the text. The objective of the chapter is to review the present legal and organizational arrangements for the management of groundwater in the light of the fundamental reorganization of the water industry which took place in 1989, to consider to what extent this reorganization has both changed and improved groundwater management, to compare the situation in the UK with arrangements in other countries, and to identify, in the context of the institutional framework, some of the challenges in groundwater management in the 1990s. The chapter will be concerned primarily with the situation in England and Wales. This is justified on two grounds. First, there has been no recent legal or organizational change in Scotland or Northern Ireland. Indeed, the Scots and Irish may say with justification that the rest of the country has now rejoined them in having independent regulation of the water environment, a situation which has existed without change there for many years. The current legal framework is established by the Water (Scotland) Act 1980 and the Water Act (Northern Ireland) 1972, which respectively places the regulatory responsibilities with the River Purification Boards and the Department of the Environment, Northern Ireland. Secondly, the proportion of groundwater used in Scotland and Northern Ireland is small both in absolute terms and relative to total consumption and therefore some of the problems found in England and Wales

are not so apparent. However, they share the important use of groundwater for private rural sources which, as in the rest of the country, present difficult problems of regulation and management.

Groundwater in UK law

The law has always had difficulty with the concept of groundwater and there exist in UK law distinctions and definitions which have no scientific basis. The problem arises because common law on this topic was largely established at a time when credible and consistent scientific advice was not available on matters relating to groundwater. As statute law on water resources and water pollution has been established, the scientific anomalies of common law have become less significant. However they still persist. In particular, a special legal status is given to the 'underground river'. UK common and statute law effectively do not recognize underground water 'flowing in defined channels' as groundwater, but regard it as if it were a surface stream. The law, therefore, completely fails to recognize that in many hydrogeological environments there is free and regular interchange between the flow within aquifer blocks and the flow in fractures and conduits between them, of which the layman's 'underground stream' is but an extreme example. From the scientific point of view, this anomaly is, in most cases, only of curiosity value and does not significantly complicate water management.

Prior to the Water Act 1989 confusions such as these were exacerbated by the fact that the definition of groundwater was different in water resource and water pollution law. There is now a common definition (Water Resources Act 1963: Section 2(2) and Water Act 1989: Section 103 (1) and Section 124(2) (b)) as follows:

'Waters contained

– in underground strata; or

– in a well, borehole or similar work sunk into underground strata, including any adit or passage constructed in connection with the well borehole or work for facilitating the collection of water in the well, borehole or work; or

– in any excavation into underground strata where the level of water in the excavation depends wholly or mainly on water entering it from the strata.'

However, in moving to a common definition the UK has moved further away from European law which defines groundwater (EC80/68 and in the draft EC Nitrate Directive):

'All water which is below the surface of the ground in the saturation zone and in direct contact with the ground or subsoil.'

This definition is different in at least two significant respects: European Law does not recognize water in the unsaturated zone as groundwater and thus excludes from the definition soil waters and percolating water yet to reach the 'water table'; secondly European Law does not recognize as groundwater, water exposed at the surface within excavations. There seems to be no valid technical reason why water in the unsaturated zone should be included within the definition, at variance with both European law and the definition which applied under the former Control of Pollution Act 1974. The implications of the definition in pollution cases remain to be assessed. The fact that pollution may be adjudged to have taken place by the fact of discharge to unsaturated strata before any contact with natural waters will present a demanding test both upon the discharger and the hydrogeologist who has to present the scientific evidence.

Water Act 1989

The Water Act 1989 amends and brings more up to date most water resources and water pollution law and establishes a new agency for England and Wales, the National Rivers Authority (NRA), to be responsible for the regulation of the water environment and *inter alia* most aspects on groundwater regulation. Those aspects not included in the remit of the National Rivers Authority will be discussed further below.

The creation of a separate and national regulatory body is a fundamental change and terminates the arrangements which had existed between 1974 and 1989 whereby virtually all aspects of the water cycle, both operational and regulatory, had been performed by multipurpose regional authorities. These regional authorities had been organized on a catchment basis and were charged with integrated management of water within the catchment. This catchment management role now lies with the National Rivers Authority, which has preserved for its regional units the same catchment boundaries. However, the role is inevitably a different one, since the water supply and sewage treatment functions are now vested in private companies which are responsible for their own utility plans. The private water companies are subject to a financial regulator, the Director General of Water Services, in addition to the regulation of their abstractions and discharges by the National Rivers Authority. The complex interrelation between water resource management and protection options and their financial implications will require co-ordination and information exchange between these two regulatory agencies. Typical of the issues involving choice and possible conflict of technical and financial interest between the utilities and the regulators and which are particularly relevant to groundwater are:

(a) possible alternative water resource strategies based on catchment or multi-catchment provision by the NRA compared with local and exclusive developments by the companies;

(b) alternative strategies for groundwater quality management based on the contrasting options of protection at source or treatment in supply.

The 1989 Water Act also places wider and more extensive duties on the National Rivers Authority in respect of conservation. Under previous legislation it was uncertain to what extent it was legitimate to limit rights to abstract groundwater if there was to be an impact upon natural wetland habitats.

The new legislation makes this duty clear, although it poses technical problems as to how monitoring can be practically undertaken to allow the necessary judgement to be made. Nothing in the new legislation changes the situation whereby there is no statutory right to groundwater being preserved at a given level, and thus there is no legal protection from the lowering of groundwater levels which cause, for example, a reduction in soil moisture availability to crops or a settlement in foundations.

The new Act also empowers the National Rivers Authority, where it considers it appropriate, to identify and to so manage water resources so as to sustain 'minimum acceptable flow' in designated watercourses. As explained later many of the situations where river flows in dry weather are perceived to be below an acceptable minimum arise because of the indirect effect of groundwater abstraction within their catchments. The concept of minimum acceptable flows (MAFs) was introduced in the 1963 Water Resources Act, but, although various proxy measures have been devised to help manage low flows, no formal MAFs have ever been designated. The concept has proved particularly difficult to operate in groundwater fed rivers because of the lack of precision in many catchments in the extent and timing of the effects of groundwater abstractions on river flows.

Issues in groundwater resource management

Most abstractions of groundwater require authorization by licence under the 1963 Water Resources Act. The powers to issue and enforce licences are now vested in the National Rivers Authority. A few new classes of abstractions, most notably all fish farms, came under control in the 1989 Water Act, but otherwise these provisions are substantially unchanged. There was extensive but unsuccessful lobbying by conservation and environmental interests during the passage of the Water Bill through Parliament to give the National Rivers Authority powers to revoke or vary downwards licences to abstract, without liability for compensation, where it could be shown that the quantity of water authorized to be abstracted exceeded the available water resources. This action was stimulated by situations where groundwater

fed streams, mainly in southern England, have become significantly depleted because of excessive groundwater abstraction. In many of these cases it is doubtful if reduction of licence is the best solution, taking into account the National Rivers Authority's general duty to conserve, enhance, and redistribute water resources. Alternative options, such as flow augmentation and channel lining are now under investigation to try and reconcile the conflicting objective of sustaining water resource provision and environmental benefit. The National Rivers Authority has identified this issue as one to be given early and detailed attention.

One of the features of water resource management in the UK during the period 1974–89 was the considerable development in catchment-wide integrated resource systems (see Chapter 3), often involving new facilities for river flow augmentation to enable river abstractions to be sustained during dry weather periods. Flow augmentation may be provided from reservoirs or from groundwater. A number of dedicated flow augmentation schemes were developed in the 1970s and 1980s. Despite their similar function surface and groundwater flow augmentation schemes have been treated differently in the new arrangements. Reservoirs used for flow augmentation to increase water resource provision remain the property of the private water companies, although their use for the benefit of water resources is governed by statutory agreements with the National Rivers Authority, which also meets the operating costs. Groundwater augmentation schemes, however, are directly owned and controlled by the National Rivers Authority, which will be responsible for their continued development as demand requires.

Groundwater resources are already at a high level of development in England and Wales and there are few areas where there is substantial scope for major enhancements in yield. Where new resources are developed, there are well established technical procedures and statutory controls to ensure that the rights of existing abstractors are protected. The protection of environmental interests is both less sure legally and more difficult technically. In the future, it will be advantageous that the major abstractor is no longer the water resource regulator, because clarity of purpose and objective are going to be necessary in addressing some of the potential conflicting interests.

Issues in groundwater resource protection

The relevant parts of the Water Act 1989 broadly follow Part II of the Control of Pollution Act 1974, but the implied concept of the latter Act that some groundwaters may be unworthy of protection is now dropped. All groundwaters, including waters in the unsaturated zone, are defined as controlled waters. A consent from the National Rivers Authority is necessary before a discharge to groundwater can be made and to cause pollution of groundwaters by an unconsented discharge is an offence. Unfortunately, the more readily controllable direct discharges to groundwaters do not often pose a significant quality risk. Perhaps the greatest threats to groundwater quality come from accidental or otherwise uncontrolled seepages and discharges (e.g. spillage of industrial or agricultural chemicals and solvents, discharges from contaminated land), leachate from older, uncontained waste disposal sites, and from diffuse pollution, especially from agriculture. In many of these situations the act cannot easily be applied or enforced.

Groundwater quality objectives

The new act contains provision for the definition of quality objectives for groundwater. This concept is already firmly established in the management of river water quality and the objectives provide the target for river quality improvement. The concept does not directly transfer to groundwater where opportunities for quality enhancement are rare and the highest objective is often for there to be no deterioration in quality. None the less the development of a robust methodology for defining quality objectives for groundwater is an essential prerequisite for developing other policies. This task has never been undertaken on a national basis before, but the 1989 Act now provides a suitable framework.

Aquifer protection

Unlike many European countries the UK has no established statutory system of land-use control around water sources. The National Rivers Authority can act to control potentially polluting practices using statutory powers under the Water Act and the Control of Pollution Act and by non-statutory consultation under the Town and Country Planning Acts. There are also opportunities to influence specific developments, for example cross-country oil pipelines under the Pipelines Act. Many of the former Water Authorities established Aquifer Protection Policies to provide a co-ordinated basis for their responses under the various legislative procedures. These have evolved independently and policies, practices, and standards vary from region to region. The National Rivers Authority now has the task of drawing up a policy for England and Wales. This policy will need to deal also with questions of diffuse pollution, rehabilitation of contaminated aquifers, and the identification of advantages of the new powers provided in the Water Act to establish Water Source Protection Zones.

Waste disposal

No one agency is responsible for the protection of groundwater from the impact of waste disposal to land. The present legislation, the Control of Pollution Act 1974 (Part I), places the responsibility for waste disposal regulation with County Councils and Metropolitan District Councils in England and District Councils in Wales, which control the location and nature of wastes deposited by a system of licensing. The UK Government has recently introduced the Environmental Protection Act 1990, which modifies and, in some areas, considerably enhances the powers in the 1974 Act, but the basic arrangement whereby local government bodies regulate waste disposal will not be changed. The National Rivers Authority will continue to have statutory rights to specify conditions in waste disposal licences to protect water quality, and, in future, the right to veto the suspension of a licence once waste disposal has ceased. However, the responsibility for the monitoring and enforcement of licences to protect water resources rests entirely with the local authorities. The lack of clarity of control on the impact of waste materials on water quality is demonstrated by the designation by the UK Government of three bodies with joint 'competency' to administer the EC Groundwater Directive: the Waste Disposal Authority and the

National Rivers Authority in respect of the responsibilities outlined above and the Planning Authorities in respect of their responsibilities for potentially polluting mineral and mining wastes which are not 'controlled wastes' under UK legislation. The development responsibilities of the Planning Authorities are also relevant in relation to restoration of contaminated land, where there is no secure process to ensure that such activities are undertaken in ways that will preserve groundwater quality.

Industrial pollution

Groundwater in many urban areas is polluted from a variety of industrial sources, two of the most common being acidified groundwaters from old industrial sites, such as gas works, and pollution from chlorinated hydrocarbon solvents (Chapter 9). The pollution is often detected at random locations and it is often difficult to identify the source. Even if the general area of the source can be identified it may not be possible to show how, by whom, and when pollution was caused with sufficient certainty to prove an offence. Proof of an offence does not mean that all groundwater pollution at one location can be attributed to one offender, and therefore legal proceedings will not be a sure means of getting the pollution cleaned up. For many of the worst urban pollution cases, no reliable clean-up techniques exist. This catalogue of legal and practical obstacles means that the problem of urban groundwater pollution will not be solved quickly. The present emphasis is on regulating current practice to prevent any further deterioration and documenting the extent of the problem wherever possible. The latter task is hampered by the lack of legal requirement on any party to maintain registers of contaminated land sites. A recommendation to establish such registers was recently made by the Environment Committee of the House of Commons following an enquiry into problems of contaminated land (Anon 1990).

Other new legislation relevant to this problem is the new power in the Water Act for regulations to be made to control the way substances are used or stored or to prevent the use of substances in defined situations. In addition, the Environmental Protection Act extends the responsibilities under the existing Control of Pollution Act to require regulation and monitoring by the new Waste Regulatory Authorities of disused landfills.

Diffuse agricultural pollution

In terms of the threat to future water resources diffuse pollution by nitrates and pesticides (the latter not all from agriculture) is the most significant groundwater quality problem. The nitrate problem is specifically addressed in legislation for the first time in the 1989 Water Act. It enables regulations to be made to create 'nitrate sensitive areas', within which farmers would agree to manage land to specified programmes designed to reduce nitrate leaching. The act provides for compensation to be paid to farmers in redress for loss of production below that which would otherwise have been achieved. It also provides for the nitrate sensitive areas scheme to be made compulsory if adequate uptake is not achieved by voluntary means. The scheme is run by the Ministry of Agriculture but the initial recommendation for areas to be designated is made by the National Rivers Authority which is also responsible for monitoring the impact of the scheme on water quality. The first of the nitrate sensitive areas, which is being undertaken as a pilot phase, was created in 1990. This scheme follows that of the state of Baden-Würtemburg in Germany as one of the first in Europe to pay compensation to farmers who undertake financially punitive changes in land management to reduce nitrate leaching. The UK scheme is, in many ways, more radical than the German scheme. The latter is based upon a percentage reduction in fertilizer use; the UK scheme incorporates a range of possibilities for change of land use including conversion from arable to low intensity grass husbandry and trees.

Areas which are vulnerable to nitrate leaching are also likely to be vulnerable to pesticide leaching. There is not yet enough research data to assess the long term risk to groundwater from pesticides and to identify any necessary control measures. Provisions exist in the Water Act to designate 'water source protection zones' but unlike nitrate sensitive areas they would be imposed purely as pollution control measures without provision for compensation.

Mineral waters

Mineral waters in law are natural waters in respect of their origin and thus fall within the scope of water legislation. As products for consumption, they are regarded as foods rather than drinking waters and are subject to their own legislation which is administered by the Ministry of Agriculture and the District Councils. Mineral waters, by definition, cannot be subject to any form of artificial sterilization and thus assurance of quality and fitness for use can only come from resource protection. The recent trend to greater use of bottled table water has greatly increased the volume of abstraction and number of sources used for commercial mineral waters. Most are deep groundwaters or springs and although the volumes abstracted are very modest compared with that for public supply, the use of groundwater for mineral waters presents a particular challenge for groundwater protection, requiring collaboration between different agencies for effective implementation.

Issues for the future

Groundwater resource planning

Demand for groundwater is increasing but, in some areas, the identified available resource may be diminishing. The reasons for a reduction in resources may be either the abandonment of sources because of increased levels of pollution, for example by nitrate, or reduction of available abstractable resource because of the need to preserve more groundwater for base-flow support to surface waters to improve or restore environmental benefit. In the past, the exhaustion of local water resources like groundwater has been met by major basin-scale developments, and it was the ability to promote such schemes which was one of the advantages of the former multifunctional structure of the water industry in England and Wales. It is not yet clear how the water resources role of the National Rivers Authority and the water supply role of the

new private companies will develop to meet these challenges. The financial structure of the National Rivers Authority will constrain its ability to invest significantly in new water resource developments and yet it seems inevitable that major expenditure will be necessary, either to undertake augmentation schemes to sustain abstractions without environmental detriment or to develop new basin-wide resources to replace those resources surrendered for quality or environmental reasons.

Legislation for resource protection

As explained above, until now aquifer protection policies and zones of groundwater protection have not had any formal basis in law. The development by the National Rivers Authority of a national aquifer protection policy must take into account the availability under the Water Act of powers to designate Water Source Protection Zones and the expected requirements of the forthcoming EC Nitrate Directive for the designation of 'Vulnerable areas' in relation to leaching of nitrate. The concept of a protection zone is well established in the legislation of many other European countries and a requirement for a 'Well head protection zone' has now been introduced by the Environmental Protection Agency in the USA. These zones are generally defined by time-variant solutions and are not best suited for catchment protection against diffuse pollutants. The challenge is, therefore, to devise a system, within the existing legal framework, which is capable of meeting the characteristics of both conservative and degradable, and point source and diffuse pollutants. It should, however, be recorded that the existing non-statutory protection policies operated by the UK water industry have been effective over the years in highlighting and allowing control of many potential pollutants to aquifers and it remains to be seen to what extent it is felt desirable or necessary to use the water source protection zone legislation for general resource protection. It seems likely that it will be used, at least initially, in special locations or for selective pollution threats, if only because, as with nitrate sensitive areas, the time and technical effort required to define zones satisfactorily for statutory purposes are considerable.

Co-operation between agencies

It will be apparent from the description of organizational and legal relationships that very many agencies besides the National Rivers Authority have a role in the management and protection of groundwater. The nature of groundwater problems, often out of sight, operating over a long time-scale and usually requiring specialist technical knowledge to understand their significance, means that many of the agencies have only a limited comprehension of the issues for which they are responsible. Even when the task is accepted and understood many agencies do not have the skilled resources to perform their function adequately. This places a considerable educational and liaison burden upon the primary agency, the National Rivers Authority. Table 2.1 lists, for England and Wales only, those organizations which have statutory or quasi-statutory responsibilities relevant to

Table 2.1   Agencies with responsibility for groundwater in England and Wales

Agency	Area of responsibility	Relevant legislation
National Rivers Authority	Primary body for groundwater resource management and protection	Water Resources Act 1963; Water Act 1989
Department of the Environment	Sponsoring ministry of NRA	Water Act 1989
	Making of regulations under Water Act	Water Act 1989
	Determination of appeals on licences to abstract from and consents to discharge to groundwater	Water Act 1989
	Determination of appeals on waste disposal licences.	Control of Pollution Act Part I 1974
	EC Groundwater Directive	EC80/68
	EC Nitrate Directive (draft)	
	Mineral extraction policy	
	Contaminated land policy	
Her Majesty's Inspectorate of Pollution	Co-ordinating body for waste regulation	Environmental Protection Act 1990
	Discharge to groundwater of 'Red List' substances	Environmental Protection Act 1990
	Discharge of radioactive substances to groundwater	Radioactive Substances Act 1960
Ministry of Agriculture	Nitrate sensitive areas Pesticide regulations	Water Act 1989
	Code of good agricultural practice	Water Act 1989
	Mineral water regulations	SI 1985 No.71

Table 2.1 *(Cont.)*

Agency	Area of responsibility	Relevant legislation
Department of Energy	Licensing of oil pipelines	Pipelines Act 1962
Property Services Agency	Crown properties are exempt from regulation on abstraction and discharging	Various legislation
Office of Water Services	Regulation of charges and levels of service of water companies	Water Act 1989
County Councils and Metropolitan Borough Councils and Welsh District Councils	Waste disposal regulation	Control of Pollution Act 1974
	Competent Authority	EC Groundwater directive
County Councils	Mineral extraction regulation	County Planning (Minerals) Act 1981
District Councils	Development planning	Town and Country Planning Acts
	Competent authority	EC Mineral Water Directive (80/777)
Nature Conservancy Council	Consultee on issue of licences to abstract and consents to discharge with conservation implications	Water Act 1989

the management and protection of groundwater. The proper management and protection of groundwater is increasingly the subject of public concern and lobbying by environmental pressure groups. The extent to which these issues are properly met, although undoubtedly considerably assisted by the institutional changes brought about by the 1989 Water Act, will also depend upon clarity of objective, effective co-operation between the many agencies involved and adequate provision of resources to those agencies.

Acknowledgements

The author would like to recognize the help of his legal and technical colleagues for the advice they have given in the preparation of this chapter. The views expressed are those of the author and do not necessarily reflect the views of his employer.

References

Ackroyd, D. S. (1986). The law relating to groundwater in the United Kingdom, in *Groundwater: occurrence, development and protection*, (ed. T. W. Brandon), pp. 591–607. Instn. Water Engrs. Sci., London.

Anon (1990). *Contaminated land*. Report of the Environment Committee of the House of Commons (Report 170–1). HMSO, London.

Howarth W. (1990). *The law of the National Rivers Authority*. Centre for Law in Rural Areas, Aberystwyth.

3. Groundwater in basin management

M. Owen, H. G. Headworth, and M. Morgan-Jones

The use of aquifer storage

The pattern of groundwater use in Britain is strongly influenced by the general high quality of the rivers and streams and their consequential value for water supply, fisheries, wildlife, and amenity. It is partly because of the good quality of the rivers that the proportion of groundwater abstracted in Britain for water supply is less than half that of Europe as a whole. The proportion of groundwater abstracted may also be a reflection of the small size of many rivers in Britain resulting in the effects of abstraction being more noticeable than on the larger continental rivers.

After a hundred or more years of conventional groundwater development for direct supply, many aquifers are highly exploited. During the 1950s and 1960s there was increasing concern that further direct supply boreholes would cause yet more depletion of stream flows. This led to the consideration of development of groundwater resources in ways which would not affect low summer flows. The introduction of the Water Resources Act 1963 provided the legislative and organizational framework to facilitate regional management of both groundwater and surface water resources in the context of river basins.

New ideas were examined which sought to overcome many of the environmental objections associated with traditional direct supply groundwater abstraction (Ineson and Downing 1964). Central to all such methods is the use of groundwater storage on a seasonal basis. There are various techniques which achieve this end, the principal ones being river augmentation using groundwater, artificial recharge, and alternate seasonal use of sources. The first two of these techniques have been under active investigation and development for the last twenty years or so. Progress on river augmentation schemes was fully reviewed by

Downing *et al.* (1981), and Headworth *et al.* (1983), and on artificial recharge by Edworthy *et al.* (1981).

Britain's varied geology has produced a wide range of recognized aquifers spanning the stratigraphic column from the Devonian to the Recent periods. However, 80 per cent of the abstraction is from only two aquifers, the Chalk (54 per cent) of southern and eastern England, and the Permo-Triassic sandstones (26 per cent) of central and northern England. This situation is reflected in the distribution of the groundwater management schemes discussed here which account for about 15 per cent of the total amount of groundwater used in the country. Details of the breakdown of use of groundwater by aquifer and region are given elsewhere (Commission of the European Communities 1982; Department of the Environment 1988a).

Depletion of low river flows

For the last thirty years or so the effect of groundwater abstraction on river flows has been widely recognized and taken fully into account in new groundwater resource developments. During this time most developments were permitted only where the effects on river flow were insignificant or where the mode of operation or deliberate compensation discharges protected flows.

Before the Water Act 1945 came into force the development of groundwater resources was largely unrestricted. The 1945 Act imposed some degree of control but it was not until the advent of the Water Resources Act 1963 that total control could be exercised. For example major expansion of the residential areas on the outskirts of London in the first half of the twentieth century was supplied largely by extensive use of groundwater from the chalk valleys of the Chiltern Hills and North

Downs. Rights were obtained by water undertakers through Parliamentary Orders, to take quantities often greatly in excess of needs at the time. This was of little consequence whilst actual abstraction was well short of entitlement. The rights to abstract were automatically confirmed as licences under the Water Resources Act of 1963 but by then actual abstraction had begun to rise steeply in response to rapidly increasing demand and has continued to do so.

The consequence for river flows is well illustrated by the case of the River Ver in Hertfordshire (Fig. 3.1.) where the upper 10 km of the originally perennial or regular bourne stream has become a valley which is now normally dry, and the present perennial section suffers from very low and inadequate flows. The environmental consequences

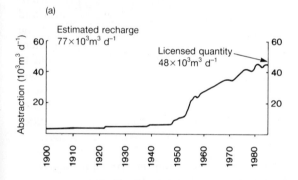

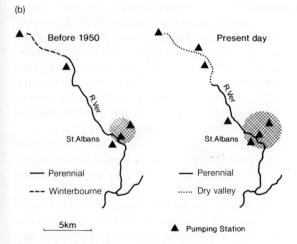

Fig. 3.1. Groundwater abstraction in the catchment of the River Ver, Hertfordshire (a) History of abstraction (b) Effect on the river.

have been major shifts in habitat, the loss of naturally sustained fisheries, and reductions in the general amenity value of the river as an integral part of the rural and urban landscapes of the catchment.

The Ver is a particularly serious example of this phenomenon. Approximately 75 per cent of the average annual recharge to the Chalk aquifer in the catchment is licensed for abstraction and is now almost totally taken up. The situation is exacerbated by the direct export of some of the abstracted water to supply areas outside the catchment and by the export of effluent generated from water supplied within the catchment.

This condition is not unique to the Ver or the Chalk streams around London. A recent preliminary survey by the National Rivers Authority (1990) has identified similar problems in up to forty catchments spread over the groundwater bearing areas of the south-eastern half of England. Licences should not have been granted for new sources at many of the sites in question but the Authority is now faced with a legacy of the past. In view of the rising interest in protecting and restoring the environment, the Authority has decided that the situation will be investigated in greater depth with a view to adopting a programme of alleviation schemes in justified cases.

Proposals for such alleviation schemes for six cases around the London Basin (Halcrow 1988) have already been made. The solutions proposed in all cases involve, in combination with other measures, changes to the use of groundwater. Improvements could clearly be brought about by closing the wells or pumping stations concerned but in the Thames catchment alone these sources provide approximately 180×10^3 m³ d⁻¹. Clearly, their closure would have major financial repercussions because of the need to provide alternative sources of supply. It is preferable to use various compensation measures to avoid this but ultimately it may be necessary to cease abstracting from some wells or markedly reduce abstraction at very sensitive locations and face the costs involved.

The River Ver is a tributary of the River Colne, a major tributary of the River Thames. The effects of groundwater abstraction in relation to rivers are well illustrated by following the river downstream. Major groundwater abstraction starting at the confluence of the Ver and Colne and extending downstream resulted in marked reduction of flow in the River Colne at times of low flows. The condition became so severe that in the early 1970s a sewage treatment works was installed purely to augment flows; action to alleviate low flows and improve the environment is not a new idea. Further downstream inputs from tributaries and from the major sewage treatment works to which the bulk of the Colne catchment is drained, produce much higher base flows. In this zone abstractions totalling 80×10^3 m^3 d^{-1} over a distance of 9 km produce no unacceptable environmental effects, even at times of low flows.

The depletion effects of non-returned abstraction are most noticeable during periods of low flow. Abstractions, which when compared to average annual resources may be quite modest, will have more marked effects at low flows, particularly if located in the headwaters of catchments where flows are low. In general, rivers on the Chalk are susceptible to low flow depletion as the effects of abstraction on river flow are direct. This can be contrasted with some locations on the Permo-Triassic sandstones where the interaction between aquifer and river is indirect due to a much stronger component of vertical flow induced by frequent marl bands. In the Bromsgrove area of the West Midlands, Salmon (1990) has shown that the effects of abstraction spread much further with widespread lowering of groundwater levels but limited direct effect on river flow. In comparison, Chalk catchments show little regional drawdown of water levels but marked effects on river flows.

River augmentation using groundwater

The idea of using groundwater to augment rivers at times of low flow was first propounded by a well-driller, Guthrie Allsebrook, in 1934. While pumped storage reservoirs are normally filled from rivers in winter to meet water demands in the summer, river augmentation using groundwater adopts the concept of increasing summer flows in dry years by pumping from boreholes. These then support increased abstraction from the river downstream, for treatment and supply. In many cases the seasonal discharge of groundwater into the river does not support a downstream river abstraction but instead compensates the river when flows

are low as a result of conventional groundwater abstraction that is put directly into supply elsewhere in the catchment. Examples of both these types of augmentation have been developed quite extensively in Britain. The locations of the main schemes are shown on Fig. 3.2. and details are given in Table 3.1. Several of these schemes were in various states of development more than a decade ago (Downing *et al.* 1981).

The principle of net gain

The principle of augmentation is straight forward. Water is pumped from wells into a river whenever its flow falls below that required to satisfy all the demands on the river, whether these are for water supply or the maintenance of adequate flows for irrigation, navigation, fisheries, wildlife, general amenity, or the dilution of effluents. The aim is to provide a high net gain in flow, this being the amount by which natural flows are augmented by the pumped discharge expressed as a percentage of the pumped quantity:

$$\text{Net gain} = \frac{G - (I + R)}{G} \times 100$$

where G is groundwater discharged to the river, I is natural groundwater flow intercepted by the wells and R is the recirculation of water from the river into the aquifer by river-bed leakage.

The net gain depends upon the transmissivity and storativity of the aquifer, the permeability of the river bed, and the position of the wells relative to the river and the aquifer boundaries (Oakes and Wilkinson 1972). In aquifers with a high transmissivity and low storativity (such as the Chalk and other limestones) the aquifer response to pumping is rapid and wells have to be sited at a considerable distance from the river to obtain adequate net gains. If the transmissivity is relatively low and the storativity is high, as is the case with many sandstones (such as the Permo-Triassic sandstones) then the wells can be sited closer to the river. However, if the river bed is impermeable then wells can be sited alongside the river and water levels deliberately lowered below the river bed to develop groundwater storage.

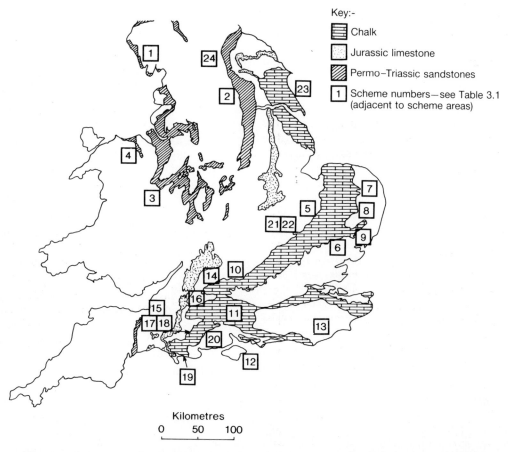

Fig. 3.2. Locations of schemes using groundwater for river augmentation in Britain.

Net gain generally decreases with time (Fig. 3.3.). However, when groundwater abstraction has completely intercepted the natural base flow, or if the river bed is impermeable, high net gains can be sustained. A water balance is always maintained. If the pumped boreholes are unable to draw on river flows, either by recirculation or interception of groundwater flow to the river, then the volume of groundwater abstracted will be wholly derived from groundwater storage and the resultant area over which drawdown occurs can be very large indeed.

To assess the efficiency of a river augmentation scheme using groundwater and its consequences both on river flows and groundwater levels, natural flows and levels have to be estimated either by correlation with rivers and wells which are outside the pumping area (Wright 1975) or from models

(Oakes and Pontin 1976; Birtles and Reeves 1977). Adequate forecasts of natural conditions require data collected over several years before pumping begins.

The effect of augmentation pumping on river flows can only be measured if the pumping rate is high enough to affect the flow to a measurable extent. If regression equations are used to assess the effects, then the abstraction rate should be over four times the standard error of estimate of the natural river flows (Wright 1975).

Schemes to support downstream river abstraction

Since the late 1960s thirteen schemes for augmenting river flows to support downstream river

Table 3.1   River augmentation schemes using groundwater

Number on Figure 3.2	Scheme	Aquifer	Number of abstraction boreholes	Gross yield ($10^3 m^3 d^{-1}$)	Total length of pipeline (km)	Stage of development
1	Calder	Permo-Triassic	3	10	0.3	commissioned 1984
2	Vale of York	Permo-Triassic	69	380	80	not yet developed
3	Shropshire Stage I	Permo-Triassic	11	40	15	commissioned 1985
	Shropshire Remainder	Permo-Triassic	59	185	58	under development
4	Clwyd	Permo-Triassic	5	14	0.1	commissioned 1979
5	Little Ouse/Thet	Chalk	52	122	67	under development
6	Stour	Chalk	15	80	0	partly operating
7	Bure	Chalk	3	11	0	to be commissioned
8	Waveney	Chalk/Crag	5	38	0	now licenced
9	Deben	Chalk/Crag	2	1	0	awaiting testing
10	West Berkshire	Chalk	33	115	80	commissioned 1975
11	Itchen Stage I	Chalk	6	31	13	commissioned 1985
	Itchen Stage II	Chalk	4	56	20	tested 1989
12	Isle of Wight Stage I	Ferruginous Sands	6	14	2	tested 1989
	Isle of Wight Stage II	Ferruginous Sands	2	4	1	not yet developed
13	Wallers Haven	Hastings Beds	10	18	0	commissioned 1988
23	Hull	Chalk	1	13	4	pilot scheme only
	Schemes for environmental protection					
14	Malmesbury	Gt/Inf. Oolite	13	66	10	commissioned 1980
15	Winterbourne Abbas	Chalk	1	5	2	commissioned 1980
16	Wylye	Chalk/U. Greensand	11	43	6	commissioned 1975
17	Devil's Brook	Chalk/U. Greensand	5	9	0	commissioned 1974
18	Blandford and Pimperne	Chalk	6	12	1	commissioned 1987
19	Tadnol/Empool	Chalk	15	35	0	commissioned 1986
20	Allen	Chalk	4	55	4	commissioned 1978
21	Cam/Rhee	Chalk	13	33	70	commissioned 1986
22	Lodes Granta	Chalk	13	36	39	being promoted
24	Thornton Steward	Millstone Grit	1	5	3	exploration

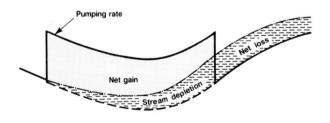

— Observed flow
—·— Estimated natural flow
– – Depleted flow (flow less augmentation)

Fig. 3.3. River augmentation using groundwater—the principle of net gain.

abstractions have been or are being developed. They vary greatly in scale and in the aquifers which are utilized. The two most important aquifers in Britain, the Chalk (Upper Cretaceous) and the Permo-Triassic sandstones, are also those most used for augmentation schemes, but other aquifers have also been developed. These are the Hastings Beds (Lower Cretaceous) in Sussex, the Ferruginous Beds (also Lower Cretaceous) in the Isle of Wight, and the Crag (Pleistocene) in East Anglia.

The largest scheme is the Shropshire groundwater scheme, which, with the commissioning of its second stage in 1991, has 70 boreholes in a catchment of 830 km² with a total pumped output of 225×10^3 m³ d⁻¹ (Severn Trent Water Authority 1977). The Permo-Triassic sandstones are also used in a small scheme by British Nuclear Fuels Ltd in the Calder catchment and are the subject of a large scheme in Yorkshire which has not yet been developed.

The largest completed scheme using the Chalk is the West Berkshire (formerly Thames) groundwater scheme (Hardcastle 1978; Owen 1981). This was the first scheme to be developed in Britain and comprises 33 boreholes with a total pumped output of 115×10^3 m³ d⁻¹. Some 80 km of pipelines conduct the water to four principal river outfalls and several minor ones. In many respects this scheme provided the test case for augmentation schemes in Britain and was authorized following a prolonged public inquiry in 1972.

Several Chalk schemes were developed by the former Anglian Water Authority (Great Ouse River Authority 1972). The first stage of the Great Ouse Scheme, in the Little Ouse and Thet catchments, is licensed and under construction. This scheme has 52 boreholes with a gross yield of 122×10^3 m³ d⁻¹. Several smaller schemes have been or are being developed in East Anglia to utilize the Chalk or the Pleistocene Crag to support downstream abstractions.

Schemes for environmental protection

In groundwater catchments where water supplies are obtained from traditional direct supply boreholes and not from river sources, the abstraction of additional supplies can lead to a further decline in low river flows to environmentally unacceptable levels unless the flows are supported by augmentation. Eleven schemes providing such environmental protection have been or are being developed in Britain.

Six of these schemes have been developed by Wessex Water. The Wylye Scheme in Dorset was the first (Avon and Dorset River Authority 1973) and utilises the Upper Greensand and Chalk aquifers (Cretaceous). This scheme, commissioned in 1975 following extensive field studies, comprises 11 boreholes with a total output of 43×10^3 m³ d⁻¹ of which 19×10^3 m³ d⁻¹ are used to support river flows.

The largest of these schemes in terms of yield is the Malmesbury scheme which uses the Inferior Oolite and Great Oolite aquifers of the Middle Jurassic. It comprises 13 boreholes with a total output of 66×10^3 m³ d⁻¹ of which 26×10^3 m³ d⁻¹ are used to support river flow. The Great Oolite is used for direct supply and the Inferior Oolite for river support as locally the latter aquifer is not directly in contact with the river system.

A third major scheme develops the Chalk in the Empool and Tadnol catchments. It comprises 15 boreholes with a total output of 35×10^3 m³ d⁻¹ of which 12×10^3 m³ d⁻¹ are abstracted for direct supply while up to 23×10^3 m³ d⁻¹ support river flows when these are low.

Groundwater has been used conjunctively with surface water in the Lancashire Conjunctive Use Scheme for many years (Walsh *et al.* 1988). The scheme does not fit easily into any of the three main types identified here. It is operated through a complex set of control rules which generally give a preference to the use of surface water on a cost basis. However, when groundwater is used, there is provision for compensation discharges of groundwater to certain rivers based on prescribed flows.

Principal characteristics of aquifers used for augmentation

Although confined and unconfined aquifers are being used, unconfined aquifers predominate. In southern England, where the Chalk outcrops are free of glacial drift, the unconfined conditions require boreholes at least 5 km from rivers. With the Chalk's high transmissivity (typically 500–2000 m² d⁻¹) and low storativity (0.5 to 2.0 per cent) such large distances are needed if net gains of 50 per cent or more are to be achieved at the end of a pumping period of four to six months. These chalk schemes generally employ boreholes which are 450–750 mm in diameter, 70–130 m deep, and are unlined below the water table. Borehole yields vary greatly but are generally 2–9 × 10^3 m^3 d^{-1} after acidization. In many respects, the Chalk is not ideal for this type of development because the field studies have shown that frequently only the top 5–40 m, of whatever part of the formation is being used, is well fissured and yields water freely.

In contrast, the Permo-Triassic sandstones of central and northern England yield water from their entire thickness of up to 250 m and possess both fissure and intergranular flow. Transmissivities of 100–500 m^2 d^{-1} are lower than for the Chalk whereas storativities of 5–20 per cent are higher. Boreholes are typically 600 mm in diameter, up to 200 m deep and are unlined below the water table. Yields may be 3–7 × 10^3 m^3 d^{-1}.

The contrasting characteristics of the Chalk and the Permo-Triassic sandstones largely determine how they are used for river augmentation. The effects of abstraction from the Chalk develop and dissipate quickly and schemes can be operated relatively frequently without cumulative effects developing. In contrast, the much greater storage of the Permo-Triassic sandstones means that net gains can be very high, but the after effects persist for a considerable period. Consequently, schemes need to be operated less frequently if high net gains are to be obtained.

Environmental effects

Because of the diversity of the areas in which schemes are undertaken the effects of augmentation vary greatly. The first schemes tested twenty years ago led to anxiety amongst farmers over the effects on agriculture of widespread lowering of the water-table. Studies were made which showed that no harmful effects would result and in fact positive benefits could occur in low-lying and waterlogged ground. Nevertheless, in the Great Ouse and Shropshire schemes soil moisture monitoring networks were established to allay this concern.

The potential detriment to fisheries has been an important issue in several schemes, notably the Clywd scheme in north Wales, the Calder scheme in Cumbria, and the Thames and Itchen schemes in southern England. In the last two, extensive fisheries and biological studies were carried out which showed that the effects of low flow augmentation in dry years was likely to be beneficial to the rivers concerned (Southern Water Authority 1978).

Stage 1 of the Great Ouse scheme provides protection to Sites of Special Scientific Interest (SSSI) involving wetlands while similar concern centred on withy beds in spring-fed boggy ground on the Isle of Wight. Extensive commercial watercress beds relying on overflowing artesian boreholes were affected by the Itchen scheme in Hampshire and have required the provision of borehole pumps for their protection.

The diversity of augmentation schemes, in size, aquifer, and environmental setting, means that there is no standard methodology which can be followed in implementing them. Each needs to be studied and planned with great care, with river flow, groundwater level, and environmental monitoring instituted several years before testing is undertaken. Moreover, the local community and interested groups need to be kept informed of progress if major opposition is to be avoided.

Artificial recharge

The use of artificial recharge to replenish groundwater storage for the benefit of enhancing resources can be seen as the ultimate step in aquifer management. Several trials were carried out between ten and twenty years ago (Edworthy *et al.* 1981) but the technique has not received such widespread application as the river augmentation schemes. The continental European practice of using the technique largely as a means of improving water quality has not been adopted in Britain.

Of the four experimental schemes reported by Edworthy *et al.*, only two have become operational, those at Hardham in Sussex and in the London Basin. In addition, borehole recharge trials have been carried out at a single site at Stourbridge in the Permo-Triassic sandstones in the Midlands but as yet there are no plans to go beyond the experimental stage.

London Basin

The most extensive area of interest and activity in Britain in the technique of artificial recharge has been and still is in the London Basin. Detailed hydrogeological desk and field studies (Water Resources Board 1972, 1974a,b) identified considerable potential for the technique over areas totalling several hundred square kilometres; an operational scheme was developed by Thames Water in the Lee Valley between 1975 and 1980 (Hawnt *et al.* 1981). Further investigative work carried out since has led to plans for major expansion into the adjacent Enfield–Haringey area and for detailed appraisal of the potential of a further area in south London (see Fig. 3.4(a) and Connorton 1988a,b).

The two essential requirements for artificial recharge for resource enhancement are a plentiful supply of recharge water and a geological structure that will trap a large proportion of the water once it has been recharged. The London Basin has both these attributes. In addition, in north London, heavy abstraction in the past has lowered levels to the extent that a substantial volume of the confined aquifer has already been dewatered thus providing a natural reservoir ready for filling. In south London the aquifer is full and recharge would follow an initial abstraction phase. Over the basin as a whole, levels are now rising again at a rate of up to 1 m per year (Simpson *et al.* 1989) and are thus adding to the available resource. However, the rate of rise is not fast enough for full use to be made of the resource (Connorton 1988 *b*); hence the need to recharge the aquifer artificially as well.

Geologically, the structure of the London Basin is an asymmetric synclinorium trending WSW–ENE. Over most of the area that will be used the aquifer is a two-layer system. The main element is the top 50 metres or so of the Chalk but with an upward extension into the so-called 'Basal Sands' of the overlying Eocene strata. The Basal Sands have two important effects: they provide an increase in the volume of rechargeable storage and have a significant bearing on water quality. In general, there is a leaky relationship between the Chalk and the Basal Sands and the combined aquifer is concealed except where erosion of anticlinal structures or the presence of scour hollows (Berry 1979, Simpson *et al.* 1989) has resulted in

(a)

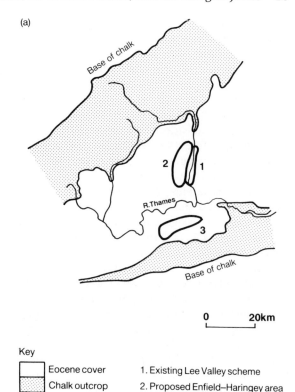

0   20km

Key

	Eocene cover	1. Existing Lee Valley scheme
	Chalk outcrop	2. Proposed Enfield–Haringey area
		3. Proposed south London area

(b)

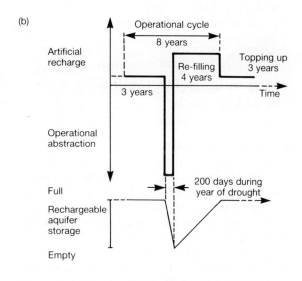

Fig. 3.4. Artificial recharge in the London Basin (a) Location of existing and proposed schemes (b) Schematic representation of idealized operational cycle (after Connorton 1988 *a, b*).

its connection with the surface with important hydrogeological consequences.

The availability of a supply of recharge water is an integral part of the conjunctive use of the scheme in combination with the river-derived sources which provide the greater part of London's supply. During times of off-peak demand, surplus capacity in the distribution system will provide adequate quantities for recharging. In use, the artificially recharged water is seen in principle as a drought resource. Water will be recharged gradually at times of surplus to build up a resource which will be pumped out in years when the stored quantity of river water is inadequate to overcome shortfalls (see Fig. 3.4 (b)). At present the frequency of operation is seen as averaging one year in eight.

In more recent years, the emphasis of work has been on investigation. Temporary limitations on the availability of water for recharge, pending completion of the London Water Ring Main (Keane and Kerslake 1988), has meant that operational recharge has not been possible, illustrating again the interdependence of artificial recharge and other elements of the water supply system. Attention has been given to four major areas of uncertainty that were perceived when the decision was made in 1985 to expand the use of the technique (Connorton 1988a). These are:

- Quantity—potential yields and the response of levels to recharge and abstraction.

- Quality—changes in groundwater quality particularly in relation to the movement of water into and out of the Basal Sands.

- Rising groundwater levels—the impact of artificial recharge on existing rates of rise and the repercussions for the integrity of subsurface structures (investigated as an independent project by the Construction Industry Research and Information Association (Simpson *et al.* 1989)—see Chapter 4).

- Water treatment—the impact of water quality changes on the specification of treatment for water abstracted after recharge.

These issues are to varying degrees interdependent and an integrated programme of work comprising

two main elements was implemented to deal with them, namely a programme of hydrogeological field studies and a set of four interrelated research projects.

The hydrogeological field studies comprised a reappraisal of the basic data on the characteristics of the aquifer and its history of development, and, as a consequence, improvement of the monitoring network, and also testing of exploratory dual-purpose recharge/abstraction sites. These site-specific investigations have been based on a combination of old well stations and recent boreholes and include an assessment of the viability of composite sites consisting of several boreholes a hundred metres or so apart. Such sites are being considered where yields from individual boreholes are expected to be low and large numbers of separate sites many not be practical due to the heavily urbanized nature of much of north London.

One of the characteristics of the old well stations is that when they were constructed groundwater levels were higher than now and their yields were much greater. From work on flow horizons in Chalk boreholes (Owen and Robinson 1978; Younger 1989), the most productive part of the aquifer is the top 30 m. Much of this zone has been dewatered over large areas of north and central London and consequently yields from boreholes and wells are substantially reduced. It is anticipated that when the aquifer has been resaturated by recharge, yields will improve markedly. The analysis of the field investigations is not yet complete but there are strong indications that good yields will be obtained through this mechanism. A contrast between the use of the Chalk aquifer for river augmentation and for artificial recharge is that the former technique uses aquifer storage as the transmissivity and storativity decline vertically whereas the latter technique uses storage as the aquifer properties improve vertically.

Each of the four research projects covers one of the areas of uncertainty noted above. Extensive groundwater flow modelling studies using the finite element method have been carried out (Water Research Centre 1990a,b) to improve understanding of regional groundwater flow and to predict the effects of various recharge and abstraction regimes. The results from this modelling have played an essential part in the research project into rising groundwater levels (Simpson *et al.* 1989) This process is happening slowly irrespective of artificial recharge and the engineering implications are dealt with in greater detail in Chapter 4. However, studying the implications of artificial recharge for this process shows the interdependence of these two research projects.

The other two research projects are also closely interrelated. Earlier work (Edworthy *et al.* 1978; Flavin and Joseph 1983) had identified the possibility of changes to groundwater chemistry, principally high concentrations of sulphate, calcium, and iron under certain circumstances. The Water Research Centre was commissioned to undertake further quality surveys and leaching experiments (Water Research Centre 1987). This work has confirmed that the 'flushing' process of alternately draining and resaturating the Basal Sands could change the quality of recharged and resident water in the aquifer. These changes have implications for the treatment of water that is subsequently abstracted and hence the fourth research project, to look at the specification for this treatment. The use of ultraviolet radiation is being investigated as an alternative to conventional super-chlorination to reduce costs and save land. The presence of iron particles in the water interferes with the effectiveness of UV; thus the quantity of iron present must be assessed as must other parameters which may need to be removed to produce a potable water.

Partly due to the implications of changes in groundwater chemistry for water treatment and partly through concern about changes in the aquifer itself a further major research project in groundwater quality is currently in progress. This research is dealing with:

(1) the mineralogy of the Chalk, Basal Sands, and other Eocene strata and related chemical conditions;

(2) the role of bacteria in reactions;

(3) the consequences of recharging chlorinated mains water;

(4) the development of a geochemical model to describe and predict quality arising from the artificial recharge operation.

If expectations are realized, the use of artifical recharge in north London and south London together could secure some $250–300 \times 10^3 \text{ m}^3 \text{ d}^{-1}$ of water resources for London on the basis of use for approximately 200 days in years of drought. This represents about 10 per cent of current demand in the London area. As such it will be one of the largest groundwater development schemes undertaken in Britain and will have contributed substantially to the body of hydrogeological knowledge on the Chalk aquifer. Using this knowledge, an overall management scheme for the aquifer will be devised once the full potential for artificial recharge is known.

Hardham

Hardham is one of the most thoroughly researched sites for artificial recharge in Britain. The Hardham Basin in Sussex comprises an elongate closed basin of Folkestone Beds (Lower Greensand) 9 km by 3 km, underlain by impermeable strata and filled in the centre by over-consolidated Gault Clay. The Folkestone Beds comprise 40–70 m of clean, medium-grained sands possessing a high transmissivity and an average storativity of 20 per cent. When full, the Basin has a storage of $158 \times 10^3 \text{ m}^3$. Natural recharge to the Basin of $9.9 \times 10^3 \text{ m}^3 \text{ d}^{-1}$ can be enhanced by induced recharge of $15 \times 10^3 \text{ m}^3 \text{ d}^{-1}$ from the Rivers Rother and Arun which flow across the Basin.

Following some early lagoon recharge experiments in the late 1960s, more intensive experiments were conducted between 1972 and 1974 using untreated river water. Three lagoons measuring 160 m by 14 m were used and maximum rates of recharge were found to be $2.6–5.9 \times 10^3 \text{ m}^3 \text{ d}^{-1}$ for each lagoon, with a maximum of $6.8 \times 10^3 \text{ m}^3 \text{ d}^{-1}$ when they were combined (Izatt *et al*, 1979). Regular cleaning of the lagoons was required because of siltation.

In 1976, a full-sized combined abstraction/recharge borehole was constructed and recharge experiments carried out using fully-treated water. The borehole was drilled using reverse circulation to a depth of 63 m below ground level through 15 m of Gault Clay. Drilling was at a nominal diameter of 915 mm but the use of clean water instead of mud resulted in enlargement of the borehole by scouring to a final diameter of 1050 mm. The finished lined diameter was 610 mm incorporating 35 m of Johnson wire-wound screen; 10 m of blank screen were installed to house the abstraction pump and recharge valve outlet. Six backwash tubes and four tremie tubes were inserted in the gravel-packed annulus, the tremie tubes to allow the pack to be topped up.

The recharge experiments carried out in 1980 and 1981 (O'Shea 1984) showed that recharge rates of almost 4.3×10^3 m³ d⁻¹ could be achieved although equilibrium was not attained in the 30 day continuous test. The recharge water displaced the groundwater over a wide front with little mixing taking place. No clogging occurred, although small turbidity peaks were recorded during periods of abstraction. Borehole recharge into the Folkestone Beds was found, therefore, to be easy and no problems were encountered.

Alternate seasonal use of sources

South Downs

While river augmentation from pumped boreholes constitutes an important development in the use of regional groundwater resources in Britain, the use of aquifer storage by seasonal abstraction has also been used to great effect. Probably the best example of this lies in the aquifer management policy adopted in the South Downs of Sussex (Headworth and Fox 1986).

The Chalk is the main aquifer in Sussex and has been exploited for 150 years to provide public water supplies to the major coastal towns of which Brighton is the largest with a population of 250 000. The Chalk outcrop of the South Downs extends for a distance of 90 km yet it is rarely more than 12 km wide. At its eastern end it abuts directly on the sea and forms spectacular coastal cliff scenery.

Saline intrusion in coastal granular aquifers commonly exhibits the classic wedge-shaped form, described by the Ghyben–Herzberg equation, in which a brackish zone deflects freshwater outflows to the sea. In fissured aquifers such as the Chalk, there is no simple saline interface and over-abstraction can induce seawater intrusion into the aquifer along discrete fissure zones, often for considerable distances (see Fig. 3.5.).

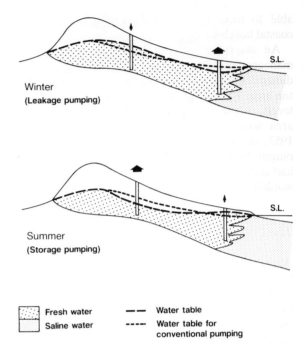

Fig. 3.5. Alternate use of groundwater sources in the Brighton area to control saline intrusion.

By 1957, the volume of groundwater abstracted in the Brighton area had reached 63×10^3 m³ d⁻¹ and had more than doubled since the beginning of the century. Increasing salinities in several of its sources led the then Brighton Corporation to introduce an abstraction policy which sought to conserve inland aquifer storage and place greater emphasis on intercepting coastal outflows which would otherwise be lost to the sea. Prior to that time, abstraction had been largely piecemeal and not integrated.

The aquifer management policy centres on making the maximum use of boreholes located along the coastal margin during the winter months when aquifer recharge and outflows occur. At the same time, reduced water demand permits pumping from inland sources to be cut back so as to allow aquifer storage to recover. As the spring and summer progress, abstraction from the coastal boreholes is reduced and more is pumped from the inland sources where groundwater levels have been allowed to build up. In drought years, following winters of below-average recharge, coastal outflows decline but inland aquifer storage is better able to meet the shortfall in abstraction from coastal boreholes.

An assessment of the benefit of the aquifer management policy has been made using a two-dimensional finite-difference model of the Brighton area. For steady-state conditions, groundwater levels, when averaged over the whole 193 km^2 area, were 0.6 m higher in the early 1980s than in 1953, despite the 33 per cent increase in total output during that time. If no increase in pumping had occurred the difference in mean water levels would have been 1.6 m.

The Cotswolds

Alternate use of different groundwater sources has been practised in the Jurassic limestones in the Cotswolds around Cirencester since 1980, the objective in this case being to limit depletion of river flow. The area covers several tributaries of the upper Thames and is immediately adjacent to the upper reaches of the Bristol Avon where the Malmesbury Scheme, mentioned above, is located. The limestones are the source of supply for the rural area of the North Cotswolds and also part of the supply for the nearby expanding town of Swindon. At the same time the Cotswolds are recognized as one of the outstanding scenic areas in Britain, one in which rivers and streams are an essential part of the landscape. The abstraction of groundwater must therefore be carefully managed to maintain the river environment.

The Jurassic limestones of the North Cotswolds consist of two aquifers, the Inferior Oolite and the Great Oolite separated by the Fullers Earth Clay. There is negligible diffuse connection between the aquifers through the clays but occasional discrete connections through fault systems are now strongly suspected in the upper Thames catchment.

In 1981, groundwater abstraction in the area was increased by 18 × 10^3 m^3 d^{-1} based on four individual sources, two in each aquifer. From pumping tests it was clear that the two Inferior Oolite sources, Baunton and Meysey Hampton (IO), had an immediate depletion effect on the flows of the rivers Churn and Coln respectively. Consequently increased abstraction was permitted only when flows were above those prescribed for the relevant river. The Great Oolite sources (Meysey Hampton

(GO) and Ashton Keynes) on the other hand did not affect river flow within the duration of pumping tests and it was concluded that in the short term they drew on storage. Hence increased abstraction was allowed from these sources when the Inferior Oolite sources were restricted by river flow constraints. When the Inferior Oolite sources were not restricted those in the Great Oolite were, to allow recovery of storage.

The area is hydrogeologically complex and at the time of authorizing the increases it was considered that there was inadequate knowledge to predict with certainty the long-term effects of this mode of operation on the Great Oolite aquifer. The increases from that aquifer were therefore initially permitted only for a period of seven years. During this time, two major pieces of investigatory work were carried out. The geology of some 300 km^2 was remapped to modern standards to provide more detailed information on the lateral and vertical distribution of strata. This work then formed part of the basis of a groundwater flow model for the Great Oolite (Rushton *et al.*, in press).

These two pieces of work modified interpretation of the behaviour of faults and recharge through the Forest Marble formation overlying the Great Oolite aquifer. In relation to the alternate use scheme, the modelling specifically showed that the effective groundwater catchment to the Ashton Keynes source is far smaller than previously believed and cannot sustain the increased abstraction without unacceptable environmental effects. This element of the scheme has therefore been dropped. The other Great Oolite source, Meysey Hampton, continues on a further trial basis; modelling indicates that the yield is available but that significant river flow depletion will occur.

As limitations on the yield of the groundwater sources in the Cotswolds became apparent, the area has been linked by trunk main to pre-existing downstream surface storage near Oxford. Here the main component of flow is groundwater discharge from the limestones and winter surpluses can be stored. Conjunctive use of the resources is developing whereby cheaper, better quality groundwater is used for direct supply in the Cotswolds when local river flows permit, otherwise the downstream storage is used, with the benefit of recirculation through effluent return.

Groundwater quality and resource development

Introduction

Two aspects of groundwater quality are crucial to the regional development of resources. These are:

(a) the processes that determine the natural or indigenous quality;

(b) the impact of diffuse pollution.

Quality is often an important tool in the investigation and design of resource development schemes and in their ultimate management to protect or improve existing quality.

Groundwater quality plays an important part in identifying groundwater flow regimes either those which exist naturally or those which are induced by abstraction. Quality measurements carried out during test pumping are an essential adjunct to the physical measurements needed to identify yield characteristics. On a more regional basis, long-term measurements of some trace constituents of groundwater quality will give some indication of regional flow patterns induced by abstraction. The use of the radioactive isotopes tritium and carbon, and the stable isotopes of oxygen and hydrogen have proved particularly useful in this context (Downing *et al*. 1979). Their time-based concentrations allow an estimation of flow times and this in turn can be used to predict potential derogation problems either to other groundwater sources or to groundwater fed rivers and springs.

Geochemical processes

The establishment of natural groundwater quality is primarily a function of soil type, aquifer type, mineralogy, hydraulic character, residence time, and whether the aquifer is confined or not.

The many differing aquifer types in Britain are overlain by an even larger variety of soil types. It is the soil zone which imprints the initial quality on recharge water. Except for thin soils on some chalk or limestone areas, most soils in Britain are acidic in nature and even those which contain an appreciable amount of alkaline material will be gradually neutralized by the generally acidic nature of rainfall. The major British aquifers, that is the Carboniferous and Jurassic limestones, the Chalk, the Permo-Triassic and Cretaceous sandstones, and the Quaternary fluvial gravel deposits, contain variable amounts of carbonate material. This carbonate, normally present as calcium carbonate, effects a significant buffering on any acidic recharge percolating through the overlying soils. This is particularly true for the limestones and the Chalk which consist predominantly of carbonate. The resultant water quality in these strata is therefore usually alkaline with a pH of 7–8 and a high bicarbonate content, that is hard water. Typical analyses of limestone and Chalk waters are shown in Table 3.2 (samples 3 and 5).

However, some sandstones and gravels may not contain high proportions of carbonate and hence will not offer the same degree of buffering control. In the Permo-Triassic sandstones and Cretaceous Lower Greensand some waters exhibit low pH values and are soft. These areas in particular are susceptible to the effects of acid rain (Kinniburgh and Edmunds 1984) and may as a result contain undesirable concentrations of dissolved species such as iron or potentially toxic metals. Any local water undertaker using these aquifers must budget for the appropriate treatment costs. Analyses of low alkalinity waters taken from these strata are also given in Table 3.2 (samples 1, 2, and 7).

Significant changes can take place to water quality upon confinement beneath impermeable strata or with increasing depth. Geochemical processes related to the removal of oxygen from the water cause the pH to change thus altering the solubility controls on many mineral species. Such processes involve ion exchange reactions and nitrate and sulphate reduction. It is not uncommon to note high concentrations of dissolved iron and manganese in these situations. Their abundance will be related to their natural presence within the rock matrix. An example of this is an unusually high concentration of manganese which has caused many problems with discoloration and deposits in the supply system (Morgan-Jones 1983).

At greater depth in the confined aquifers the presence of connate water, that is water entrapped since the rocks were laid down and subjected to slow geochemical evolution, has been identified. This water is usually saline and may contain high

Table 3.2 Chemical analyses of various groundwaters

		pH	Eh	O$_2$ %	Ca	Mg	Na	K	HCO$_3$	SO$_4$	Cl	NO$_3$-N
					\multicolumn{8}{c}{Major constituents mg l^{-1}}							
Permo-Triassic	1. Bourne Vale	7.0	—	—	69	7	15	5.0	89	69	34	12
sandstones	2. Clumber Park	6.5	—	—	35	16	10	5.6	88	64	24	2.7
Chalk	3. Brightwalton	7.0	390	70	116	2.0	7.0	1.3	328	4.9	9	4.8
	4. Mortimer	7.5	41	0	52	10.3	80	5.1	268	34	72	0.5
Jurassic	5. Baunton	7.6	305	73	100	4.3	8.1	1.2	244	38	14	6.9
limestones	6. Lechlade Mill	8.5	−140	1	12	5.0	270	4.6	380	7.2	178	0.5
Lower	7. Tilford	5.9	442	64	30	1.5	9.5	1.7	34	12	17	1.3
Greensand	8. Slough Estate	7.9	105	0	26	3.5	124	4.3	239	61	60	0.1

concentrations of minerals and gases which make it unusable for supply purposes. The geochemical processes taking place downgradient in confined aquifers have been identified by a number of authors (Ineson and Downing 1963; Edmunds 1973; Morgan-Jones and Eggboro 1981; Morgan-Jones 1985; Edmunds *et al.* 1987). Good examples of these reactions occur in the Chalk and Lower Greensand of the London Basin and in the Jurassic limestones of Lincolnshire and Gloucestershire (see samples 4, 6, and 8 in Table 3.2). A typical downgradient geochemical profile in a limestone overlain by clay is shown in Fig. 3.6.

Saline intrusion

Large tracts of the coastline of Britain are made up of permeable strata in direct contact with the marine or estuarine environments. Major groundwater abstraction points located close to the coast are susceptible to saline intrusion if they are overpumped. This has occurred along the Thames estuary, the Mersey estuary, and in the Lincolnshire, Kent, and Sussex coastal areas (Lloyd 1981).

Saline water is generally described as having a chloride content in excess of 300 mg l^{-1} (related to acceptable potability levels) and hence it does not

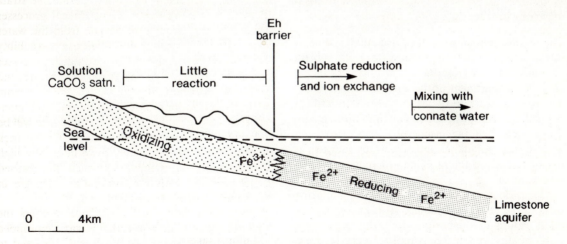

Fig. 3.6. Schematic diagram showing relationship between oxidizing and reducing conditions in confined and unconfined limestones (after Edmunds 1973).

take much intrusion of sea water (with about 19 000 mg l⁻¹ Cl) to cause deterioration. The actual movement of saline water into the freshwater environment is difficult to predict due to the complexity of mixing related to pumping rates and density variations between the two water types. Freshwater, being less dense than saline, will constitute the upper body of water. Provided the density difference can be maintained there will be little upward movement of saline water. However, overpumping may destroy this balance, causing upconing and a resultant deterioration in quality. Time based profiles of such deterioration have been produced for abstractions from the Chalk on the south coast (see Fig. 3.5.) and in the sandstones adjacent to the Mersey estuary.

Clearly pumping regimes must be established in such areas which produce a balance between recharge and abstraction rates. Any short term demands for overpumping must be weighed against the long-term damage that may ensue. The aquifer could be written off as a supply source for many years if sensible control is not applied. An established control scheme in the Chalk of the Sussex coast has already been mentioned. Major investigations, including substantial use of modelling, have been carried out on the Chalk of Lincolnshire to develop a comprehensive management scheme for the aquifer including control of saline intrusion (University of Birmingham 1987).

Diffuse pollution

The relationship between land use and leaching of nitrate and pesticides to groundwater is the subject of much concern (Kendrick *et al.* 1985; Foster *et al.* 1986; Department of the Environment 1988*b*). The need to work to rigorous quality standards as defined by European legislation has thrown into sharp focus the increasing problems associated with diffuse pollutants and the water industry is having to address the problem as a matter of long-term control.

The various aquifer types respond in differing ways to the effects of land use. It is known that in many well-fissured limestone areas the response of water quality to recharge is rapid. Hence land use controls applied at the surface should have an effect in a reasonably short time. However, in

some areas of the Chalk and in sandstones, responses to changes at the surface may take many years to manifest themselves at the water table. In these areas, in particular, the water undertakers must consider either on-site treatment, such as nitrate removal, or resource blending with better quality water if they are to avoid losing the source.

The reduction of nitrate downgradient below confining beds can be of critical relevance in some areas. It is possible to use a proportion of this low nitrate water and blend it with a high nitrate water at outcrop thus ensuring the supply standards are maintained whilst continuing to optimize the yield of the aquifer. However, great care must be taken not to induce the downgradient movement of water with high nitrate contents by overpumping the confined zone as this will destroy the concept. A management scheme following this principle, and which also deals with elevated concentrations of iron and manganese, is being developed in South Yorkshire (R. J. Aldrick, pers. comm.).

The same principles of control apply to other diffuse pollutants, particularly pesticides, which are of considerable concern because of their potentially carcinogenic effects. It has been considered that commonly used pesticides, such as simazine, atrazine, and isoproturon, either degrade rapidly in the soil zone soon after application or are bound up by adsorptive processes and become immovable. However, analyses of groundwaters from a diversity of areas show that this is clearly not the case and it is therefore necessary to effect a change in land use practice or pesticide type to ensure that the situation does not become unmanageable. A list of pesticides found in some Chalk groundwaters is given in Table 3.3.

Another potential source of pollution of aquifers constraining use of the resource is contaminated land. In the industrialized areas of Britain the soil zone has been seriously contaminated by inorganic and organic residuals from such activities as gas works, plating works, refineries, mining, and other industrial processes, many of which ceased activity a long time ago. Once these areas are disturbed by development and are opened up to the atmosphere, the possibility of downward leaching of contaminants by infiltration of rainfall exists. Many old industrial sites are located in river valleys and are in close proximity to both surface water courses and groundwater supplies. If the surface and

Table 3.3 Pesticides identified in Chalk ground-water at concentrations above drinking water limits

---

* Chlortoluron
* Atrazine
  Carbetamide
* Simazine
  Propyzamide
  Tridimefon
* Isoproturon
  2,4-D
  Prochloraz
  Diuron
  Phosalone

---

* = common occurrence

groundwater systems are in hydraulic continuity, as is often the case, there may be two pathways by which a supply could become contaminated, firstly, by direct leaching into the aquifer or, secondly, via polluted run off into the adjacent watercourse.

There are areas of Britain where the extent of industrial contamination has been so severe that the aquifer is already seriously contaminated. Such a situation of predominantly organic pollution by chlorinated hydrocarbons has been described in the Permo-Triassic sandstones under Birmingham (Rivett *et al.* 1990; see Chapter 9). The level of pollution has resulted in the effective abandonment of part of the aquifer for potable supply since the remedial measures which would be required to treat the water, such as air stripping or carbon activation, would be prohibitively expensive in terms of the benefit to water supply.

Aquifer protection.

A knowledge of the processes controlling water quality and the situations which can lead to deterioration are clearly a vital part of the overall management of the aquifer system. The increasing awareness of the vulnerability of groundwaters to pollution has resulted in the concept of the aquifer protection policy. These policies or guidance documents have been used by the regulatory authorities in Britain for the past few years.

In essence the policies employ a two tier approach to the problem of resource protection.

On a local scale the immediate catchment of each public supply abstraction is identified and this then acts as the protection zone for that source. Certain known potentially polluting activities are identified in the policy and it is made clear that objections will be raised to their presence within the zone.

On a regional basis the areas of aquifer outcrop and recharge are identified and risk factors ascribed to them. Detailed scrutiny of potential polluting activities are undertaken in order to protect the resource. However, on a regional basis the approach must of necessity be more pragmatic than that adopted for the local zonal control. Recent legislation, as enacted in the Water Act (1989), has given strength to this means of groundwater source protection by providing for the establishment of mandatory water protection zones. The National Rivers Authority in England and Wales will be able to define a nationally enforceable approach to aquifer protection in which certain activities may well be proscribed within protection zones.

A special case of a water protection zone is a nitrate sensitive area. In this case, a problem of diffuse pollution, land use control is applied within a designated catchment zone around a source which either exceeds or may soon exceed the European Standard for nitrate in drinking water (that is 50 mg l^{-1} as NO_3). It is likely this concept may also be applied to other forms of diffuse pollution such as pesticides. There is little doubt that the tightening of controls on discharges into the ground and the more rigorous application of quality standards to groundwaters will necessitate increased monitoring in order to satisfy the needs of the new and revised European Quality Directives. A National Aquifer Protection Policy will play an important role in this process.

The future

Various types of multiple source groundwater development schemes have permitted the extension of the use of the groundwater resource over the past two decades. Several major schemes, particularly the augmentation schemes in Shropshire and East Anglia and the artificial recharge schemes in the London Basin, are currently being implemented but are not yet complete. Artificial recharge is the ultimate step in aquifer management and has the attraction of using water surpluses, a distinct advantage in this environmentally conscious era. The costs and logistics of delivering the surplus water to the recharge areas in a chemically compatible state is one of the major issues facing further use of this technique together with very thorough hydrogeological investigations.

Without multiple source schemes, further use of groundwater would have been very restricted. New direct supply abstractions in small catchments during this period have been very limited in number and will continue to be so unless linked with compensation measures for environmental protection. Indeed there is a strong move now through the National Rivers Authority to investigate such catchments where long-standing abstractions have caused river flow depletion and to implement schemes to alleviate the environmental damage that has resulted. These alleviation schemes will attempt to maintain existing abstractions but in some cases the only solution may be to reduce them and seek new resources.

However, there have been, and will continue to be, some opportunities for direct supply abstractions where they are close to major rivers in which the flow depletion is limited by recirculation of effluents or is environmentally acceptable by virtue of the size of the river. Such abstractions are in effect indirect river abstractions and make little use of storage. Abstractions near the sea may also be acceptable (Houston *et al.* 1986) providing they do not cause saline intrusion.

A feature of increasing concern in many British aquifers is the deteriorating trend in quality in relation to diffuse contamination from nitrate and pesticides in rural areas and from trace organics in urban areas. In the short-term, and possibly in the long term also, these trends may have major cost implications through the need for treatment to remove contaminants.

Two economic advantages of regional groundwater schemes over surface water schemes have been held to be lower unit cost and the facility for staged implementation to match growth in demand. The latter attribute is still valid but unit costs, through the need for increased environmental protection and water treatment, will not be as universally low as once thought. The selection of water resources developments in the future may be influenced as much by the principle of 'best possible environmental option' as on cost alone. In this respect, groundwater has no overriding advantage over surface water except perhaps in the technique of artificial recharge.

Acknowledgements

The authors wish to thank Mr B. J. Connorton of Thames Water plc for material supplied on the artificial recharge schemes in the London Basin. The views expressed in this publication are the views of the authors and not necessarily the organizations to which they are affiliated.

References

Avon and Dorset River Authority (1973). *The Upper Wylye investigation*. Bournemouth, UK.

Berry F. G. (1979). Late Quaternary scour-hollows and related features in central London. *Q. J. Eng. Geol.*, **12**, 9–29.

Birtles, A. B. and Reeves, M. J. (1977). Computer modelling of regional groundwater systems in the confined—unconfined flow regime. *J. Hydrol.*, **31**, 91–127.

Commission of the European Communities (1982). *Groundwater resources of the United Kingdom.* (EUR 7946 EN), Brussels.

Connorton, B. J. (1988a). Artificial recharge in the London Basin. In *Special Subject No. 13, Recent developments in artificial recharge*, Int. Water Supply Assoc., Rio de Janeiro congress.

Connorton, B. J. (1988b). Water resources management and rising groundwater levels in the London Basin, (ed. E. A. Kozlovsky), Vol. 1, pp. 139–148. *Water management and the geoenvironment*, UNESCO. Paris.

Department of the Environment (1988a). *Assessment of groundwater quality in England and Wales.* HMSO, London.

Department of the Environment (1988b). *The nitrate issue*. HMSO, London.

Downing, R. A., Pearson, F. J., and Smith, D. B. (1979). The flow mechanism in the Chalk based on radioisotope analysis of groundwater in the London Basin. *J. Hydrol.*, **40**, 67–83.

Downing, R. A., Ashford, P. L., Headworth, H. G., Owen, M., and Skinner A. C. (1981). The use of groundwater for river augmentation. In *A survey of British hydrogeology 1980*, pp. 153–72. The Royal Society, London.

Edmunds, W. M. (1973). Trace element variations across an oxidation—reduction barrier in a limestone aquifer. *Proc. Symp. Hydrogeochem. Biogeochem.*, Tokyo 1970, 500–26.

Edmunds, W. M., Cook, J. M., Darling, W. G., Kinniburgh, D. G., Miles, D. L., Bath A. H., Morgan-Jones, M. J., and Andrews, J. N. (1987). Baseline geochemical conditions in the Chalk aquifer, Berkshire, UK: a basis for groundwater quality management. *App. Geochem.*, **2**, 251–74.

Edworthy, K. J., Stott, D. A., and Wilkinson, W. B. (1978). Research into the physical and chemical effects of artificial recharge in the Lea Valley, London. *Water Resources Bull.*, **14**, 554–75.

Edworthy, K. J., Headworth, H. G., and Hawnt, R. J. E. (1981). Application of artificial recharge techniques in the United Kingdom. In *A survey of British hydrogeology 1980*, pp. 141–52. The Royal Society, London.

Flavin, R. J. and Joseph, J. B. (1983). The hydrogeology of the Lee Valley and some effects of artificial recharge. *Q. J. Eng. Geol.*, **16**, 65–82.

Foster, S. S. D., Bridge, L. R., Geake, A. K., Lawrence, A. R., and Parker, J. M. (1986). *The groundwater nitrate problem*. Hydrogeol. Rep. No. 86/2. British Geological Survey, Keyworth.

Great Ouse River Authority (1972). *Great Ouse groundwater pilot scheme. Final report*. Cambridge, UK.

Halcrow (Sir William Halcrow and Partners) 1988. *Study of alleviation of low river flows resulting from groundwater abstraction*. Thames Water Authority, Reading, UK.

Hardcastle, B. J. (1978). From concept to commissioning. In *Thames groundwater scheme*. Instn Civ. Engrs, London.

Hawnt, R. J. E., Joseph, J. B., and Flavin, R. J. (1981). Experience with borehole recharge in the Lee Valley. *J. Instn Water Engrs*, **35**, 437–51.

Headworth, H. G. and Fox G. B. (1986). The South Downs Chalk aquifer: its develpment and management. *J. Instn Water Engrs Sci.*, **40**, 345–61.

Headworth, H. G., Owen, M., and Skinner, A. C. (1983). River augmentation schemes using groundwater. *British Geologist*, **9**, 50–4.

Houston, J. T. F., Eastwood, J. C., and Cosgrove, T. K. P. (1986). Locating potential borehole sites in a discordant flow regime in the Chalk aquifer at Lulworth using integrated geophysical surveys. *Q. J. Eng. Geol.*, **19**, 271-82.

Ineson, J. and Downing, R. A. (1963). Changes in

the chemistry of groundwaters of the Chalk passing beneath argillaceous strata. *Bull. Geol. Surv. Gt. Brit.*, **20**, 176–92.

Ineson, J. and Downing, R. A. (1964). The groundwater component of river discharge and its relationship to hydrogeology. *J. Instn Water Engrs*, **18**, 519–41.

Izatt, D., Fox, G. B., and Tague, M. (1979). Lagoon recharge of the Folkestone Beds at Hardham, Sussex, 1972–75. *J. Instn Water Engrs Sci.*, **33**, 217–36.

Keane, M. A. and Kerslake, J. C. (1988). The London water ring main: an optional water supply system. *J. Instn Water Env. Sci.*, **2**, 253–66.

Kendrick, M. A. P., James, H. A., Clark, L., Gibson, T. M., Baxter, K. M., Turrell, M. B., and Fleet, M. (1985). *Trace organics in British aquifers*, Lab. Rep. 823-M. Water Res. Centre. Medmenham, UK.

Kinniburgh, D. G. and Edmunds, W. M. (1984). *The susceptibility of UK groundwaters to acid deposition*. Hydrogeol. Rep. No. 86/3. British Geological Survey, Keyworth.

Lloyd, J. W. (1981). Saline groundwaters associated with fresh groundwater reserves in the United Kingdom. In *A survey of British hydrogeology 1980*, pp. 73–84. The Royal Society, London.

Morgan-Jones, M. (1983). *Investigations into high concentrations of manganese in Chalk groundwater at Taplow*. Res. Rep. N/1, Thames Water Authority, Reading, UK.

Morgan-Jones, M. J. (1985). The hydrogeochemistry of the Lower Greensand aquifers south of London. *Q. J. Eng. Geol.*, **18**, 443–58.

Morgan-Jones, M. and Eggboro, M. D. (1981). The hydrogeochemistry of the Jurassic limestones of Gloucestershire, England. *Q. J. Eng. Geol.*, **14**, 25–40.

National Rivers Authority (1990). *Extent and impact of over-abstraction upon river flows*. London.

Oakes, D. B. and Pontin, J. M. (1976). *Mathematical modelling of a Chalk aquifer*. Tech. Rep. 24. Water Res. Centre, Medmenham, UK.

Oakes, D. B. and Wilkinson, W. B. (1972). *Modelling of groundwater and surface water systems. I— Theoretical base flow*. Water Resources Board, Reading, UK.

O'Shea, M. J. (1984). Borehole recharge of the Folkestone Beds at Hardham, Sussex. 1980–81. *J. Instn Water Engrs Sci.*, **38**, 9–24.

Owen, M. (1981). Thames groundwater scheme. In *Case-studies in groundwater resources evaluation* (ed. J. W. Lloyd), pp. 186–202. Clarendon Press, Oxford.

Owen, M. and Robinson, V. K. (1978). Characteristics and yield in fissured chalk. In *Thames groundwater scheme*. Instn. Civ. Engrs., London.

Rivett, M. O., Lerner, D. N., Clark, L., and Lloyd, J.W. (1990). Organic contamination of the Birmingham aquifer, UK. *J. Hydrol.*, **113**, 307–23.

Rushton, K. R., Owen, M., and Tomlinson, L. M. (In press.) The water resources of the Great Oolite aquifer. *J. Hydrol.*

Salmon, S. (1990). The significance of vertical components of flow in groundwater with special reference to the Bromsgrove aquifer. Unpublished Ph.D. thesis. University of Birmingham.

Severn Trent Water Authority (1977). *Shropshire groundwater. Report of the investigation and proposals for development*. Birmingham, UK.

Simpson, B., Blower, T., Craig, R. N., and Wilkinson, W. B. (1989). *The engineering implications of rising groundwater levels in the deep aquifer beneath London*. CIRIA, Special publication 69, London.

Southern Water Authority (1978). *Final report of the Candover pilot scheme*. Worthing, UK.

University of Birmingham (Department of Civil Engineering) (1987). *North and south Chalk modelling study. Final report*. Anglian Water, Huntingdon, UK.

Walsh, P. D., Walker S., and Pearson, D. (1988). Derivation of operating policies for surface-water sources in North West Water. *J. Instn. Water Env. Mgmt.*, **2**, 51–9.

Water Research Centre (1987). *Groundwater quality in the London Basin*. Thames Water Authority, Reading UK.

Water Research Centre (1990a). *Groundwater model of the Enfield–Haringey Area. Final Report.* Thames Water plc, Reading UK.

Water Research Centre (1990b). *Groundwater model of the London Basin*. Thames Water plc, Reading UK.

Water Resources Board (1972). *Artificial recharge of the London Basin—I Hydrogeology*. Reading, UK.

Water Resources Board (1974a). *Artificial recharge of the London Basin—III Economic and engineering desk studies*. Reading, UK.

Water Resources Board (1974b). *Artificial recharge of the London Basin—IV Pilot recharge works in the Lea Valley*. Reading, UK.

Wright, C. E. (1975). The assessment of regional groundwater schemes by river flow regression equations. *J. Hydrol.*, **26**, 209–15.

Younger, P. L. (1989). Devensian periglacial influences on the development of spatially variable permeability of the Chalk of south-east England. *Q. J. Eng. Geol.*, **22**, 343–54.

4. Rising groundwater levels—an international problem

W. B. Wilkinson and F. C. Brassington

The fall in groundwater levels

Readily available and good quality groundwater has made a major contribution to the growth of industry and commerce in many major cities. During the nineteenth and early twentieth centuries groundwater development was rapid in those European cities below which there was such a resource. Many thousands of wells and boreholes were sunk. A similar pattern is seen in North America although the major urban growth generally occurred later than in Europe. The heavy abstraction of groundwater often exceeded natural recharge to the underground strata and this led to a substantial fall in groundwater levels. The deficit between the amount of groundwater pumped and the natural recharge was met by water drawn from aquifer storage (Wilkinson 1984, 1985, 1986).

In the United Kingdom the two principal strata that have been used for groundwater development are the Chalk and the Permo-Triassic sandstones. There are also many minor aquifers which provide valuable water supplies. The outcrop of the principal aquifers is shown in Fig 4.1. The figure shows that London is located on Chalk and Birmingham, Liverpool, Manchester, and Nottingham on the Permo-Triassic strata. These cities have drawn heavily on the underlying aquifers in the past and, indeed, all still make use of these water sources. Due to the substantial abstraction in the late 1800s and early 1900s, a situation was reached in the 1940s where the water level had fallen some tens of metres below ground level over large areas beneath several of these cities. As the groundwater pumping proceeded, natural springs dried up and boreholes which had once naturally overflowed to the surface ceased to do so. Pumps were necessary to give continued use of these 'artesian' wells. The exact sites of many of the springs were lost and indeed many were built over as the cities spread. However, their former presence is often recorded in the old place names.

As the demand for sites in the centre of the cities rose, so larger buildings were constructed with deeper basements and large raft or piled foundations below the original, natural rest water level. Transport and cable tunnels, sewers, and water pipes were all built, again often below the original rest water level. The site investigations for such structures may have revealed that the groundwater level at the time was several metres below the base of the proposed structure. Furthermore, the records may have indicated that it had been at, or close to, this level for possibly one or two decades or even longer. In some cases the assumption was made either that these were natural groundwater levels or that they would remain depressed indefinitely into the future. In any event, it was assumed that the groundwater level could be safely ignored in the geotechnical design of the structure. This assumption would have been borne out during the construction phase. Thus the long term, large scale dewatering of these aquifers below many cities led to an often unrecognized benefit in the design and construction of the foundations of some large buildings and other underground structures at city sites.

The rise in groundwater levels

Where groundwater abstraction is reduced, the water-table will begin to rise and with time a new water-table and groundwater-flow regime will be established. If pumping is stopped then the

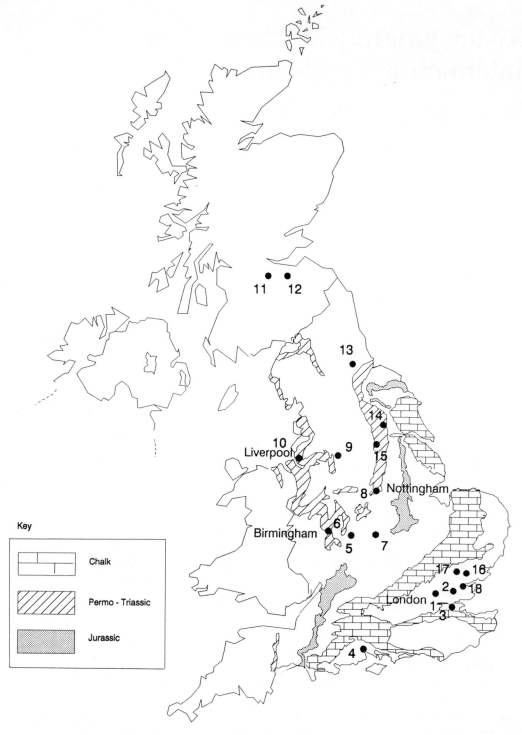

Fig. 4.1. Outcrop of the principal aquifers in the UK and locations of sites where a reduction in groundwater abstraction is leading to a rise in the water-table. (see Table 4.1).

groundwater system will move towards the natural, pre-abstraction condition. The groundwater component of river flow will increase as will spring flow and those springs that had been dry for many years and possibly forgotten will re-emerge. The rising groundwater may be of poor quality. A particular problem has arisen from the reduction or cessation of dewatering pumping following closure of coal mines. The groundwater emerging from the mine workings may often have high iron and sulphur contents resulting from the oxidation of iron pyrites in the coal-bearing rocks (Henton 1978).

Over the last ten to fifteen years a progressive rise in water levels has been observed below several of the major cities in the UK. This has already caused some problems in Birmingham (Hurst and Wilkinson 1985; Lloyd and Lerner 1986) and Liverpool (Brassington and Rushton 1987; Rushton *et al.* 1988) and the rate of rise in the aquifer below London is of concern. The principal cause for the rise is the reduction in groundwater abstraction in recent years from the central areas of the cities. The main factors contributing to this were:

- the destruction of large areas of the inner cities during World War II which resulted in the loss of many wells;

- licensing controls under the 1945 Water Act which reduced the number of new wells being constructed;

- the deterioration in the quality of groundwater due to pollution entering aquifers from the surface or by saline intrusion from an estuary.

Figure 4.2 shows the quantity of groundwater abstracted for public supply and for industry in England and Wales for the period 1974–1987 (Brassington 1990). During this period there was a decline of more than 30 per cent in the use of groundwater by industry and an increase in public supply which balanced the decline. These changes in groundwater abstraction caused a redistribution in the pattern of groundwater pumping giving a reduction in the city centres and an increase, mainly in public supply abstraction, in the suburbs and the rural areas. Over-abstraction has also occurred in a number of areas remote from the cities and this has led to the same changes in spring

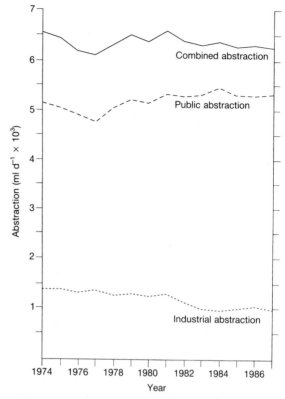

Fig. 4.2. Quantity of groundwater abstracted for public supply and for industry in England and Wales for the period 1974–1987.

and river flow and water quality that are described above.

Another cause of rising water levels in some areas may be the leakage from the many miles of sewers and water mains that criss-cross cities (Price and Reed 1989). The percentage loss of drinking water by leakage from mains over the country as a whole is about 30 per cent. It could be appreciably larger than this in the centres of the industrial cities due to the age of the mains. If the superficial strata in which the mains are located are of low permeability, i.e. clayey in nature, then this lost water will find its way into sewers or surface water drains and the recharge to the underlying water-bearing strata, and consequently, the rise in groundwater levels, will be unaffected. However, if the mains are in highly permeable surface deposits which are in good hydraulic contact with the aquifer the enhancement of recharge and the associated rise in water levels could be significant. Similar consider- ations apply to the leakage losses from sewers which are above the water-table. This loss/recharge is generally very difficult to quantify. The natural recharge, in contrast to the contribution from water mains and sewers, may have declined due to the extensive paved areas and roof drainage in cities, and much of it is collected by surface-water drainage rather than contributing, where the geol- ogy permits, to refill the aquifer.

In some cases, the rises in water level have been sufficiently dramatic to capture the attention of the media. In Birmingham the flooding of a number of basements has occurred and in Liverpool the rate of pumping from a British Rail tunnel has had to be increased. A rise in general water levels in Nottingham has also been reported. There has been a rapid and sustained rise in groundwater

levels in the aquifer below London (Marsh and Davies 1983). The maximum rate is about 1.5 m a^{-1}. The water table is still very deep in Central London but modifications to the design of the foundations of the new British Library have been made to resist an elevated water-table at some future date. An increase in the inflow of water to the basements of buildings founded in the Thames river gravels has also been observed recently and this may be due to a rise of the water- table in these deposits.

Brassington (1990) identified situations in the UK where a reduction in groundwater abstraction is leading to a rise in the water-table. Table 4.1 lists these occurrences and gives brief descriptions of the geological settings and the potential prob- lems. The locations of the sites given in Table 4.1 are shown in Fig. 4.1.

Table 4.1 Summary of UK case histories

Location of aquifer (numbers refer to Fig. 4.1)	Extent of rise	Cause	Consequences
1. *London* Chalk and London Tertiaries	Rise of 20m between 1968 and 1985	Groundwater abstraction reduced by 46 per cent (106 Ml d^{-1})	Ground levels may rise and lead to differential movement in buildings. There is the possibility of flooding of tunnels and basements and also of instability of some building foundations.
2. *Tilbury* Chalk	Minor	Cessation of de-watering when deep quarrying finished	Quarries flooded and pumping re-started to protect new industrial buildings
3. *North Fleet* Chalk	c. 1.5m over 1970–1986	Industrial abstraction reduced by 85 per cent (51 Ml d^{-1}) because of saline intrusion	Pumping increased for quarry de-watering
4. *Fawley (Hants)* Bagshot Sands	Rises between 15–45m over 1930–1985	Industrial abstraction ceased (4.5 Ml d^{-1})	None known
5. *Birmingham* Permo-Triassic sand- stone	Rises between 5–10m over 1981–1987	Groundwater abstraction reduced by 68 per cent (31 Ml d^{-1})	Basements flooding controlled by de-watering
6. *Wolverhampton* Permo-Triassic sst	18 m rise rise between 1973 and 1987	Groundwater abstraction of 20 Ml d^{-1} ceased	None known

Table 4.1   (*Cont.*)

Location of aquifer (numbers refer to Fig. 4.1)	Extent of rise	Cause	Consequences
7. *Coventry* Coal Measures	Minor	One abstraction shut down	Minor flooding in housing estate
8. *Nottingham* Permo-Triassic sandstone	3m between 1965 and 1987	Groundwater abstraction reduced by 45 per cent (12.2 Ml d^{-1})	Flooding of some city centre basements and medieval caves
9. *Trafford Park, Manchester* Permo-Triassic sandstone	15m between 1970 and 1986	Groundwater abstraction reduced by 72 per cent (21 Ml d^{-1})	None known
10. *Liverpool* Permo-Triassic sandstone	2.5m between 1975 and 1985	Groundwater abstraction reduced by 80 per cent (35 Ml d^{-1}) over 20 year period	Flooding of railway tunnels and some basements
11. *Edinburgh* Devonian/Lower Carboniferous	Not measured	Groundwater abstraction ceased after contamination by leaking sewers	Increased flows in sewers. Flooding in some basements
12. *Musselburgh* Coal Measures	Not measured	Industrial abstraction of 6 Ml d^{-1} shutdown	Minor flooding in housing estate
13. *Rusheyford* Magnesian Limestone	10m rise between 1974 and 1988	Cessation of coal mine de-watering	None known
14. *Beverley* Chalk	Not measured	Industrial abstraction reduced by 50 per cent	Minor flooding of agricultural land and some properties
15. *Doncaster* Permo-Triassic sandstone	Not measured	Significant reduction in abstraction during late 1970s	Sewer flows increased
16. *Ipswich* Chalk	Not measured	Groundwater abstraction reduced by 3 Ml d^{-1} because of saline intrusion	Minor flooding in some cellars
17. *Braintree* Chalk	15–20m between 1978 and 1988	Groundwater abstraction reduced by 78 per cent (5.4 Ml d^{-1})	None known
18. *Southend* Chalk	13.5m between 1975 and 1988	Groundwater abstraction reduced by 14 Ml d^{-1} because of saline intrusion	None known

A study of the London Basin has been recently completed in which the reasons for the rise in groundwater levels were investigated and predictions of future levels were made. The effect of high groundwater levels on structures was assessed and options for groundwater control were proposed. A description of the main findings will be given later.

An international problem

The problems associated with rising water levels are not restricted to the UK. A number of examples from other countries are given below.

France

In France, Bergeron *et al.* (1983) describe problems associated with rising water levels in a number of areas. The situation in Paris is of particular interest. Large quantities of water were pumped from the alluvium and Tertiary limestone aquifers which are in hydraulic connection with the River Seine. In the mid-1800s water levels in the aquifer were about 4 m above the river level. Due to the extensive groundwater pumping from the aquifer, water levels fell by up to 30 m near Aubervilliers in the north-west outskirts of Paris. The water was used for public supply, industry, and air conditioning. A major dewatering scheme was also installed in the 1970s during the construction of the urban expressway and other major public buildings with deep foundations. During the 1960s and 1970s many buildings with deep basements were built for use as warehouses or car parks. It appears that in a number of cases the assumption was made that it was unnecessary to design the basements against a high water level. Since 1972 there has been a progressive rise in water level due to a steady cut-back in industrial abstraction associated with factory closures, a cessation of the major dewatering activity as the major public works were completed, and a run of years with a higher than average rainfall. One multi-storey building was constructed on the Champs Elysées in the 1960s where the groundwater level was +17 m NGF (a datum level in France related to sea level). The base of the foundation was set at +14 m NGF and the multiple basements were made waterproof to +19 m NGF. The designers appeared to be totally unaware of the fact that the level of the River Seine was +26 m NGF and that the groundwater level at the site had originally been at +29 m NGF. In 1973 the rising water levels overtopped the sealed section and several basement car parks flooded and were rendered totally unusable. If the reduction in abstraction continues it is likely that the water levels will approach those in the Seine.

USA

In the USA, similar problems have been reported from New York (Soren 1976; Van der Leeden *et al.* 1980) and Louisville, Kentucky (Kernodle and Whitesides 1977). In Brooklyn, Long Island, New York, large quantities of groundwater were pumped from the underlying sand aquifer. Original groundwater levels were 3–6 m above mean sea level but due to the abstraction between 1900–1940, the water table had fallen to 4 m *below* mean sea level. As a result of this depression sea water was drawn into the aquifer reducing the quality of the pumped water. This led to a progressive cut-back until in 1976 groundwater abstraction had almost ceased. During the 1940s and 1950s when the groundwater levels were low several buildings were designed and built on the supposition that the water table was in an equilibrium state. Due to the cut back in pumping the water levels had risen to between 2.5–3 m above mean sea level in 1976. Flooding occurred, in this densely populated area of New York, in a hospital, a store, and in the basements of seven schools. The water levels are now being controlled by pumping and disposal to the sewers; there is concern that the sewers and sewage works may become overloaded.

The problem appeared to have resolved itself when another issue arose in September 1978 (Van der Leeden *et al.* 1980). A US coastguard observed a large concentration of oil in Newton Creek leading to East River. It was established, following a detailed drilling programme, that the source of this pollution was a plume of oil lying on the water-table and flowing into the creek. The plume was some 25 ha in extent and varied in thickness from a trace lying on top of the water-table to a layer 6 m thick. It was estimated that the spill contained some 77×10^3 m³ of highly combustible oil. The investigations indicated that the spill had occurred during the 1950s when the water-table was depressed and had risen with the water-table as groundwater abstraction in adjacent areas was cut back. Eventually, the situation was reached where the groundwater gradient was towards the Newton Creek and the oil began to escape. Depths of basements and sewers in the area were checked and fortunately all were found to be clear of the spill. The rehabilitation cost is estimated to be several million dollars but work has apparently been delayed due to a number of complex legal issues.

Japan

Japan also has problems with rising groundwater levels. Ohta (1987) described how the water level in the sand formations below Tokyo was reduced by 100 m due to industrial groundwater abstraction which started in the 1920s. The fall in water level led to consolidation of the clay strata in the sequence causing surface settlement to the extent that an area of sea-shore is now below sea level. In the 1960s a restriction was imposed on groundwater pumping so that groundwater levels could rise and surface settlement could be controlled. As the water levels have increased problems have arisen recently with the construction of a 35 m deep basement excavation in Tokyo. Water from poorly constructed site investigation boreholes 'spouted' into the base of the excavation just as the lean-mix concrete was about to be poured to consolidate the excavation base. Chemical grouting and deep dewatering were necessary to enable the work to be completed.

The Persian Gulf States

In the Persian Gulf States the rapid increase in living standards has been associated with the growth of public parks and private gardens in cities. In such arid countries there is an understandable enthusiasm for trees and grass. Irrigation of trees and plants using either public water supply, or increasingly, treated sewage effluent, is essential. Much of this irrigation is uncontrolled and La Dell (1986) suggested that it often greatly exceeds

the plant needs. The surplus water that does not evaporate percolates to the water-table. This, together with leakage from city mains, sewers, and septic tanks, has led to a rise in water levels and water-logging or flooding of some lowland areas. Problems have been reported in Riyadh, Jeddah, Mecca, Kuwait City, and Qatar (Watson 1990).

Germany

In Germany at least 17 situations have been identified where groundwater levels have shown a progressive rise and eight cases of flooding or other damage have been reported (Anon 1984).

Categorizing the causes and effects of rising water levels

The examples given above indicate that rising groundwater levels may be due to a number of causes. These are listed in Table 4.2. The range of

Table 4.2   Causes of rising water levels

Man made
Recovery of groundwater levels and flow following the reduction or cessation of groundwater pumping for:
– public water supply
– industrial water supply
– mine dewatering
– construction dewatering

Rise of the water-table above natural level due to:
– engineering works; dams; barriers,
– leakage of sewers or pipes,
– irrigation.

Artificial groundwater recharge through basins or wells.

Man induced settlement of the ground surface giving a relative rise in groundwater level.

Natural
Extreme variations in groundwater level due to exceptional:
– precipitation
– tidal conditions

---

problems that may develop are given in Table 4.3 and suggested counter measures in Table 4.4.

Rising groundwater levels in the London Basin

The London Basin is a large synclinal fold in Cretaceous and Tertiary strata. A simplified geo-

Table 4.3   Possible effects of a rising ground-water level

Increase in spring and river flows

Re-emergence of 'dry springs'

Flooding of basements

Increased leakage into tunnels

Surface water flooding

Spread of pollution underground

Pollution of surface waters

Reduction of slope and retaining wall stability

Reduction in bearing capacity of foundations and piles

Increased hydrostatic uplift on basement structures

Increased swelling pressures on underground foundation and structures

Swelling of clays leading to surface uplift

Chemical attack on foundations

Table 4.4   Countermeasures

Regional pumping policy

Local pumping adjacent to structure

Permeable basement and drain to sump

Relief wells draining to a sump

Waterproof structure

Anchor or ballast foundation

Drainage channels

logical map is shown in Fig. 4.3 and a cross section through the basin is given in Fig. 4.4. Cretaceous Chalk crops out in the Chiltern Hills to the north and in the North Downs to the south. The Chalk lies below younger deposits of sands and clays which are exposed at the surface in the centre of the basin.

The Chalk is a soft, fine grained, white limestone with a maximum thickness of about 240 m. Water flows through the aquifer in a complex of small joints and fissures. Its hydraulic properties are defined by a transmissivity which ranges from a low value of about 20 m^2 d^{-1} in the more deeply buried part of the aquifer to high values of 2000 m^2 d^{-1} along the major valleys of the outcrop. The Chalk has storage coefficients of about 2 per cent and 0.01 per cent in the unconfined and confined situations respectively.

The Tertiary deposits comprise a sequence of clays and sands overlain by the London Clay, which is up to 150 m thick at the centre of the basin. The lower deposits in this series are sandy and are generally referred to as the Basal Sands. They are made up of the geological sequence known as the Thanet Beds and part of the Wool-wich and Reading Beds, and Blackheath Beds. The Basal Sands reach a thickness of 35 m in the east of the Basin but are thin to the west. It is not possible to quote a representative transmissivity for the Basal Sands due to their variable nature. However, the Thanet Beds are more uniform in nature and their hydraulic conductivity is about 2.5 m d^{-1}. The unconfined storage coefficient of the Basal Sands lies in the range of 5–10 per cent. The Basal Sands rest on, and form an aquifer unit with, the Chalk. The strata between the top of the Basal Sands and the London Clay are predominantly clays with occasional layers of more coarse-grained material. These clay strata act as a confining layer to the Chalk and Basal Sands aquifers; however, the exact level of this boundary is subject to interpretation.

Original groundwater levels

A map which shows the original groundwater levels prior to any major abstraction was prepared by the Water Resources Board (1972) (Fig. 4.5). This was based on early records of water levels in

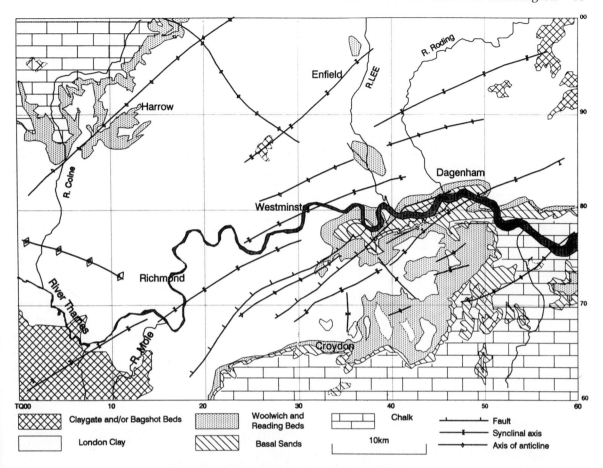

Fig. 4.3. Geological map of the London Basin.

wells and an interpretation of the way in which the hydrogeological boundaries would control the groundwater flow in the London Basin.

Groundwater abstraction

Total groundwater abstraction in the confined area of the London Basin rose steadily from about 9×10^6 m³ a⁻¹ in the early 1800s to a peak of 227×10^6 m³ a⁻¹ in 1940. From World War II onwards the quantity abstracted steadily declined to about 190×10^6 m³ a⁻¹ in 1965. More recent data supplied by the former Thames Water Authority has also shown a fall to 121×10^6 m³ a⁻¹ in 1984.

The cut-back in abstraction during the 1950s and 1960s halted the fall in water levels which were probably at their lowest during the early 1960s

(Fig. 4.6). The groundwater flow directions in Fig. 4.6 show marked changes when compared with the original flow paths (Fig. 4.5). Flow in 1965 was from the outcrops towards the depression in Central London. The River Thames to the east of the Isle of Dogs also became a source of recharge and this led to saline intrusion of the aquifer adjacent to the river.

Rise in groundwater levels

The reduction in groundwater abstraction has led to a steady rise in groundwater levels since 1965. The groundwater level contours for 1985 are shown in Fig. 4.7. Typical well hydrographs are given in Fig. 4.8. These show that in Central London the rise over the last 15 years has been 15–20 m. Along

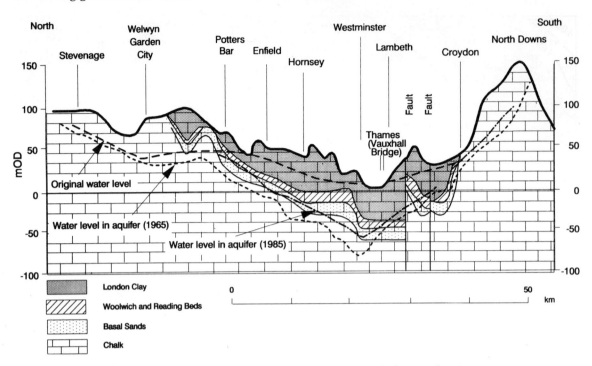

Fig. 4.4. North to south geological section through the London Basin.

the Chalk outcrop to the north and south of the basin, abstractions have been maintained or, in some cases increased, and consequently the water levels have changed little or shown a modest decline.

The CIRIA study

Wilkinson (1984) expressed concern over the rise in groundwater levels in the London area and used a simplified groundwater model of the London Basin to show that the important factors controlling the rate of rise were (a) groundwater abstraction in the confined area and (b) the position of the interface between the Basal Sands and the overlying clay. In 1985 the Thames Water Authority commissioned the Water Research Centre to prepare a more detailed model of the basin. The model findings were to be used as part of their analysis of proposals for the development of new groundwater resource projects using artificial recharge of surplus water in the London Basin (see Chapter 3). A detailed study by the Construction Industries Research and Information Association (CIRIA) of the engineering impacts of rising groundwater levels in the deep aquifer beneath London began in 1986 (Simpson *et al.* 1989). The Thames Water Authority agreed to their groundwater model being used to support the CIRIA investigations.

The objectives of the CIRIA study were to:

(1) investigate the rate of rise of groundwater levels in the London Basin and predict the final levels that would be attained;

(2) assess the effect that an elevated groundwater level would have on existing structures, buildings, foundations, tunnels, etc.;

(3) recommend monitoring procedures;

(4) recommend local and regional control measures to be adopted as necessary;

(5) draw the attention of those responsible for the design of new structures to the possibility of an elevated groundwater level.

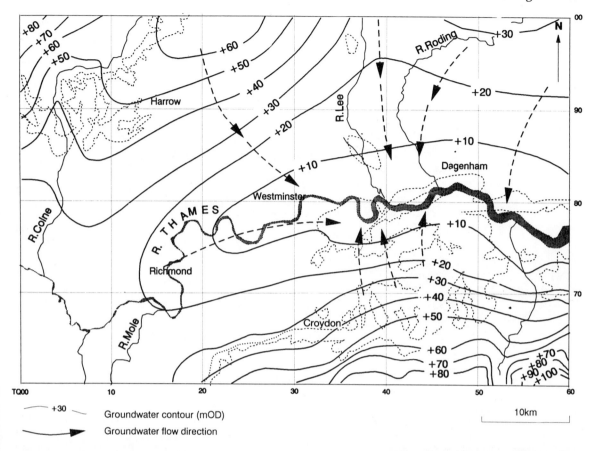

Fig. 4.5. Original groundwater levels prior to major abstractions and groundwater flow directions (after Water Resources Board 1972).

Groundwater models

The groundwater model developed by the Water Research Centre has played a central role in the CIRIA studies.

The finite element representation for the model and the hydrological boundary conditions are shown in Fig. 4.9. It covers an area of some 100 km², each triangular element having an area of about 4 km². The model computes groundwater levels at each intersection node formed by the triangles. The model was centred on the Enfield–Haringey area as the Thames Water Authority's prime objective was to use it to assess the feasibility, and determine the operating rules and effects of an artificial groundwater recharge project in this area. The model is able to represent

the groundwater flow in both the outcrop areas and in the areas below the clay cover towards the centre of the basin. For the outcrop area the monthly recharge to each triangular element is calculated by subtracting evapotranspiration from rainfall.

In the centre of the basin the Chalk is overlain by the Basal Sands and these are covered by impermeable clays. The levels of the Chalk/Basal Sands interface and Basal Sands/clay interface are defined for each node. The model assumes that the Chalk and Basal Sands aquifers are in perfect hydraulic continuity. Within a given element of the model the Chalk and Basal Sands are assumed to have a common transmissivity but the unconfined storage coefficient is the appropriate Chalk or Basal Sands value depending on whether the

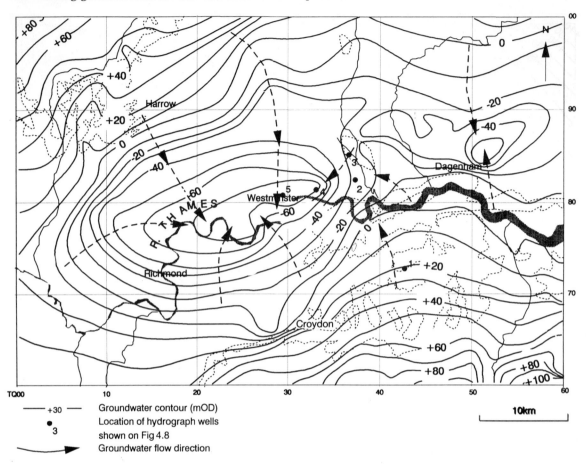

Fig. 4.6. Groundwater levels in 1965 and groundwater flow directions (after Water Resources Board 1972).

groundwater level is in the Chalk or in the Basal Sands aquifer.

In all model runs the unconfined storage coefficient in the Chalk was kept at 2 per cent at all elements, but, due to the heterogeneous nature of the Basal Sands, its unconfined storage coefficient was varied between simulations from 2–10 per cent. It was thus possible to test the sensitivity of the model predictions against this parameter. Where the water level rises above the top of the Basal Sands a hydraulic pressure develops at the base of the confining clay layer and the groundwater system becomes confined. In such circumstances a confined storage coefficient of 0.01 per cent was used in the model. For each element, at each time step, the model compares the ground-

water level with the level of the Chalk/Basal Sands or Basal Sands/clay interface and enters the appropriate storage coefficient into the calculations. Each element in the model is given a transmissivity value. This varies from 3000 m^2 d^{-1} in the outcrop area to 20 m^2 d^{-1} or less in the centre of the basin.

An average groundwater abstraction or recharge rate over each nodal element can be introduced at every time-step so that past or future pumping or artificial recharge patterns can be represented accurately.

Model calibration

The model was calibrated using groundwater level maps and abstraction or recharge sequences for the following periods:

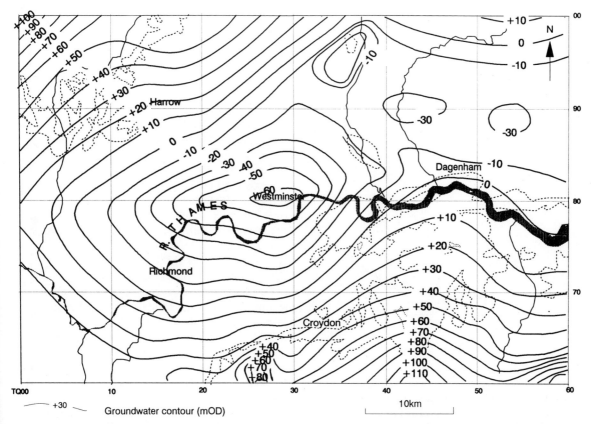

Fig. 4.7. Groundwater levels in 1985 based on information compiled by the Thames Water Authority.

(a) pre-development (*c*.1800);
(b) 1965–1975;
(c) 1975–1985.

Calibration consisted of making numerous runs of the model and between each run, adjusting the transmissivity pattern or boundary conditions until a satisfactory match was obtained between the observed water levels (case (a) above) or observed water level fluctuations (cases (b) and (c) above) and the modelled water levels.

In general the modelled and 'observed' values at the nodal points correspond to within ±5 m. This is considered a satisfactory match as the 'observed' nodal values are based on an interpretation of observation well records which are rather sparse in some areas of the model. The observation well records may also be affected by local pumping and this may not represent the mean water level for the triangular finite element in which they lie.

A problem was encountered with the calibration pre-development situation (case (a) above). The initial runs gave much higher groundwater levels from the model than were observed in a confined area of the aquifer just north of the River Thames. This is an area where a number of 'scour holes' have been observed in the London Clay. Some of these penetrate the full thickness of the London Clay and into the Lower London Tertiary deposits. The holes are filled with a mixture of disturbed clays, alluvium, sands, and gravels with a diapiric structure. Their origin is uncertain but it is suggested that some may be caused by groundwater under high artesian pressure, bursting through in areas where the London Clay was thin. In order to fit the model with the observed values it was necessary to allow upward drainage in the model from several of the elements adjacent to the River Thames. No recharge from the surface was permitted at these nodal points. Their location is shown

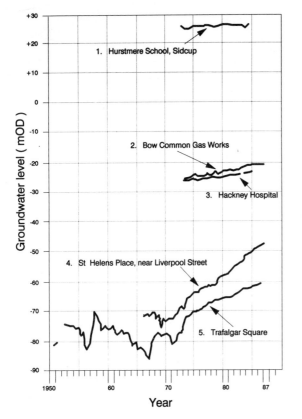

Fig. 4.8. Typical well hydrographs. The well positions are shown in Fig. 4.6.

in Fig. 4.9 where they are described as a 'one way-leakage boundary'.

Future groundwater level predictions—abstraction and artificial recharge scenarios

Once the model was calibrated it could be used to predict future groundwater levels. As a first step it was necessary to select a range of future artificial recharge and abstraction scenarios to be used in the model simulations. The abstractions that are of particular importance are those in the centre of the basin. These have declined markedly in recent years. Several scenarios were modelled, two of which are shown in Fig.4.10:

(a) to maintain the abstractions in the central London area at their 1985 values into the future and;

(b) to reduce the central London abstractions to zero by the year 2000.

Water levels are rising in the aquifer but not sufficiently rapidly in the short-term to meet water resource demands and the Thames Water Authority wished to investigate supplementation of the flow into the aquifer by artifically recharging surplus, treated, surface water. The area proposed for recharge was Enfield-Haringey in North London. The Thames Water Authority was already operating an artifical recharge scheme using wells and boreholes in the Lee Valley.

For modelling purposes the operation of both the existing Lee Valley scheme and that proposed for Enfield–Haringey was simplified to recharge over seven years followed by heavy abstraction of the recharged water in the eighth year. However, the recharge was controlled in the model so that if the water level at the recharge node rose above the top of the Basal Sands the recharge was stopped. It was only resumed once the levels had again fallen into the sands. The nodal abstraction rate in the eighth year was kept constant even though recharge at that node had been curtailed during the seven year recharge phase. The theoretical abstraction rates in the Lee Valley and Enfield–Haringey schemes were selected as 44 and 100×10^3 m³ d⁻¹ respectively.

A range of future abstraction scenarios was modelled. A typical hydrograph from the scenario which assumes that the groundwater public supply abstraction will remain at the 1985 value but that the private and industrial abstractions in Central London will steadily decline to zero over the next 10 years (case (a) above) is shown in Fig. 4.11. Under this regime the groundwater levels in Central London would rise to within 3–5 m of the pre-abstraction levels within 60 years (a rise of some 75 m from the 1965 levels in Central London).

There is some difficulty in defining the interface between the Basal Sands and the overlying clays. Indeed the interface is not a sharp one but a gradual transition with sand layers becoming thinner and less frequent higher in the sequence. A number of model simulations were made with this interface at different levels. It is an important boundary in that it controls the jump from an unconfined to a confined condition. The scenario described above was run on the model but with the thickness of the Basal Sands reduced by 30 per cent. The resulting hydrograph is shown in Fig. 4.11. The rate of rise is much more rapid and

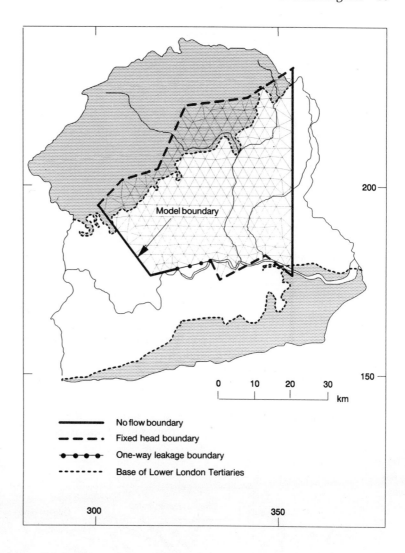

200 —

150 —

Model boundary

0   10   20   30
|___|___|___|
km

No flow boundary
Fixed head boundary
One-way leakage boundary
Base of Lower London Tertiaries

300   350

Fig. 4.9. Hydrological boundary conditions of the finite element models (after Simpson *et
al.* 1989).

equilibrium conditions are established after about
30 years.

Because of the way in which they are operated
it appears that with time the 'recharge' schemes
become net abstraction schemes and if so this will
contribute to the holding down of water levels in
Central London.

Engineering implications

It is now well established that a steady rise in
groundwater levels is taking place in Central
London. If this is maintained, the models predict

that the water levels will rise into the Basal Sands
within the next few decades. Some of the large
structures in London are founded in the Basal
Sands and future foundations for buildings are also
likely to extend down into these strata.

The consequences of an elevated water-table on
deep basements, foundations, and tunnels were
investigated as part of the CIRIA study (Simpson
et al. 1989). The effects are likely to be most
marked and rapid in the strata with high permeability such as the Basal Sands. In the London Clay
and other thick clays in the sequence differential
movement due to swelling may lead to structural
cracking. The principal effects on engineering

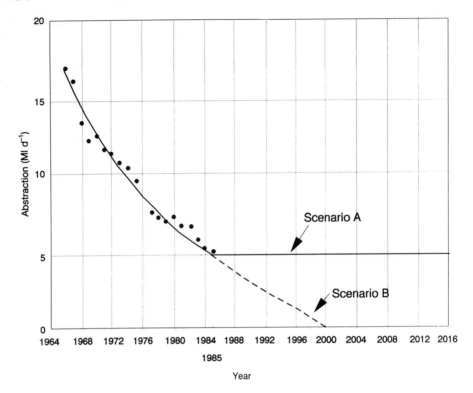

Fig. 4.10. Two of the scenarios modelled for possible future abstractions in the central London area.

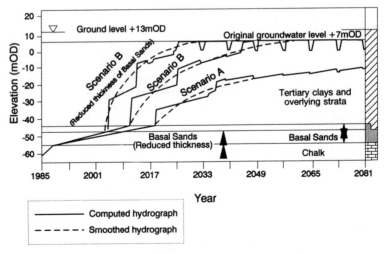

Fig. 4.11. Predicted hydrographs for Trafalgar Square (point 5 Fig. 4.6) from the groundwater model for scenarios A and B given in Fig. 4.10.

structures would be from flooding, swelling of clays, reduction in bearing capacity, hydrostatic uplift pressures, chemical attack, and difficulties during construction.

Tunnels

It is estimated that there may be some 130 km of tunnels located near the base of the London Clay or in the Lower London Tertiary sediments which are in critical locations with respect to rising water levels in Central London. If the water level continues to rise some tunnels will suffer an increase of water pressure, deformation, minor cracking, and spalling, although the possibility of structural failure is very small. High water levels could lead to a major increase in the cost of new tunnels although such problems could be avoided if the groundwater level could be controlled at about −20 m OD in Central London.

Swelling

The fall in groundwater levels and the associated increase in effective stress that occurred as pore-water drained from the London Clay and the Woolwich and Reading clays led to consolidation of the clays and surface settlement of 250–500 mm over a period of about one hundred and twenty years. With rising groundwater levels the process will be reversed as the clays swell. The amount of swelling may be less than the preceding consolidation. This could cause problems where a building has its foundations at different depths. Although movement due to differential swelling may be small and in most cases would not affect the structural integrity of the building, there could be substantial damage to the facade of important buildings which could be extremely costly to repair.

Bearing capacity and hydrostatic uplift

Calculations show that shallow and deep foundations and piles in Central London may show a loss of bearing capacity as water levels rise. Water pressures on the walls of deep basements could also increase the lateral thrust. High artesian pressures may lead to uplift on the base slabs of deep basements. In composite raft/pile foundations the increase in water pressure could cause a redistribution of load between the piles and the raft. Such problems have been considered in the design of

the new British Library building at a site close to St Pancras Station. The deepest basement extends down to 25 m below ground level, penetrating 4 m into the Woolwich and Reading clays. Large diameter piles extend about 12 m into the Woolwich and Reading sands (Basal Sands). The groundwater level at present lies in the Chalk and the original design assumed that this level would remain stable during the life of the building. Realization that the water levels were rising led to a redesign to increase the bearing capacity of the piles and to strengthen the basement retaining walls. A drainage system below the basement slab, connected to relief wells which extend into the Chalk, has been installed to prevent a rise in artesian pressure above the basement slab level.

Construction

The high artesian pressures at the base of the Woolwich and Reading clays or the London Clay could be hazardous during the excavation of deep foundations. These would need to be controlled by a dewatering system. Difficulties may also arise with artesian pressure when constructing bored pile foundations in sandy strata or in clays with more permeable sand or silt layers.

Chemical attack

During recharge experiments in the Lee Valley very high sulphate concentrations of up to 2000 mg l^{-1} were measured in the groundwater recovered from a few of the boreholes. These high sulphate concentrations are believed to result from the oxidation of pyrite in the Basal Sands following dewatering. Such high concentrations coming into contact with foundations could lead to attacks on concrete and its deterioration. Although this is not thought to be a widespread problem, research is in hand to identify the distribution of sulphate in the dewatered Basal Sands.

Concluding remarks

A number of situations exist within the UK and overseas where groundwater levels are rising appreciably. In most cases the rise is due to a reduction in groundwater pumping below cities; however, additional recharge to an aquifer from leaking sewers and water mains may be a significant factor in some places. The reasons for the cut-back in groundwater pumping are many, but are principally a decline in industrial activity and a demand for water in city centres, and a reduction in the quality of the pumped water.

Groundwater levels had been so low for so long in some cities that the assumption was made by the engineers responsible for the design of buildings and structures that a long-term equilibrium in the levels had been established. The prospect of a much higher groundwater level was either not considered or was disregarded. This lack of appreciation of the hydrogeological factors has led to flooding in basements and tunnels and some structural problems.

Within the UK, such problems have occurred (and are only likely to occur) at a limited number of locations. The legal framework relating to groundwater matters was devised as a means of managing water resources and has no provision for the continuation of pumping when supplies are no longer required. A number of dewatering schemes are operating in the UK but these are carried out by individual companies or organizations to protect their own buildings or structures, rather than by community action.

In London, the consequences of groundwater approaching its pre-abstraction levels would be very serious, and the recent CIRIA study has identified the types of structure and areas at risk. The study shows that no serious problems are likely to occur within the next thirty years, which gives time to set in place a regional groundwater-control policy. In the meantime, it is incumbent on the engineering community to consider the long-term fluctuations in groundwater levels, both in London and elsewhere, and to make provision for such fluctuations within their designs.

References

Anon, (1984) *Renewed rise of groundwater following discontinuation of draw-down measures in the Federal Republic of Germany*. Federal Institute of Geo-Sciences and Raw Materials. Archive No. 95 678. Hanover.

Bergeron, G., Dehays, M., and Pointer, T. (1983). *Remontées des nappes d'eau souterraine: cause et effects*. B.R.G.M., Orléans.

Brassington, F. C. and Rushton, K. R. (1987). A rising water table in central Liverpool. *Q. J. Eng. Geol.*, **20**, 151–8.

Brassington, F. C. (1990). Rising water levels in the United Kingdom. *Proc. Instn Civ. Engrs*, **88**, 1037–57.

Henton, M. P. (1978). Abandoned coalfields: problem of pollution. *The Surveyor*, 31 May, 9–11.

Hurst, C. W. and Wilkinson, W. (1985). Rising groundwater levels in cities. *Proc. 21st regional meeting of the Engineering Group*, Sheffield. pp. 11–20. Geol. Soc., London.

Kernodle, J. M. and Whitesides, D. V. (1977). Rising groundwater level in downtown Louisville, Kentucky, 1972–1977. *Water-Resources Investigations*, pp. 77–92. US Geological Survey.

La Dell T. (1986). *Rising groundwater levels in cities*. Contribution to a meeting of the Hydrogeological Group of the Geological Society of London.

Lloyd. J. W. and Lerner, D. N. (1986). *Aspects of rising groundwater in the Triassic sandstone aquifer under Birmingham*. Contribution to a meeting of the Hydrogeological Group of the Geological Society of London.

Marsh, T. J. and Davies, P. A. (1983). The decline and partial recovery of groundwater levels below London. *Proc. Instn Civ. Engrs*, Part 1., **74**, 263–76.

Ohta, H. (1987). Deep excavation performed after restriction of deep-well pumping in Tokyo. *Proc. 9th European Conference on Soil Mechanics and Foundation Engineering* (ECSMFE), **2**, 715–18.

Price. M. and Reed, D. W. (1989). The influence of mains leakage and urban drainage on groundwater levels beneath conurbations in the UK. *Proc. Instn Civ. Engrs*, Part 1., **86**, 31–9

Rushton, K. R., Kawechi, M. W., and Brassington, F.C. (1988). Groundwater model of conditions in Liverpool sandstone aquifer. *J. Instn Water Env. Mgmt*, **2**, 67–84.

Simpson, B., Blower, T., Craig, R. N., and Wilkinson, W. B. (1989). *The engineering implication of rising groundwater levels in the deep aquifer beneath London*. The Construction Industry Research and Information Association, Special publication 69, London.

Soren, J. (1976). Basement flooding and foundation damage from water-table rise in the east New York section of Brooklyn, Long Island, New York. *Water-Resources Investigations*, pp. 76–95, US Geological Survey.

Van der Leeden, F., Braids. O. C., and Fleishell, J. L. (1980). *The Brooklyn oil spill*, Geraghty and Miller, New York.

Water Resources Board (1972). *Artificial recharge of the London Basin. I : Hydrogeology.* Water Resources Board, Reading.

Watson, R. (1990). Saudi Arabia faces disastrous rise in groundwater. *New Civil Engineer*, 26 July, 6–7.

Wilkinson, W. B. (1984). Rising groundwater levels and geotechnical consequences. Informal discussion, Ground Engineering Group and Water Engineering Group. *Proc. Instn Civ. Engrs*, Part 1., **76**, 791–4.

Wilkinson, W. B. (1985). Rising groundwater levels in London and possible effects on engineering structures. In *Hydrogeology in the Service of Man*, 145–57. Memoirs of the 18th Congr. Int. Assoc. Hydrogeol. Cambridge.

Wilkinson, W. B. (1986). *Rising groundwater levels in cities.* Contribution to a meeting of the Hydrogeological Group of the Geological Society of London.

5. Groundwater recharge and water movement in the unsaturated zone

C. M. K. Gardner, J. P. Bell, J. D. Cooper, W. G. Darling, and C. E. Reeve

Introduction

The unsaturated zone is defined here as the zone of an unconfined aquifer above the water-table where pore-water pressures are negative; thus it comprises the soil, the capillary fringe, and the unsaturated material between these. It is characterized by vertical water movement both downward and upward; lateral potential gradients are only significant in the rooting zone of the soil.

Study of the movement of water within the unsaturated zone is important for a number of reasons:

1. A knowledge of recharge amount is a fundamental requirement for resource evaluation. Recharge generally reaches the unconfined groundwater body via the unsaturated zone. Investigation of water movement processes within this zone is thus essential for understanding the factors controlling recharge quantity and timing.

2. Concern has been rising, in Britain and the rest of the world, about pollution of groundwater resources. Understanding the processes of water transmission through the unsaturated zone assists in modelling the quantities and timing of pollutant transfer to the water-table.

This pollution can arise from both point or areal (diffuse) sources, and the unsaturated zone is involved in almost all cases. Prominent contaminants are agricultural chemicals: nitrates have been studied in the unsaturated zone of the major aquifers for some years; more recently, pesticides and their residues have also given cause for concern. Waste disposal activities are another possible

source of pollution of groundwater; landfill leachates are a well-known problem; hazardous and radioactive waste disposal represent other activities where answers are needed. Pollution from various accidental causes, for instance chemical spillages, pipeline fractures, or nuclear disasters, is another example.

3. In the case of shallow unconfined groundwater, changes to the water-table may have an important impact on the unsaturated zone itself, and particularly on the vegetation which it supports, whether natural or agricultural. The modification to the water-table may come about through water resources development, river level changes, construction, mineral extraction, or artificial drainage. The consequences of such changes depend not only on the depth by which the water-table is adjusted, but whether it is a temporary, recurring, or permanent adjustment.

4. Artificial recharge is often performed by impounding water deliberately, or accidentally, at or near to the ground surface, and relying on the unsaturated zone to conduct it to the underlying groundwater body. Examples include 'rainfall harvesting' schemes, recharge canals, and leakage from irrigation canals. These are mainly of relevance to semi-arid regions.

As a result, the unsaturated zone has been a subject of much research interest in the UK and in the rest of the world during the past decade. It has been marked by an emphasis on field studies, largely supplanting the laboratory-based work prevalent up until the early seventies, and the maturing of numerical modelling into a more practical tool, rather than an activity of mainly academic interest.

A distinction can be drawn between studies aimed at an understanding of the unsaturated zone overlying deep groundwater and that overlying shallow groundwater. In the case of deep groundwater, a significant intermediate zone of unsaturated material separates the soil from the capillary fringe. The unsaturated and saturated zones are clearly distinguishable, the zone of fluctuation of the water-table is relatively small and, below the root zone, water movement is always downwards.

In the case of shallow aquifers the intermediate zone is thin or absent. Seasonally, the capillary fringe may extend into the soil and the water-table may rise to, or above the ground surface, a large proportion of the 'unsaturated zone' becoming part of the saturated zone. Furthermore, there is the possibility of upward potential gradients developing from the water-table in response to transpirational demand, causing upward water movement. Thus the groundwater body itself may contribute towards evapotranspiration, either by extraction of water by plant roots below or just above the water-table, or via the unsaturated zone. This leads to some conceptual problems concerning the definition of both groundwater and groundwater recharge in the context of shallow aquifers.

Recognition of the importance of field studies has led to a demand for better and more convenient instruments to measure the variables of interest in the unsaturated zone. A number of these were described in the predecessor to this publication (Kitching *et al.* 1981). Here, a capacitance technique for the measurement of water content is described.

Field studies of water movement processes in the unsaturated zone and recharge measurements using both soil physical and geochemical techniques are also discussed. These have tended to focus on the English Chalk, both because of its great importance as a source of water supply and because of the challenge in attempting to understand flow processes in an unsaturated zone characterized by 'bimodal' permeability. This work has gone a long way to providing an understanding of the processes involved, but the measurement techniques employed inevitably provide point information.

One of the persistent problems in groundwater hydrology is obtaining a reliable areal value for groundwater recharge. Progress has been achieved in the last decade particularly through the use of models, and there are now a number of promising techniques although not all of them are applicable to every problem. A recent volume, edited by Simmers (1988) contains many papers reviewing the international state of the art in this subject. Here, the use of a robust inverse method is described.

The capacitance probe for measurement of soil water content

The variables of most interest in hydrological studies of the unsaturated zone are the volumetric water content and the water potential. During the past decade, an important advance in instrumentation has been the emergence of methods for water content measurement, based on the relationship between the dielectric constant of unsaturated soil and its volumetric water content. The recognition of this relationship is not new, but it is only in recent years that the electronics have become available to enable robust and stable field instrumentation to be designed. Commercial systems are now available, offering an alternative to the well established neutron probe method, for many, but not all, applications.

The neutron probe has been widely adopted as the principal method for measuring water content in soil and porous media for the past 25 years (Bell 1976; Greacen 1981), but is not ideal in several respects, difficulties arising from radiological safety aspects perhaps being the most prominent of these. In the developed world, safety regulations already constrain the use and transport of neutron probes, and these are likely to become ever more tightly drawn in the future. Furthermore, it is worrying that in developing countries, where large numbers of neutron probes already exist, often there is no proper control of the radioactive sources within them. What will happen to these over the next few thousand years, during which they will remain hazardous, should be a matter of serious concern. Considerations of safety and cost militate against the neutron probe for use as a static sensor for monitoring water content changes in the soil profile and an instrument for this purpose has long been sought. The weight and bulk of shielding needed to protect the operator also makes the neutron probe inconvenient for use in many field situations. For some research applications, there has been a long-term requirement for a method capable of providing a high degree of depth resolution.

Two methods for water content measurement have been developed which utilize the dielectric constant principle: the so-called 'Time domain reflectometry' method (Topp and Davis 1985), developed mainly in Canada, and the capacitance method. In the capacitance method, the fringe capacitance of the soil surrounding a pair of electrodes is determined. Capacitance depends on the dielectric constant of the three-phase mineral/air/water system comprising the soil. The geometry and configuration of these electrodes varies according to the design philosophy, but there are three principal variants: a depth probe system (operating within a plastic access tube) for determination of water content profiles, a 'push-in' probe for topsoil use only, and static sensors for monitoring water content changes over a period of time at a specific position in the soil. For hydrogeological purposes we are concerned primarily with the depth probe.

Principles, design, and operation

The dielectric constant of free water at frequencies of less than 1000 MHz is around 80 and that of the mineral components of most soils is typically 4–6. Measurement of soil dielectric constant therefore has the potential to provide a very sensitive method for determining water content. Early work was carried out in the kHz range and gave unreliable results (Smith-Rose 1933). In the 1960s the importance of interfacial polarization effects in heterogeneous materials (i.e. soil) was recognized (Hoekstra and Delaney 1974), and hence the need to operate at much higher frequencies. Thomas (1966) used a bridge method operating at 30 MHz to measure soil dielectric constant by capacitance, employing a wedge-shaped electrode system which could be pushed into the upper layer of the soil.

For hydrological studies, several workers have developed designs in which the electrodes comprise part of a probe which operates within a vertically installed plastic access tube (Kuraz and Matousek 1977; Malicki 1983; Galfy 1984; Dean *et al*. 1987). Several systems are available commercially, but it is beyond the scope of this account to describe them individually. The Institute of Hydrology (IH) Soil Water Capacitance Probe, a British system, is therefore cited to illustrate the general principles and to highlight features specific to that design.

The probe body and access tube are both made of plastic, a material which is electrically non-conductive, as the physics demand. Mounted within or on the surface of the probe is a pair of metal electrodes arranged so that the soil outside the access tube acts as the primary dielectric material. Different electrode geometries are favoured by different designs, and influence depth resolution and field penetration. There are two preferred geometries: either a parallel pair of annular rings oriented co-axially with the probe, spaced about 30 mm apart (the IH design), or two metal strips mounted vertically on the outside or just within the probe.

The IH probe also contains the electronic circuitry, a miniature lithium cell power source and an infra-red photo-emitter. This latter feature is unique to the IH design. The signal is transmitted via a fibre-optic cable to a hand held frequency meter, de-coupling the probe electrically and thus avoiding problems associated with standing waves in cables and interaction with external objects near the cable. In the case of the IH probe, the handle is formed by sectional plastic tubes forming a continuation of the probe body, extendible to any required length up to 2 m. This handle enables the probe to be located precisely and centrally within the access tube, at the required depths, by means of a click-stop device. Other probes are lowered by cable; this makes for a more compact system but offers room for inaccuracies due to the necessary loose fit in the access tube.

Access tubes are made from any non-conductive plastic but require a high specification on wall thickness; they are usually 38–50 mm in diameter. The tube must be installed with extreme care because the sensitive radius around the probe is small. Dean *et al*. (1987) showed that 90 per cent of the response is created within a diameter of 130 mm. i.e. a radius of 40 mm in the soil around a 50 mm access tube. However, within this zone the response is heavily weighted to the soil immediately adjacent to the access tube wall. Hence, it is essential that air gaps are not created during the installation procedure and special rigs are necessary to install access tubes to the required precision.

Claims have been made that the capacitance method is largely independent of soil type, but there is theoretical and experimental evidence that this is not so (Ansoult *et al.* 1985; Dean *et al.* 1987; Bell *et al.* 1987). The problem is inherent in the physics of the system because each component—mineral, air, and water—contributes to the dielectric constant but these are not directly additive (Ansoult *et al.* 1985). Figure 5.1 shows the general curve, with best fit lines for four individual soils. The IH access tube installation method incidentally provides calibration samples for each tube and so a calibration specific to the access tube can be derived. The working calibration curve for each soil is relatively short, and hence may be regarded as linear.

It should be noted that comparisons with data derived by neutron probe should be treated with caution (Bell *et al.* 1987). Both instruments are calibrated against thermogravimetric data (oven drying at 105°C) but the 'water' driven off at this temperature does not necessarily correspond to the definition of 'water' as sensed by either system.

In the case of the capacitance probe, it is conceivable that a proportion of water driven off at 105°C may not be free enough within the soil to respond to the field reversals. Conversely, it is equally conceivable that the opposite applies—water which can be sensed by the probe may not be driven off at normal oven temperature; research is required to clarify this. In contrast, the neutron probe works on entirely different physical principles. It responds to hydrogen, not all of which is present in the form of free water. This hydrogen may be chemically combined in many different organic and inorganic forms which will have differing stability in relation to the thermogravimetric process. It is therefore probable that any relationship in the responses of the two systems will be specific to each soil.

Notwithstanding these theoretical considerations, good agreement has been achieved in the general form of profiles derived by the two methods. Figure 5.2 shows that the neutron probe produces a smoothed water content profile while the capacitance probe has better resolution and corresponds more closely with the gravimetric data (each point representing a 40 mm layer).

The capacitance method offers a viable alternative to the neutron probe and has the advantages of

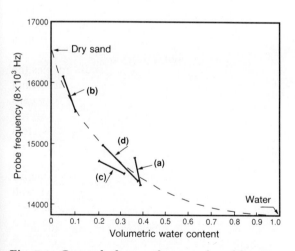

Fig. 5.1. General form of general calibration curve with best fit calibration lines for four different soils: (a) a chalk soil, (b) medium-fine sandy soil, (c) sandy-clay drift soil, (d) silty drift overlying gravel (after Bell *et al.* 1987).

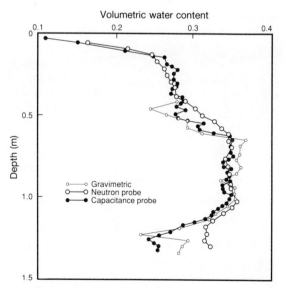

Fig. 5.2. Comparison of a soil water profile determined by three different methods (after Bell *et al.* 1987).

being cheaper, non-radioactive, and quicker to use. It also has better depth resolution and is easily logged automatically. Its disadvantages are the requirement for precise access tube installation and complete absence of air gaps. This precludes use in shrink-swell soils and causes difficulties in stony soils. Present methods of access tube installation limit the method to the upper 2 m, but further work is expected to extend this.

Hydrology of shallow unsaturated zones

In Britain, shallow aquifers occur mainly in coastal and river flood plain areas including, for example, the Lincolnshire fenland, the Somerset Levels, and the Thames Valley. There has been little research in Britain on the unsaturated zone of such areas until recently; more work has been conducted in the Netherlands, where shallow aquifers dominate the hydrology of a substantial part of the country.

From a hydrogeological viewpoint, the main interest in the unsaturated zone over a shallow aquifer is its control of the timing, rates and quantities of recharge, and discharge through it. However, plants in a shallow water table soil often make use of the groundwater; indeed the character of a natural habitat may depend upon its presence. Understanding plant use of groundwater, and hence groundwater discharge through the unsaturated zone, is therefore of interest from two perspectives: that of the hydrogeologist whose principal concern is the saturated zone, and that of the farmer or conservationist interested in the welfare of crops or a nature reserve, respectively. In fact, independent measurement of plant use of groundwater from shallow aquifers has been problematic until recently (see below). Water balances and models of the shallow unsaturated zone have had to rely on use of the Penman potential evapotranspiration estimate for well watered vegetation (Penman 1948).

Changes to the water-table above a shallow aquifer may be imposed due to, for example, groundwater exploitation or localized pumping associated with sand and gravel extraction. Modification of the water-table can have a significant impact on the vegetation: lowering may deprive plants of part of their water supply; raising the level can cause excessive soil water logging with

attendant problems. The groundwater models necessary to simulate the effects of changes of the water-table, have to incorporate some representation of unsaturated zone processes which will also change. A difficulty arises in coupling the one-dimensional descriptions of vertical water flux developed by soil physicists, with two- or three-dimensional saturated zone models.

Youngs *et al.* (1989) addressed the three-dimensional problem of a spatially-varying groundwater depth controlling evaporation by coupling a one-dimensional, steady-state description of the unsaturated zone processes controlling evaporation, to a two-dimensional groundwater model, derived from drainage theory. This allowed them to model the shape of the water-table for areas drained on all sides by ditches. Although the behaviour of the unsaturated zone was considerably simplified, this represents a good example of a model of the unsaturated zone being linked to a shallow groundwater model.

Such models vary in their degree of complexity but rely on the same fundamental assumption, i.e. that the upward flux from the water-table equals the smaller of either the potential evapotranspiration rate, or the maximum steady flux which can be conducted through the unsaturated zone from a water-table at the prevailing depth (Gardner 1958). Invariably such models require a considerable amount of detailed information regarding the hydraulic properties of the unsaturated zone, namely water release characteristics and unsaturated hydraulic conductivity functions for all soil layers, information which is difficult to supply.

An aim of recent work by the Institute of Hydrology has been to study the detail of the annual water regime of soils developed over a shallow aquifer and to evaluate the success of various models. Water potentials, contents, and levels have been monitored at several sites on the Thames flood plain to the north-west of Oxford. Periods when discharge via the unsaturated zone is occurring can be identified using the measurements of potential (Fig. 5.3). At sites where the water-table is very shallow throughout the year, water potentials fluctuate about the equipotential profile determined by the water-table depth. During dry summer periods, an upward potential gradient develops above the water-table, along which there

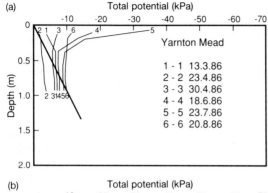

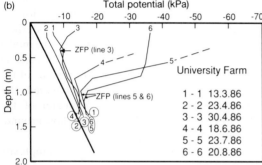

Fig. 5.3. Sequence of hydraulic potential profiles recorded at two shallow water table sites near Oxford in spring and summer 1986: (a) water table falls to 0.8 m, (b) water table fluctuates between 1.1 and 1.9 m.

is an upward flux of water (Fig.5.3(a)). Quantification of discharge, via the unsaturated zone, in these circumstances, requires the use of an estimate of evaporation rate; this was assumed to equal the Penman potential rate in the absence of any direct measure of plant water use.

At another site where the soil was developed on coarser sediments and the water-table was deeper, the lower part of the potential profile tended to equilibrate with the water-table (Fig. 5.3(b)). However, the form of the upper profile was influenced less by the shallow water-table; a zero flux plane was regularly observed during the summer months, indicating that there was no discharge from the groundwater. Table 5.1 shows the water balances which were calculated for these sites, and a third where the water-table fluctuation was intermediate between the two. The sites are within a kilometre of one another and so subject to very similar meteorological conditions, yet considerable differences in recharge and discharge quantities occur due to the physical properties of the soil and sediments comprising the unsaturated zone, and the depth of the water-table. Caution needs to be exercised in interpreting these figures, since evapotranspiration has been assumed to equal the potential rate at two of the sites, whereas it was estimated independently from soil water measurements at the other.

These datasets have also been used to evaluate MUST (Model for Unsaturated flow over Shallow water Tables) (De Laat 1985). It was found to simulate conditions in the unsaturated zone well in a normal, and a wet summer, 1985 and 1986 respectively (Gardner and Lumadjeng 1989).

The work described is ongoing. A new innovation is the use of a Hydra, a newly developed instrument which measures bulk transpiration rate over an area extending approximately 200 m upwind (Shuttleworth *et al.* 1988). A Hydra was first employed in the summer of 1989. It was found that evaporation from a river meadow which had been cut for hay was only 70 per cent of the Penman potential rate during July and August, then increased to near the Penman rate in early October. These data have, for the first time, provided a direct measure of water use by plants from soils over a shallow aquifer. The information will be used to test the basic assumptions of the models

Table 5.1   Water balance results for sites in the Thames Valley near Oxford, August 1985–August 1986

Site	Water table fluctuation (m)	Recharge (mm a^{-1})	Discharge (mm a^{-1})
Yarnton Mead	0.2–0.8	266	210
Long Pond	0.6–1.5	241	102
University Farm	1.1–1.9	323	11

referred to above and to calculate recharge to, and discharge from, shallow aquifers.

Hydrology of the deep unsaturated zone

Physical studies of the unsaturated zone of the English Chalk have produced results leading to a better understanding of both the processes of water flow, and the rates, timing, and magnitude of fluxes through the unsaturated zone to the water-table. This work has been conducted at several sites on the Middle and Upper Chalk (Wellings and Bell 1980; Wellings 1984*a*,*b*; Gardner *et al.* 1990; Cooper *et al.* 1990).

The approach taken has involved long term, field monitoring of the hydraulic behaviour of the chalk augmented by *in situ* measurements of unsaturated hydraulic conductivities. The results have provided detail as to the relative contributions of the chalk fissure system and its matrix in transporting water and solutes through the unsaturated zone, complementing other studies based on chemical profiling (e.g. Geake and Foster 1989). The methods which have been employed are applicable to other geological formations and work has recently commenced on the Triassic sandstones in south-west England (Cooper *et al.* 1990).

Sites and methodology

Seven sites have been studied. Four of these have shallow, rendzina type soils developed on weathered chalk. The depth of the base of the weathered chalk layer varies from 0.8–2.1 m. Such soils are characteristic of much of the chalk outcrop. At another site, the thin soil had been removed and replaced by a 0.2 m layer of gravel. The remaining two sites have a shallow cover of permeable clayey drift, varying in depth from 1–2 m, over the chalk. These drift sites were paired with nearby sites which have no drift. Grass grew on all but the gravel covered experiment, but at two sites additional plots were established in adjacent arable fields.

A similar approach was applied at each experimental site. Neutron probes and mercury manometer tensiometers were used to monitor soil water content and hydraulic potential changes on a weekly or more frequent basis, over a period of two or more years. The neutron probe access tubes and tensiometers were installed to a depth of 3 m or more at each site. At most sites, there was in addition a set of pressure transducer tensiometers and borehole pressure transducer tensiometers were installed to enable hydraulic potential measurements to several metres. Arrays of gypsum resistance blocks (Wellings *et al.* 1985) were employed to provide measurements when potentials fell below the operating range of tensiometers, (−80 kPa). An on-site tipping bucket rain gauge monitored rainfall hourly.

Hydraulic properties of the *in situ chalk*

Unsaturated hydraulic conductivity was measured at one site using the instantaneous profile method (Watson 1966; Hillel *et al.* 1972), and at three others using a combination of this and the constant rate infiltration method, as described by Poulovassilis *et al.* (1974) and Cooper (1979). The extra equipment required was that necessary to irrigate the site according to the method used, and a cover to prevent rainfall infiltration or evapotranspiration losses.

The unsaturated hydraulic conductivity-pressure potential relationship measured in the hydraulic conductivity experiments at four sites is shown in Fig. 5.4. The saturated conductivity values range from 50–1000 mm d^{-1} but for each site the curve falls steeply to 1–6 mm d^{-1} at a pressure potential of −3 to −5 kPa. At each site measurements continued until the pressure potentials fell below tensiometer range. However, below −5 kPa very little decrease in conductivity was measured in all cases.

The low conductivity values i.e. 1–6 mm d^{-1}, are in the same range as laboratory measurements of saturated conductivity of chalk sampled from both the Upper and Middle Chalk (Price *et al.* 1976). The near constancy of these values, despite falling potential, is consistent with *in situ* measurements of the pressure potential–water content relationship of the unweathered chalk at all the sites. They show virtually no change in chalk water content between potentials of −5 to −80 kPa; an example is illustrated in Fig. 5.5. The combination of these results indicates that the almost constant unsaturated conductivity value measured in the field

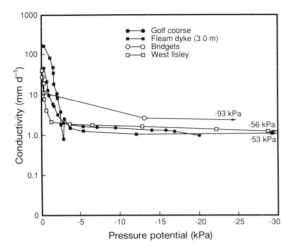

Fig. 5.4. Unsaturated hydraulic conductivity as a function of pressure potential, at 2.1 m depth, at four sites on chalk (after Wellings and Cooper 1983) The figures on the graph refer to the lowest pressure potential at which conductivity was measured. The lines shown would extend to these values.

represents the saturated conductivity of the chalk matrix. Supporting evidence that the chalk blocks of the matrix have almost no specific capacity until potentials below −50 kPa and much lower are attained, is provided by laboratory measurements of water release (Croney and Coleman 1954) and of chalk pore and pore throat sizes (Price *et al.* 1976; Price 1987).

The steep increase in conductivity at potentials

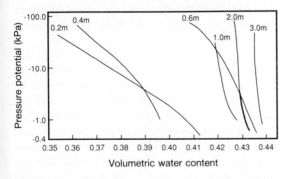

Fig. 5.5. *In situ* measurements at West Ilsley, Berkshire, of the relationship between hydraulic potential and water content, as the chalk profile drains (after Cooper *et al.* 1990).

above −3 kPa to saturation is attributed to the conductivity of the fissure system. As the graph demonstrates, the Chalk does not have to be saturated for flow in fissures to occur; the unsaturated conductivity increases rapidly as the fissure system fills. However, probably only a small part of the fissure system need be filled to increase the conductivity significantly above that of the chalk matrix.

Hydrological behaviour of the Chalk's unsaturated zone

The measurement of water potential profiles enabled assessment of the likely contribution of fissure flow at the various sites. Observation of pressure potentials above −3 to −5 kPa at several adjacent depths was interpreted as evidence of fissure flow in that zone of the Chalk. Figure 5.6(a) shows an example of this occurring in the top 3 m in two successive weeks early in the year. During three winters of measurement at this site, profiles of high potentials similar to those illustrated were observed on about 15 per cent of measuring occasions. At the other sites the evidence for fissure flow, or otherwise, was clearer. At some, water potentials persisted at values well above the −5 kPa threshold during the winter months. At others, they remained below it throughout most of the winter periods observed, indicating that fissure flow did not occur. Exceptional rainfall might cause potentials to increase temporarily to above −5 kPa, creating an isolated pulse of fissure flow, as for example the single event which was observed in five years of measurement at a site near Winchester (Wellings 1984*a*).

At those sites where evidence of fissure flow was not observed, the chalk matrix possesses sufficient conductivity to transmit drainage from the soils without the potentials being forced to rise into the 0 to −5 kPa fissure flow range. The presence of a soil, a layer of weathered chalk, and in some cases drift, provides a capacity to buffer the effects of heavy rainfall periods, soaking up the water and then releasing it slowly later. However, there seems, at present, no ready method to characterize a site's propensity to fissure flow on the basis of the nature of the soil, drift, or weathered chalk, or their respective depths. Indeed, the results from a site in Sussex serve to demonstrate that antecedent

(a)

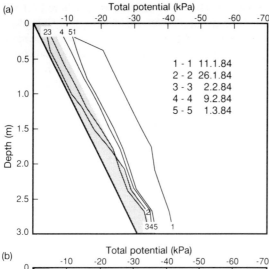

(b)

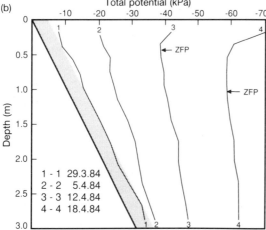

Fig. 5.6. Sequences of hydraulic potential profiles recorded at West Ilsley, Berkshire in 1984: (a) Mid-January to early March, (b) March to mid-April. ZFP denotes the zero flux plane depth and the shaded area indicates pressure potentials above −5kPa (after Gardner *et al.* 1990).

weather conditions can be an important factor in determining the form of the winter potential profiles; one year high potentials were frequently recorded whilst in the following drier winter, potentials remained below −5 kPa.

Figure 5.6(b) shows a sequence of potential profiles within which there is a change from a wholly draining profile to one where the potential gradient in the upper part induces an upward flux of water to the plant roots and soil surface.

Between the two parts of the profile i.e. that where the water potentials indicate that the flux is upward and that where it is downward, is a plane across which there is no water movement, a divergent 'zero flux plane' (ZFP). Measured water content changes above and below this ZFP can be ascribed to evapotranspiration and drainage respectively. This is the basis of the zero flux plane method of recharge estimation described in the following section.

Recharge estimation by soil physical methods

Soil drainage is the downward water flux at any point below both the maximum rooting depth and the maximum depth that the ZFP is likely to reach. Drainage from the soils developed above a deep aquifer will eventually recharge the aquifer. Storage changes within the unsaturated zone can modify the timing of the recharge. However, over a long period, the total quantity of soil drainage equates with recharge at the water-table. Thus, if soil drainage can be quantified, so can aquifer recharge.

An obvious approach is to use Darcy's law to calculate the downward water flux directly:

$$q = -K\,(\psi_m, \theta)\,\mathrm{d}\psi\,/\,\mathrm{d}z$$

where q is the water flux density in the vertical (downwards) direction;

K is the unsaturated hydraulic conductivity of the formation; at pressure potential ψ_m and volumetric water content θ;

ψ is the hydraulic potential of water;

and z is depth beneath the ground surface.

Unfortunately, hydraulic conductivity varies by several orders of magnitude over the seasonal range of water content found in most field soils and is also spatially variable. It is also very difficult to measure, and so this method is impracticable to apply in most operational situations. Soil water drainage models are frequently subject to similar limitations; much data, including unsaturated hydraulic conductivity functions, must be supplied to allow confidence to be placed in the results.

The zero flux plane method

One solution to these problems is the application of the so-called zero flux plane method. A zero flux plane is a level in the soil profile at which the total water potential gradient is zero. Above a divergent zero flux plane (ZFP) the potential gradient is upward, moving water towards the rooting zone. Below the ZFP the gradient is downward, inducing drainage. A ZFP is usually present at times when, on average, evapotranspiration exceeds rainfall. Figures 5.6(a) and (b) illustrate the development of a ZFP in a chalk profile; the zero flux plane concept has been discussed in detail by Wellings and Bell (1982).

The soil water balance equation:

$$\Delta S_z = P - R - E_A - D_z$$

where ΔS_z is water content change above depth z;

P is precipitation;

R is surface runoff;

E_A is actual evapotranspiration;

and D_z is drainage below depth z;

belies the problem of separating the evapotranspiration and drainage components of that balance. Rainfall quantities are readily measurable, as are soil water content changes, but partitioning the residual water content change into evapotranspiration and drainage requires information about the direction of movement of water within the soil, in order to distinguish upward from downward fluxes.

However, water content changes due to evapotranspiration and drainage, can be separated from one another when a ZFP is present. Since no water can flow through the ZFP any measured changes of water content in the profile above it will be due to exchange of water at the surface, either rainfall infiltration or evapotranspiration. Similarly, any change of water content below the ZFP will be due to drainage out of the base of the profile, which will appear eventually as groundwater recharge. The ZFP depth can be identified with tensiometers, which are simple and cheap devices, whilst water content can be measured readily by neutron probe or capacitance probe.

The zero flux plane method for measuring soil drainage is not a new idea, but studies at a variety of sites in the UK, principally on the Chalk (Cooper 1980; Wellings and Cooper 1983; Wellings 1984a; Gardner et al. 1990; Cooper et al. 1990) have demonstrated that it is a robust approach with an inherent precision of about 20 mm a^{-1}.

The soil water balance method

In Britain, ZFPs are rarely observed during the winter half of the year, and may be absent for short periods at other times, after heavy rainfall. In the absence of a ZFP another approach to obtaining soil drainage quantities is required. If no ZFP can be identified it is most often because the profile is wet and it is reasonable to assume that evapotranspiration losses will be close to the potential rate as calculated from meteorological data (Penman 1948) or supplied by the Meteorological Office MORECS Service (Thompson et al. 1981).

Drainage from the profile, and hence the groundwater recharge, can be calculated from a simple water balance of the soil profile, knowing rainfall, evapotranspiration, and measured soil water content changes. In temperate climates evapotranspiration rates are normally low during such periods, so that the method is unlikely to be unduly sensitive to errors in estimation of evapotranspiration.

Application of the combined ZFP and water balance approach

Evapotranspiration and drainage on a plot scale can be calculated by combining the water balance approach with the zero flux plane method. The method gives not only gross annual recharge, but also its seasonal distribution, and the changing water flux with depth. It has been applied at the Middle and Upper Chalk sites referred to earlier and Table 5.2 summarizes the recharge measurements. Mean annual recharge totals are given for several periods for some sites to permit inter-site comparisons. Figure 5.7 illustrates the cumulative values of the components of the annual water balance for 1981, for the Fleam Dyke site, close to Cambridge.

Greater recharge amounts were measured at the Hampshire site than in Cambridgeshire, and attributed to the much higher rainfall there. The measurements at the Berkshire sites were made over

Table 5.2 Recharge rates to the English Chalk aquifer determined by the combined zero flux plane and water balance method

Aquifer	Average annual rainfall (mm a⁻¹)	Soil	Crop	Years	Mean annual recharge (mm a⁻¹)
Upper Chalk Hampshire	798	Thin	Grassland	1976–81	348
				1979–81	258
		Thin	Arable	1979–81	266
Upper Chalk Berkshire	760	Thin	Grassland	1983–4	160
Upper Chalk Berkshire	724	Shallow drift	Grassland	1983–4	264
Middle Chalk Cambridgeshire	550	Thin	Grassland	1979–82	147
				1979–81	132
				1980–82	126
		Thin	Arable	1980–82	132
Middle Chalk Cambridgeshire	550	Thin	Grassland	1980–82	130
Middle Chalk Cambridgeshire	550	Shallow drift	Grassland	1979–81	162

(Figures from Wellings 1984*a*; Cooper *et al*. 1990)

different years so are not directly comparable. However, despite the relatively high average annual rainfall, recharge at the thin soil site in Berkshire was similar to the Cambridgeshire figures. This small value is, at least in part, due to the large atmospheric evaporative demand at this particularly exposed and so very windy site. An automatic weather station was used there and calculations of Penman potential evapotranspiration rate, using data from it, were significantly higher than at the nearby site.

This presents problems when comparing the two Berkshire sites. The fact that more recharge occurred at the drift covered site, must largely be due to the different meteorological conditions there. However, more recharge was also measured at the drift covered Cambridgeshire site, relative to its neighbours, where the meteorological conditions were similar. Also, a fairly even distribution of recharge throughout the year was found at both drift covered sites. This contrasts with the thin soil sites from which drainage ceased during the main summer ZFP period, as in Fig. 5.7. Evidently the presence of a shallow cover of drift (1–2 m depth) over the Chalk affects the timing of recharge, and may increase the amount relative to that from areas having thin soils (Gardner *et al.* 1990; Cooper *et al.* 1990).

Arable cultivation was found to have little effect on recharge from the sites in question, as indicated in Table 5.2. However, what is not apparent from Table 5.2 is just how variable recharge to the Chalk was, year to year, at each site. Though the variations relate to differences in rainfall amount, no simple relationship between rainfall and drainage can be derived as both the timing and quantity of the rainfall are significant (Gardner *et al.* 1990).

Geochemical techniques applied to the unsaturated zone

The chemistry of water in the unsaturated zone has been much studied in the developed world

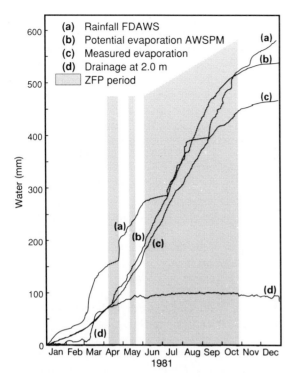

Fig. 5.7. Cumulative values of the components of the 1981 water balance for Fleam Dyke, Cambridge (after Cooper *et al.* 1990).

since the 1960s, primarily to gain information about sources and rates of groundwater pollution; recent advances in that area are covered in Chapters 8 and 10. However, chemical techniques can also be used to give more basic information about water movement in the unsaturated zone and rates of groundwater recharge. This brief review considers how recent developments in geochemistry in the wider sense (including isotopic techniques) are being used for these purposes in the United Kingdom and to assist the developing world.

Estimation of recharge rate by solute balance

This technique is based on the concept that the increase in concentration of a particular solute in pore-water in the unsaturated zone, over its concentration in rainfall, is a direct result of concentration by evapotranspiration. Provided that average amounts of rainfall and its concentration of solutes

are known, an average groundwater recharge figure can be calculated. Certain assumptions must be made to use the method with confidence, the chief of which is that the solute is conservative, being neither added to nor subtracted from the unsaturated zone by any organic or inorganic agency. The chloride ion is usually selected for measurement, because it is normally the least-affected species in water in the unsaturated zone. To obtain a satisfactory average concentration for unsaturated zone waters, it is necessary to make a series of pore-water measurements from the surface downwards via a suitable sampling technique (hand augering or dry drilling). Variations in near-surface concentrations are due to fluctuations in the direction of water movement, but beneath the zero flux plane (Wellings and Bell 1980) the depth profile settles to a 'steady state' (Fig. 5.8) and the average concentration should be similar to the

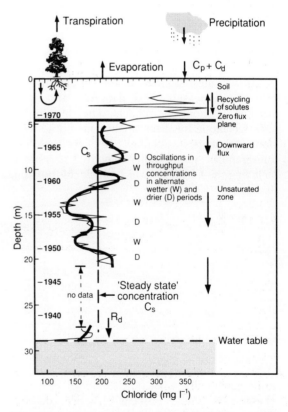

Fig. 5.8. The solute balance method of recharge estimation illustrated by an unsaturated zone chloride profile sampled in Cyprus.

concentration finally entering the aquifer. The method is summarized by the following expression:

$$R_d = P \frac{C_p + C_d}{C_s}$$

where R_d is recharge,

  P is amount of precipitation,

  C_p is average chloride concentration in rainwater,

  C_d is a term covering any dry depositional sources of chloride (dust, aerosols),

and C_s is the average chloride concentration of deep pore-water.

The chloride balance technique is usually most appropriate to 'natural sites' where there have been no anthropogenic influences such as the use of fertilizers.

Application in Britain

Many years of data for rainfall exist for the UK, and reliable averages can be calculated. Long-term information about chloride in rainfall is less satisfactory both with respect to quantity and quality, though, because of improvements in the precision of low-level measurements, data from at least the last decade should be reliable. The chloride balance technique has been applied mainly to sandstone lithologies, although any material with intergranular permeability would be suitable, including the Chalk for which much chloride data has already been gathered during diffuse pollution studies. Table 5.3 gives recharge rates calculated for various sandstone aquifers (Edmunds *et al.* 1989; Moss and Edmunds 1989). The results are broadly in line with general recharge rates estimated from water balance studies, but have the advantage that they can be applied very locally, to illustrate, for example, the difference in recharge rate beneath open ground and woodland.

Application overseas

In developing countries, particularly those with semi-arid climates, calculation of recharge rates by conventional water balance methods is often not feasible because of cost and logistical difficulties. The chloride balance technique largely circumvents these problems, and has been applied in Cyprus (Edmunds and Walton 1980), where calibration with a lysimeter was also carried out (Kitching *et al.* 1980), northern Sudan (Edmunds *et al.* 1988) and Senegal (Edmunds 1989). Estimated recharge rates from individual sites cover a range up to 94 mm a^{-1}; Table 5.4 shows that the method gives credible average recharge amounts for each of the areas. However, the accuracy of the estimates is constrained by the adequacy of historical data for chloride in rainfall and estimates of dry chloride deposition, the term C_d in the equation, becoming more significant with increasing aridity. There is therefore a need for increased effort to be devoted to the measurement of dry deposition and rainfall chemistry in these areas.

Table 5.3 Recharge rates in UK aquifers calculated by chloride balance

Aquifer	Rainfall (mm a^{-1})	Rainfall Cl$^-$ (mg l^{-1})	Land-type	Pore-water Cl$^-$ (mg l^{-1})	Recharge (mm a^{-1})
Tertiary sands, Surrey	750	3.0	Open	8	280
Cretaceous sands, Surrey	830	3.0	Open	15	165
Cretaceous sands, Surrey	830	3.0	Open	13	190
			Wooded	39	65
Triassic sandstone, W Midlands	700	2.3	Open	4.5	360
Triassic sandstone, W Midlands	700	2.3	Wooded	15	105

(Figures from Edmunds *et al.* 1989; Moss and Edmunds 1989)

Table 5.4   Recharge rates in semi-arid climates calculated by chloride balance

Country	Rainfall (mm a^{-1})	Total Cl$^-$ deposition equiv (mg l$^-$)	Pore-water Cl$^-$ range (mg l^{-1})	Pore-water Cl$^-$ mean (mg l^{-1})	Mean recharge (mm a^{-1})
Cyprus	406	26	83–650 (8 sites)	183	57
Senegal	325	2	26–290 (9 sites)	131	4.9
Sudan	200	5	173–3936 (14 sites)	1629	1.0

(Figures from Edmunds *et al.* 1988; Edmunds 1989)

Isotopic techniques

Oxygen and hydrogen isotope techniques possess a major advantage over chemical solutes as a means of studying water movement, because they are part of the water molecule and therefore not subject to processes which may change the concentration of chemical species in the unsaturated zone. (Even rain-derived chloride, normally highly conservative, may in certain instances be supplemented, for example from evaporite deposits). Studies have been based on the natural variation of the stable isotopes ^{2}H and ^{18}O and the radioisotope ^{3}H, and on the use of ^{2}H-enriched water as a tracer.

Application in Britain

Cyclic variations of isotope ratios with depth in the soil zone at a site in Hampshire were linked to seasonal variations in rainfall input and used to infer a downward water movement through Upper Chalk of approximately 1 m a^{-1}, which was consistent with movements of nitrate and chloride at the same site (Wellings and Cooper 1983). Following detailed studies of stable isotopes at a site on Middle Chalk in Cambridgeshire, Bath *et al.* (1982) and Darling and Bath (1988) came to a number of conclusions: 'steady state' isotope values of pore-waters were similar to those of weighted average annual rainfall, implying a contribution from rainfall at all times of the year; there were differences in the isotopic content of matrix and macropore

water which varied with land use, although bulk recharge compositions remained the same; the 'piston displacement' model of flow in the unsaturated zone, developed from results from southern England, was inappropriate to the Cambridgeshire site. This was attributed to a lower recharge rate in eastern England, which was the combined result of geological differences and a lower rainfall.

Partly to confirm the apparent annual nature of cyclic variation of stable isotope depth profiles obtained in southern England, an experiment using ^{2}H-labelled water was carried out on Upper Chalk in Hampshire (Wellings 1984*b*). This supported the earlier views of Wellings and Cooper (1983) by demonstrating downward movement of a ^{2}H peak by just over 1 m a^{-1}, and also showed that nitrate, chloride, and water all moved at a similar rate through the Chalk. Further tracer work on the Upper Chalk in Berkshire by Gardner *et al.* (1990) showed a ^{2}H movement rate of 0.8 m a^{-1}, which was paralleled by that of a chloride tracer applied at the same time.

The sharpness of the thermonuclear tritium (^{3}H) peak resulting from 1963–5 rainfall continued to decline in the Chalk's unsaturated zone during the 1980s, owing to the combined effects of dispersion and radioactive decay. However, the peak remained easily detectable, and tritium follow-up studies were carried out in Cambridgeshire and west Norfolk to supplement work started in the late 1970s (Foster and Smith-Carington 1980; Foster and Bath 1983). These studies found rates of water movement in the unsaturated zone that were similar to those inferred from previous environmental ³H and tracer ²H work at other sites, for example a rate of 0.7 m a⁻¹ for west Norfolk (Geake and Foster 1989). Differences in the shapes of depth profiles between sites were consistent with the stable isotopic evidence that modes of infiltration to the Chalk vary as a function of geographical and geological positions (Darling and Bath 1988).

Application overseas

Despite the promise of quantitative assessment of evaporation in arid areas held out by stable isotopic techniques in the last decade (summarized by Fontes and Edmunds 1989), application has proved difficult in developing countries because of a lack of supplementary soil physical and meteorological information. Nevertheless, isotopic techniques have become an important qualitative tool for research into flow processes in the unsaturated zone in arid areas.

Difficulties in sampling the unsaturated zone's profile in arid terrain are both logistical and methodological. Various techniques have been tried: dry percussion drilling, augering, and sampling during dug well construction (Edmunds *et al.* 1988). Further problems are encountered with the extraction of water from soil and rock with low water contents: the process used must not cause isotopic fractionation. Although centrifugation techniques such as the immiscible liquid displacement method of Kinniburgh and Miles (1983) are adequate for water contents in sand above 10–15 per cent by weight, other methods have had to be developed for lower water contents. Extraction by vacuum has proved a suitable technique when $\delta^{18}O$ is required, but direct reduction, which reduces handling steps to a minimum has been developed to provide rapid δ^2H profiling (Darling and Talbot 1989).

An example of the application of isotopes to infiltration in arid terrain is provided by an integrated study of rainfall, pore-water in the unsaturated zone, and groundwater undertaken in northern Sudan and summarized in diagrammatic form in Fig. 5.9 (Darling *et al.* 1987). Near-surface soil water has extremely enriched heavy isotope values as a consequence of high surface evaporation, and from these values amounts of evaporation can be calculated, given certain assumptions

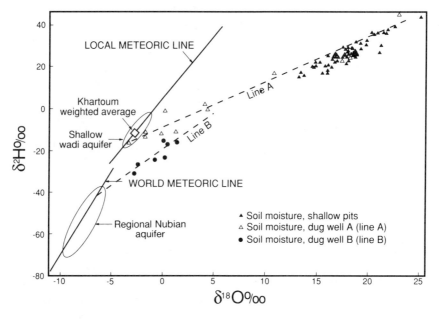

Fig. 5.9. Delta plot of unsaturated zone water from the Butana region of northern Sudan. Also shown are rainfall and groundwater compositions, and the local and world-wide rainfall lines.

about soil physical conditions and atmospheric relative humidity. A dug well (A) at the same site shows a smooth progression from near-surface soil isotope values to a composition close to the underlying local wadi aquifer, whereas another well (B), some 80 km away, shows a progression towards the regional aquifer composition. If significant recharge were occurring, these profiles would be similar, but instead they appear to be the product of diffusion between the soil zone and the water in whichever aquifer lies beneath, with the implication that the water in the unsaturated zone is essentially static. The chloride balance work of Edmunds *et al.* (1988), cited above, confirms that recharge rates are close to zero. Additional indirect evidence that recharge is negligible away from wadi beds is obtained from the isotopic composition of regional groundwater in the Nubian formation. This was recharged during a postglacial pluvial episode recognized elsewhere in northern Africa (Bath 1983; Fontes and Edmunds 1989) and retains a distinctive composition separate from that of modern shallow groundwater.

Comparatively little use has been made of thermonuclear tritium for measuring water movement in the unsaturated zone in arid areas. A reasonably well-defined ^{3}H peak was found in semi-arid Cyprus (Edmunds and Walton 1980), but in general little work was carried out in the 1980s, and with increasing time and southerly latitude, the technique may not be easily applicable in Africa during the 1990s. The main reason that tritium has not been used more in infiltration studies of arid areas is the logistical problem of transporting large, low water content soil and rock samples back to the laboratory.

Directions of future research

Whereas in the United Kingdom physical data and techniques are applied to studies of the unsaturated zone, rather than geochemical methods, in developing countries, especially those with arid climates, such data and techniques are not generally available. The application of geochemistry to problems of water movement in the unsaturated zone in dry areas has, therefore, become established during the last decade as an important element in understanding the water balance, and

during the next decade should embrace an increasingly well-defined set of techniques for estimation of infiltration and evaporation. The two categories of geochemical investigation considered here—solute balance methods and isotopic techniques—are not, of course, mutually exclusive, and indeed have been applied together whenever practicable. Particularly in developing countries, the solute balance method is cost-effective and should form the main thrust for assessing amounts of recharge, constrained where necessary by isotopic evidence. For studies where a knowledge of evaporation rate is important, isotopic methods hold great promise for the future.

Developments in the inverse method for estimation of groundwater recharge

Important problems in the evaluation of groundwater resources include the estimation of aquifer properties (storage coefficient and transmissivity) and of aquifer recharge. These are difficult to measure directly in the field; it is therefore natural to try to develop indirect methods for their estimation. One such method, generally known as the inverse method, operates through the use of mathematical models, which may then subsequently be used to predict the behaviour of the system. Much research has been undertaken into the estimation of aquifer parameters, but less has been carried out on the equally important problem of aquifer recharge estimation.

Previous studies

The problem of recharge estimation has been approached in two ways. Krishnamurthi *et al.* (1977) and Rushton and Ward (1979), have tackled this problem by considering the processes of infiltration. Venetis (1971), Gelhar (1974), Besbes *et al.* (1978), and Gelhar *et al.* (1979) have used water-table data as a starting point but proceeded in diverse ways. Gelhar used a stochastic approach to identify the recharge mechanism and hydrogeological parameters, whilst Venetis, although working within a deterministic framework, restricted his field of application to a problem with one spatial dimension and one particular form of boundary condition. Besbes and collegues were concerned with predicting the recharge from specific flood events.

Recent studies

In short-term arid zone studies of groundwater, it is often found that the least satisfactory aspect is the quantification of recharge, although qualitative information concerning its distribution is often available. Research has therefore been directed at developing methods which make use of this qualitative information, together with field estimates of aquifer parameters, to provide initial quantitative estimates of recharge. In arid and semi-arid environments, recharge, although variable, follows an essentially cyclical pattern, thus allowing the use of a steady state formulation (Nutbrown 1975).

Smith and Wikramaratna (1981) outlined a method, which they called the inferred recharge method, which is particularly applicable to semi-arid zones. The partial differential equation describing steady state groundwater flow is approximated by a set of simultaneous equations with recharge as the dependent variable, water levels as the independent variable, and the parameter of transmissivity as the coefficient matrix linking the two. The obvious method of estimating the recharge by evaluating these matrix equations, commonly known as the implied recharge method, was shown to be undesirable even for small errors in water level and transmissivity data. The inferred recharge method uses information about the location of recharge areas to obtain a more robust numerical method for estimating recharge. The linearity of the formulation allows the solution to be written in closed form and therefore avoids the need for an iterative solution. It is also possible to obtain measures concerning the reliability of the results with reference to errors in the data.

By applying their algorithm to a typical problem, Smith and Wikramaratna drew the following conclusions:

(a) The inferred recharge method is an improvement on the more obvious implied recharge method.

(b) An error in the transmissivity distribution has less effect than the same proportional error in the water level observations. This means that the method can be applied in the early stages of modelling studies when only a crude assessment of the hydrogeological parameters is available. These early results can be used to modify collection of field data, resulting in a better description of the hydrogeological processes. In this case, the inferred recharge should not be viewed as the definite recharge distribution.

(c) The sensitivity of the model to data errors increases with the proportion of the modelled area which receives recharge; the limiting case where recharge occurs at all locations corresponds to the implied recharge model. The advantages of the inferred recharge model depend on being able to restrict the recharge to a small proportion of the modelled area. In many practical problems this can be a serious limitation.

This inferred recharge model has been applied successfully during consultancy projects in Jordan, Botswana, Somalia, and Qatar.

Wikramaratna and Reeve (1984) proposed modifications to the method of Smith and Wikramaratna (1981). It was assumed that the recharge can be divided into a number of different types, each of which is confined to a known area within the region of interest. Within each area, it is assumed that the distribution of the corresponding recharge type is known, although its quantity is unknown. At any given location there could be more than one recharge type. For example, it could be assumed that recharge from rainfall occurs at all nodes of the model, and is distributed in proportion to the nodal areas. On the other hand, recharge from a river might be restricted to nodal areas beneath the river, and proportioned according to the length of river within each nodal area. This modification means that only a small number of types of recharge need to be estimated, even though recharge could be occurring at a large proportion of the nodes. Thus the modification overcomes the most serious limitation to the practical application of the original inferred recharge method, which is that it depends on being able to restrict the recharge to a small proportion of the modelled area. It has the additional advantage that it uses recharge types, such as rainfall and river recharge, and thus parallels the approach used in deriving a water balance.

The inferred recharge model, as proposed by Smith and Wikramaratna (1981), requires water level data at the model nodes and it calculates values which give the best least squares fit to these nodal water levels. In practice, data points do not correspond exactly to model nodes, and there may be considerable variation in the reliability of the data. Interpolation between data points to give nodal values presents a real problem, especially when water level data are sparse or unevenly distributed. A further refinement, introduced by Wikramaratna and Reeve (1984), allows recharge values to be found which give the best weighted least squares fit to the water levels at the data points, the weights being chosen to reflect the reliability of the data.

The modified inferred recharge model is more versatile than the original, since it can be applied to problems where recharge occurs at a large proportion of the model nodes. It is also more objective since it uses water level data at the locations where this is measured thus removing the need for subjective interpolation between data point and nodal values. The modifications proposed make the inferred recharge model a substantially more useful technique for application to practical problems.

The method must be used with caution. Its application depends on the formulation of a conceptual model of the recharge processes. The reliability of the recharge estimates will depend on the quality of the hydrogeological interpretation. The method will always produce a solution, but clearly if the wrong conceptual model has been applied, this solution may be meaningless and misleading. The agreement between the observed and best fit water levels provides an indication of the quality of the conceptual model and allows comparison between different, hydrogeologically reasonable models of the system.

The modified inferred recharge model has been applied during consultancy work both overseas and in the UK to provide an initial estimate of the magnitude of recharge from different sources.

The modified inferred recharge method

Description

The two-dimensional steady state flow of groundwater through an aquifer can be described by the equation:

$$\nabla . [T(x,y) \nabla \phi (x,y)] = r(x,y)$$

where (x,y) is a coordinate vector,
∇ is the differential operator,
T is the transmissivity,
ϕ is the hydraulic head,
and r is a source/sink term per unit area,
$r > 0$ representing abstraction and
$r < 0$ representing recharge.

This equation is defined in a region and is subject to Dirichlet and Neumann boundary conditions.

Using the finite difference method, equations can be approximated by a set of simultaneous equations of the form:

$$M\phi = b + q$$

where M is a matrix of coefficients involving the parameters of transmissivity and the constant head boundary conditions,
b are the components of recharge whose location and magnitude are known, for example abstraction from wells,
and q are the recharges to be estimated.

If it is assumed that there are several recharge types for which the distributions are known but the relative magnitudes unknown, q can be written as:

$$q = Bt$$

where t are the magnitudes of the unknown recharge types,
and B is a matrix containing coefficients which relate the amount of each recharge type to the recharge at the individual model nodes.

Within the study area the hydraulic head will be known at a certain number of data points; in general these will not correspond with the model nodes. To take account of this, a matrix, Ω, is introduced which interpolates between the nodal heads, ϕ, to give the predicted heads at the data points.

For any set of different types of recharge, t_0, the weighted sum of squared errors at the data points can be defined as :

$$S(t_0) =$$
$$\cdot(h - \Omega M^{-1}Bt_0 - \Omega M^{-1}b)\, 'w'w\, (h - \Omega M^{-1}Bt_0 - \Omega M^{-1}b)$$

where Ω is a diagonal weighting matrix and the primes denote the matrix transpose; the diagonal elements of w represent the reliability of the water level observations at the data points. For example, water levels measured at boreholes which have been accurately surveyed would be more reliable and therefore given a higher weight than those from a borehole levelled by altimeter. The best estimate of the magnitude of the different recharge types is found by minimizing $S(t_0)$ and is given by:

$$B'(M^{-1})\,'\Omega'w'w\Omega M^{-1}Bt_0 =$$
$$B'(M^{-1})\,'\Omega'w'wh - B'(M^{-1})\,'\Omega'w'w\Omega M^{-1}b$$

Application

To investigate the performance of the modified inferred recharge method, Wikramaratna and Reeve (1984) applied it to a problem based on the coastal aquifer of northern Oman. This was the same problem as that used by Smith and Wikramaratna (1981) when they first suggested the inferred recharge method. The exact solution to the problem has equal amounts of recharge at six nodes corresponding to recharge from minor wadis only, at one node there is recharge from both major and minor wadis, and three further nodes have major wadi recharge only.

The true solution to the test problem is known so the effects of different water level and transmissivity errors on the recharge estimates can be investigated by adding pseudo-random errors to the true values of these parameters at the nodes. In order to investigate statistically the effects that different water level and transmissivity errors have on the recharge a number of replications were carried out.

Figure 5.10 compares the error variance of the modified inferred recharge method with the original inferred recharge method and the implied recharge technique. Figures 5.10(a), (b), and (c)

compare the results with exact transmissivities and different amounts of added water level error. It is clear that the modified method produces substantially smaller recharge errors than the original inferred recharge method. Both techniques are a significant improvement on the implied recharge technique.

Figures 5.10(d), (e), and (f) illustrate the results for the cases when the transmissivity error varies for a number of water level errors. In general, the modified method produces a smaller recharge error variance. The exceptions to this occur where there are large transmissivity errors and very small water level errors; these points are circled in Fig. 5.10(d). Both of the inferred recharge methods are significantly better than the implied recharge technique. As well as resulting in a lower overall error variance, the modified inferred recharge method gives a lower error variance at each of the model nodes which receive recharge.

Concluding comments

During the past decade progress in studies of the unsaturated zone and the estimation of groundwater recharge has consolidated work started mainly in the 1970s. Many techniques have been brought to a state where they may be used in practical situations, and their strengths and weaknesses have been highlighted.

Soil physical and geochemical investigations of the deep unsaturated zone of the English Chalk have continued to the point where there is now a reasonably good understanding of the processes of water movement through it, and attention is turning to the unconfined sandstone aquifers. Serious study of the hydrology of the unsaturated zone above shallow aquifers has only recently commenced and the detail of the processes has yet to be fully clarified.

The effect of stratal variation on hydraulic properties within the unsaturated zone still requires investigation. Vertical variation in the unsaturated hydraulic conductivity function will influence the timing (and in some cases the quantity) of recharge to an aquifer. Water residence times within the unsaturated zone are very important in pollutant transfer and a better understanding of vertical variation of flow processes within the unsaturated

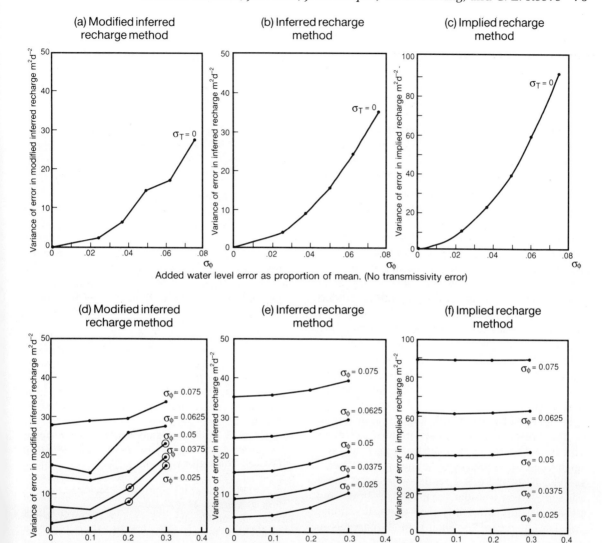

Fig. 5.10. Error variance for modified inferred recharge, original inferred recharge, and implied recharge methods with different combinations of added error.

zone will facilitate improvements to models of pollution.

The problem of obtaining areal values from *in situ* measurements to represent the unsaturated zone above both deep and shallow aquifers has yet to be addressed successfully. There is more small scale areal variability of processes within the unsaturated zone above a shallow water-table than over a deep aquifer, which poses particular difficulties

for measurement and representation. The soil physical and geochemical approaches which have been described are inherently 'point' methods. This makes them very suitable for investigating processes but less appropriate for describing water movement through the unsaturated zone on an areal basis. Geostatistical techniques may offer a solution to the problem of quantifying the variability of the unsaturated zone, although the cost implications of attempting to acquire the necessary data using some of the methods described here are considerable. However, even if spatial variability can be described, another problem is to understand the relationship between it, and its aggregate effect upon water movement to the aquifer below.

Research effort is required to look specifically at the link between point measurements and areal averages of recharge, possibly by interfacing point data with modelling techniques. In addition, practical methods for obtaining measurements representative of larger areas should be sought and investigated.

It can be argued that whilst modelling of the saturated zone is in advance of the understanding of saturated zone processes, the reverse applies to the unsaturated zone. Knowledge of processes has not been incorporated into models to any great extent, and the use of models to improve understanding of the processes involved has been very limited. This rift between modelling and processes needs to be bridged. The frequently complex and data-hungry models which are developed for research purposes will have to be supplanted by models capable of application in practical situations.

The present trend for hydrological modelling to be the preserve of computer-oriented scientists with little experience of field hydrology militates against the development of sound, process based models. Also, there is a tendency amongst those using models to consider them as black boxes and not to question sufficiently their physical validity. This lack of regard for what actually happens in the field needs to be redressed somewhat if models are to be applied to best effect, particularly in extrapolative applications to predict the consequences of land use and climate changes.

Present comprehension of water flow through the unsaturated zone assumes persistence of the prevailing climatic conditions. Given the prospect of climatic change, the development of models to forecast its probable impact on groundwater resources and quality should be urgently addressed. Processes now dominant may not maintain their importance under a slightly different climatic regime but be replaced by others which are currently less important. The consequent effects on recharge quantity and pollution susceptibility should be amenable to estimation.

References

Ansoult, M., De Backer, L. W., and Declercq, M. (1985). Statistical relationship between apparent dielectric constant and water content in porous media. *Soil Sci. Soc. Am. J.*, **49**, 47–50.

Bath, A. H. (1983). Stable isotopic evidence for palaeo-recharge conditions of groundwater. In *Palaeoclimates and palaeowaters*, pp. 169–86. I.A.E.A., Vienna.

Bath, A. H., Darling, W. G., and Brunsdon, A. P. (1982). The stable isotopic composition of infiltration moisture in the unsaturated zone of the English Chalk. In *Stable isotopes* (ed. H. Schmidt, H. Fröstel, and K. Heinzinger), pp. 161–6. Elsevier, Amsterdam.

Bell, J. P. (1976). *Neutron probe practice*. Rpt 19. Inst. Hydrol. Wallingford, Oxon.

Bell, J. P., Dean, T. J., and Hodnett, M. G. (1987). Soil moisture measurement by an improved capacitance technique: II. Field techniques, evaluation, and calibration. *J. Hydrol.*, **93**, 79–90.

Besbes, M., Delhomme, J. P., and de Marsily, G. (1978). Estimating recharge from ephemeral streams in arid regions: a case study at Kairouan, Tunisia. *Water Resour. Res.*, **14**, 281–90.

Cooper, J. D. (1979). Water use of a tea estate from soil moisture measurements. *E. Afr. Agric. For. J.*, **43**, 102–21.

Cooper, J. D. (1980). *Measurement of water fluxes in unsaturated soil in Thetford Forest*. Rpt 66. Inst. Hydrol. Wallingford, Oxon.

Cooper, J. D., Gardner, C. M. K., and Mackenzie, N. (1990). Soil water controls on recharge to aquifers. *J. Soil Sci.* **41**, 613–30.

Croney, J. and Coleman, J. D. (1954) Soil structure in relation to soil suction (pF). *J. Soil Sci.*, **5**, 75–84.

Darling, W. G. and Bath, A. H. (1988). A stable isotope study of recharge processes in the English Chalk. *J. Hydrol.*, **101**, 31–46.

Darling, W. G. and Talbot, J. C. (1989). Extraction methods for the isotopic analysis of porewaters. Report WD/89/66. British Geological Survey, Keyworth.

Darling, W. G., Edmunds, W. M., Kinniburgh, D. G., and Kotoub, S. (1987). Sources of recharge to the basal Nubian sandstone aquifer, Butana region, Sudan. In *Isotope techniques in water resources development*, pp. 205–24. I.A.E.A., Vienna.

De Laat, P. J. M. (1985). *MUST—a simulation model for unsaturated flow*. Int. Inst. Hydraul. Env. Eng. Rpt. 16, Delft.

Dean, T. J., Bell, J. P., and Baty, A. J. (1987). Soil

moisture measurement by an improved capacitance technique: I. Sensor design and performance. *J. Hydrol.*, **93**, 67–78.

Edmunds, W. M., (1989). Groundwater recharge project (Senegal): report of progress to March 1989. Report WD/89/5R. British Geological Survey, Keyworth.

Edmunds, W. M. and Walton, N. R. G. (1980). A geochemical and isotopic approach to recharge evaluation in semi-arid zones—past and present. In *Arid zone hydrology*. pp. 47–68. I.A.E.A., Vienna.

Edmunds, W. M., Darling, W. G., and Kinniburgh, D. G. (1988). Solute profile techniques for recharge estimation in semi-arid and arid terrain. In *Estimation of natural groundwater recharge* (ed. I. Simmers), pp. 139–57. Reidel, Dordrecht.

Edmunds, W. M., Cook, J. M., Kinniburgh, D. G., Miles, D. L., and Trafford, J. M. (1989). Trace element occurrence in British groundwaters. Report SD/89/3. British Geological Survey, Keyworth.

Fontes, J-C. and Edmunds, W. M. (1989). *The use of environmental techniques in arid zone hydrology.* UNESCO, Paris.

Foster, S. S. D. and Bath, A. H. (1983). The distribution of agricultural soil leachates in the unsaturated zone of the English Chalk. *Env. Geol.*, **5**, 53–9.

Foster, S. S. D. and Smith-Carington, A. K. (1980). The interpretation of tritium in the Chalk unsaturated zone. *J. Hydrol.*, **46**, 343–64.

Galfy, J. (1984). *High frequency moisture probe.* Commercial brochure, Vituki Res. Centre, Budapest.

Gardner, C. M. K. and Lumadjeng, H. S. (1989). Responses of soil moisture regimes to changes in groundwater level. In *Flow regimes from experimental and network data (FREND)*, Vol. 1, pp. 118–222. Inst. Hydrol., Wallingford, Oxon.

Gardner, C. M. K., Cooper, J. D., Wellings, S. R., Bell. J. P., Hodnett, M. G., Boyle, S. A., and Howard, M. J. (1990). Hydrology of the unsaturated zone of the Chalk of south-east England. In *Chalk*, pp. 611–18, *Proc. International Chalk Symposium, Brighton 1989.* Thomas Telford, London.

Gardner, W. R. (1958). Some steady-state solutions of the moisture flow equation with application to evaporation from a water table. *Soil Sci.*, **85**, 228–32.

Geake, A. K. and Foster, S. S. D. (1989). Sequential isotope and solute profiling in the unsaturated zone of the British Chalk. *Hydrol. Sci. J.*, **34**, 79–95.

Gelhar, L. W. (1974). Stochastic analysis of phreatic aquifers. *Water Resour. Res.*, **10**, 539–45.

Gelhar, L. W., Gross, G. W., and Duffy, C. J. (1979). Stochastic methods of analysing groundwater recharge. In *The hydrology of areas of low precipitation* (Proc. Canberra Symp., December 1979), pp. 313–21. Publ. no. 128. IAHS.

Greacen, E. L. (ed.) (1981). *Soil water assessment by the neutron method.* CSIRO, Australia.

Hillel, D., Krentos, V. D., and Stylianou, Y. (1972). Procedure and test of an internal drainage method for measuring hydraulic characteristics *in situ. Soil Sci.*, **114**, 395–400.

Hoekstra, P. and Delaney, A. (1974). Dielectric properties of soils at UHF and microwave frequencies. *J. Geophys. Res.*, **79**, 1699–708.

Kinniburgh, D. G. and Miles, D. L. (1983). Extraction and chemical analysis of interstitial water from soils and rocks. *Env. Sci. Tech.*, **17**, 362–8.

Kitching, R., Edmunds, W. M., Shearer, T. R., Walton, N. R. G., and Jacovides, J. (1980). Assessment of recharge to aquifers. *Hydrol. Sci. Bull.*, **25**, 217–35.

Kitching, R., Bell, J. P., Edworthy, K. J., and Tate, T. K. (1981). New instrumentation in hydrogeology. In *A survey of British hydrogeology 1980*, pp. 113–23. Royal Society, London.

Krishnamurthi, N., Sunada, D. P., and Longenbaugh, R. A. (1977). Mathematical modelling of natural groundwater recharge. *Water Resour. Res.*, **13**, 720–4.

Kuraz, V. and Matousek, J. (1977). A new dielectric soil moisture meter for field measurement of soil moisture. *Int. Comm. Irrig. Drain. Bull.*, **26**, 76–9.

Malicki, M. (1983). A capacity meter for the investigation of soil dynamics. *Zesz. Probl. Postepow Nauk Roln*, 201–13.

Moss, P. D. and Edmunds, W. M. (1989). Interstitial water–rock interaction in the Permo-Triassic sandstone aquifer. In *Water–rock interaction*, (ed. D. L. Miles), pp. 495–9. Balkema, Rotterdam.

Nutbrown, D. A. (1975). Identification of parameters in a linear equation of groundwater flow. *Water Resour. Res.*, **11**, 581–8.

Penman, H. L. (1948). Natural evaporation from open water, bare soil and grass. *Proc. R. Soc. A*, **193**, 120–45.

Poulovassilis, A., Krentos, V. D., Stylianou, Y., and Metochis, Ch. (1974). Soil water properties of layered soil determined *in situ*. In *Isotope and radiation techniques in soil physics and irrigation studies 1973*, pp. 205–24. I.A.E.A., Vienna.

Price, M. (1987). Fluid flow in the Chalk of England. In *Fluid flow in sedimentary basins and aquifers* (ed. J. C. Goff and B. P. Williams), pp. 141–56. Geol. Soc. Spec. Pub. 34, London.

Price, M., Bird, M. J., and Foster, S. S. D. (1976).

Chalk pore size measurements and their significance. *Wat. Serv.*, **80**, 596–600.

Rushton, K. R. and Ward, C. (1979). The estimation of groundwater recharge. *J. Hydrol.*, **41**, 345–61.

Shuttleworth, W. J., Gash, J. H. C., Lloyd, C. R., McNeil, D. D., Moore, C. J., and Wallace, J. S. (1988). An integrated micrometeorological system for evaporation measurement. *Agric. For. Met.*, **43**, 295–317.

Simmers, I. (ed.) (1988). *Estimation of natural groundwater recharge*. Reidel, Dordrecht.

Smith, P. J. and Wikramaratna, R. S. (1981). A method for estimating recharge and boundary flux from groundwater level observations. *Hydrol. Sci. Bull.*, **26**, 113–36.

Smith-Rose, R. L. (1933). The electrical properties of soils for alternating currents at radio frequencies. *Proc. R. Soc. A.*, **140**, 359.

Thomas, A. M. (1966). *In situ* measurement of moisture in soil and similar substances by fringe capacitance. *J. Sci. Instrum.*, **43**, 21–7.

Thompson, N., Barrie, I. A., and Ayles, M. (1981). *The Meteorological Office rainfall and evaporation calculation system: MORECS* (July 1981). Hydrol. Mem. No. 45, Meterorological Office, Bracknell, Berks.

Topp, G. C. and Davis, J. L. (1985). Measurement of soil water content using time-domain reflectometry (TDR): a field evaluation. *Soil Sci. Soc. Am. J.*, **49**, 19–24.

Venetis, C. (1971). Estimating infiltration and/or the parameters of unconfined aquifers from groundwater level observations. *J. Hydrol.*, **12**, 161–9.

Watson, K. K. (1966). An instantaneous profile method for determining the hydraulic conductivity of unsaturated porous media. *Water Resour. Res.*, **2**, 709–15.

Wellings, S. R. (1984a) Recharge of the Upper Chalk aquifer at a site in Hampshire, England: I. Water balance and unsaturated flow. *J. Hydrol.*, **69**, 259–73.

Wellings, S. R. (1984b). Recharge of the Upper Chalk aquifer at a site in Hampshire, England: II. Solute movement. *J. Hydrol.*, **69**, 275–85.

Wellings, S. R. and Bell, J. P. (1980). Movement of water and nitrate in the unsaturated zone of Upper Chalk. *J. Hydrol.*, **48**, 119–36.

Wellings, S. R. and Bell, J. P. (1982). Physical controls of water movement in the unsaturated zone. *Q. J. Eng. Geol.*, **15**, 235–41.

Wellings, S. R. and Cooper, J. D. (1983). The variability of recharge of the English Chalk aquifer. *Agric. Wat. Mgmt.*, **6**, 243–53.

Wellings, S. R., Bell, J. P., and Raynor, R. J. (1985). *The use of gypsum blocks for measuring soil water potential in the field*. Inst. Hydrol. Rpt. 92. Wallingford, Oxon.

Wikramaratna, R. S. and Reeve, C. E. (1984). A modelling approach to estimating aquifer recharge on a regional scale. *J. Hydrol. Sci.*, **29**, 327–37.

Youngs, E. G., Leeds-Harrison, P. B., and Chapman, J. M. (1989). Modelling water-table movement in flat low-lying lands. *Hydrol. Process.*, **3**, 301–15.

6. Shallow groundwater systems

C. Neal, D. G. Kinniburgh, and P. G. Whitehead

Introduction

The need to understand the problems caused by 'acid rain' and the development of predictive models of acidification has required an integrated approach looking at both upland surface water and groundwater quality. The development of such integrated models is likely to continue as the need to quantify the impact of pollution on both surface water and groundwater increases.

Much of the chemical character of surface water and groundwater is established in the unsaturated zone. This zone, especially the soil, is a region of rapid change in water chemistry and it provides an important buffer protecting the groundwater from surface-derived pollutants. Investigations of the unsaturated zone have already proven useful in tracing the movement of agriculturally-derived nitrate to the water-table and there is scope for applying the same techniques to the investigation of other pollutants.

As is evident from the most recent generation of water quality models, the flow of water in the near surface environment is critical in trying to understand the dynamics of surface water quality. There is mixing of water from different depths and the flow paths within catchments are complex. Water moves up and down through the soil profile in response to seasonal changes in soil water potential and this movement can even be upwards from depths greater than the rooting area. On a catchment scale, topographic differences in the landscape mean that groundwater can also emerge as a baseflow component in streams and rivers leading to mixing on a three dimensional scale.

Shallow groundwaters play an important part in determining the quality of both surface waters and deeper groundwaters in the UK.

Hydrogeochemical processes

Chemical inputs into catchments

Surface waters and shallow groundwaters derive their solutes from a variety of sources: atmospheric inputs in the form of wet and dry deposition is the minimum input; weathering of minerals in soils can either dominate the water chemistry (e.g. in calcareous soils) or play a rather minor role (e.g. in acid-sensitive soils especially where rainfall is high); the breakdown of soil organic matter can provide large and rapid inputs at certain times (e.g. of K and NO_3 during clearfelling, Adamson and Hornung 1990); fertilizers can be major sources of some solutes, including minor ones (e.g. Cd in phosphate fertilizers). Atmospheric deposition consists of components derived principally from the sea (e.g. Na, Cl, and Mg), the land (e.g. Si, alkalinity and many trace metals) and pollutant sources (e.g. various forms of N and acidity), or major contributions from all three (e.g. SO_4, F, and Ca). Volcanic gases are also a source of some volatiles including sulphur. The oxides of sulphur and nitrogen can travel hundreds of kilometres and can undergo various chemical interactions before being deposited on vegetation and land surfaces (Fowler and Irwin 1989; Derwent *et al.* 1989; Georgii 1988).

Vegetation has the capacity to scavenge solid particles, mists, and gases. These scavenging processes are extremely complex and depend on many factors: meteorological conditions, type of vegetation, leaf shape, degree of leaf wetness, and even the presence or absence of daylight can affect the scavenging ability by orders of magnitude (Cryer 1986). Vegetation has the ability to modify the inputs in several ways. Components such as NO_3 and K can be either scavenged by the vegetation or leached from it depending upon the time of year and vegetation stress. Furthermore, stem flow and throughfall can be modified both by direct chemical reaction with scavenged particles, such as the dissolution of wind blown soil minerals, and by biologically mediated reactions, e.g. I absorption by lichens and moss, and N transformations by bacteria.

In areas near the sea, the rainfall has a highly variable load of the major sea-salt components such as Cl, Na, and Mg; further inland, both the variability and the mean concentration of these components decline. Rainfall is comprised of drops of water with varying particle sizes and chemistries: in general the finer particles, as in occult precipitation, tend to have higher salt concentrations (Ferrier *et al.* 1990). Vegetation has the ability to scavenge these fine particles to different degrees (Mayer and Ulrich 1974; Cryer 1986; Miller 1989). Consequently in upland areas, for example, chloride concentrations are higher in the stem flow and throughfall compared to rainfall, even allowing for evaporative concentration on the vegetation (Reynolds and Pomeroy 1988).

This scavenging is reflected by the difference in chloride catch in a standard rainfall collector from that in a similar collector covered with mesh or a stranded covering (Cape *et al.* 1987; Ferrier *et al.* 1990). Concentration factors of two or more are often found and orders of magnitude in the case of the hydrogen ion. However, the large concentrations in throughfall also result from the cycling of Cl from the soil water zone back to the leaf surface where it is leached during rainfall. This recycled flux is difficult to measure directly and it is uncertain how well the mesh-covered collectors act as surrogates for vegetation.

Comparisons of input-output budgets for upland conditions suggest that the mist and dry deposition scavenging can be as high as 30 per cent of the Cl input for spruce forest and about half this value for moorland (Neal *et al.* 1988a).

Hydrological pathways

Solutes are transferred to groundwaters and streams from atmospheric and terrestrial sources by complex routes which involve a series of hydrological, chemical, and biological processes. These are both poorly understood and interactive (Christophersen and Neal 1990). At the catchment scale, these processes are integrated to produce stream and groundwater chemistries that often vary in deceptively simple ways.

Soil chemical processes

Ion exchange reactions

The organic matter component of topsoil plays an important part in modifying the chemistry of the water passing through it. Unlike permanent charge clay minerals such as montmorillonite, the effective cation exchange capacity of organic matter depends to a large extent on the pH of the soil solution and the type and concentrations of cations present. The cation exchange capacity tends to increase as the solution concentration of cations increases and so, unlike permanent charge clays, organic matter has the ability to alter the total cation concentration in the soil solution with the result that simple cation exchange reactions do not describe the ion exchange processes well (Neal *et al.* 1990b). This has the effect of minimizing fluctuations in the water chemistry passing through the soil. The high moisture retention capacity of soil organic matter also tends to dampen out the flow of water passing through the soil. Since organic matter binds multivalent cations much more strongly than monovalent ions, it is often an important store for Ca^{2+}, Mg^{2+}, and Al^{3+} rather than Na^+ or K^+ (Reuss and Johnson 1986; Neal *et al.* 1989; Reynolds *et al.* 1988, 1989; Tipping and Hurley 1988). Of course, H^+ is an exception to this valence rule.

In acid soils, the exchangeable H^+ and Al^{3+} content of organic matter is an important source of soil acidity. However, unlike clay minerals where high concentrations of H^+ tend to decompose the mineral, organic matter stores most of its acidity in the form of highly stable weak acid functional groups such as carboxylate. These release (or adsorb) H^+ over a wide range of pH and provide the principal mechanism for buffering the pH of soil solutions, especially in acid soils. Titration curves of organic matter and acid soils (Kinniburgh 1986) tend to be approximately linear over a broad pH range.

An important example of cation exchange reactions is the way that these reactions modify the rainfall inputs within the soil in acid-sensitive areas. For example, when rainfall with a high sea salt content enters an acid soil, the high concentration of sea salts partially displaces the soil exchangeable cations (H^+, Ca^{2+}, Mg^{2+}, and Al^{3+}) into the soil solution. In some cases, this may lead to the stream waters containing enhanced concentrations of these soil-derived cations rather than those from the marine components (Langan 1986; Reuss and Johnson 1986; Neal *et al.* 1989). Hence, episodes of high sea salt deposition can result in acid flushes within streams.

Biological reactions

Within the organic zone, a variety of biological processes actively break down the organic materials and supply additional solutes to the soil solution (Wild 1988). However, these biologically mediated processes can also lead to the immobilization of components as well as their release (e.g. SO_4 and NO_3) depending upon temperature and soil moisture conditions (Swank 1986; Adamson and Hornung 1990). When the vegetation is fully developed, there is a net balance between uptake and release of solutes from the organic matter. The main problem in demonstrating such steady state conditions is the large store of solutes in the soil which results in considerable insensitivity of the solute mass balances to annual fluxes of the aqueous components.

Man's impact on the ecosystem is often to increase the flux of components from certain parts of the cycle and this perturbation tends to upset the steady state: forest development results in a rapid increase in the vegetation biomass and an associated uptake in components such as base cations: forest harvesting results in a release of components such as SO_4, NO_3, and K (Adamson and Hornung 1990). Atmospheric pollution not only tends to increase the total salt content of the atmospheric input, but, since the range of solutes associated with atmospheric pollution is quite different from most soil solutions, such inputs tend to change the soil chemistry in the long term. The increased input of strong acid anions, such as SO_4 and NO_3, is not usually balanced by the release of base cations from weathering and so the soil tends to become depleted in base cations. This is mitigated to some extent by the sorption of SO_4 by the soil. Loss of base cations continues until a new equilibrium is attained. Liming results in an increased input of Ca and leads to an increase in biological activity which in turn can increase the mineralization of organic matter.

Carbon dioxide production in the soil

Biological activity in the soil provides a high partial pressure of carbon dioxide in the soil atmosphere. These high partial pressures result in elevated concentrations of dissolved carbon dioxide and in near neutral and calcareous soils, carbonic acid is a principal source of H^+ for the weathering of soil minerals. The bicarbonate content of soil solutions and shallow groundwaters reflects the extent of these reactions.

Weathering of minerals

Within the lower soil horizons, the contribution of organic matter is less than in the upper soils and inorganic materials (primary minerals from the bedrock/drift, clay minerals, etc.) predominate volumetrically. In these lower soils, when weatherable minerals are present, acidity (H^+ and H_2CO_3) deposited or generated in the upper soil is partially consumed with the release of base cations (Reynolds *et al.* 1989). Under acidic conditions aluminium may also be released to solution (Reynolds *et al.* 1989; MacMahon and Neal 1990; Neal *et al.* 1989). Under less acidic conditions the aluminium remains on the solid phase as a secondary clay or oxide/hydroxide components and carbonic acid is converted to bicarbonate.

In the lower soil horizons, organic matter in the form of surface coatings may still have a strong influence on the ion exchange properties of the soil (Mulder *et al.* 1989) but clay minerals also become increasingly important as ion exchange materials. Biological reactions may also be important indirectly in controlling the weathering rates of the inorganic as well as the organic constituents (Eckhardt 1985; Krumbein and Dyer 1985).

Stream and lake water chemistry

The principal sources of Cl in shallow groundwaters are atmospheric deposition, fertilizers, and various forms of pollution including road salt. Although Cl is an important biological solute and is cycled along with other nutrients, it is generally not derived in significant quantities from the weathering of minerals nor is it adsorbed strongly by organic matter or minerals. It should therefore be one of the simpler solutes to understand since many of the complexities of chemical reaction are absent. In particular, it should help to refine the hydrological models (Neal *et al.* 1988*a*; Christophersen and Neal 1990).

Even though the incident rainfall shows a large variation in Cl its concentration often shows remarkably little variation in the stream water of areas with thin soils and impermeable bedrock (Sklash and Farvolden 1979; Neal *et al.* 1988*a*). Similar damping has been observed in lysimeter drainage (Darling and Bath 1988). Sometimes, however, no such damping occurs and the stream response directly reflects the rainfall Cl (Langan 1986). This illustrates an important feature of catchment response: different catchments can behave in different ways even when they have apparently similar geomorphological and flow hydrograph characteristics.

Having passed through the vegetation cover, the atmospheric inputs are transferred, via the soil, to the groundwater and stream. In the soil zone mixing processes occur which dampen the variation in the input while evaporation of water increases concentrations. During summer months, water moves upwards as well as downwards through the unsaturated zone further enhancing dispersion. There is evidence for the upward movement of water and solutes from a depth of at least 3 m in the unsaturated zone of the Chalk (Wellings and Bell 1980).

Transport of water through the soil zone is complex and both macro and micropore movement are very influential. In the uplands, natural pipe networks can allow rapid flow downslope, and in fissured aquifers such as the Chalk, movement of water and pollutants through fissures can reach velocities greatly exceeding 1 md^{-1}. Within streams, mixing of different parcels of water means that variations in water chemistry can be damped even further than observed in the soil (Reynolds and Pomeroy 1988).

Stream and lake water chemistry represents a mixture of water from various soil horizons, groundwater baseflow, and atmospheric inputs and is closely linked with the hydrograph response (Walling and Webb 1986). Degassing of CO_2 when soil and groundwater enters a stream usually

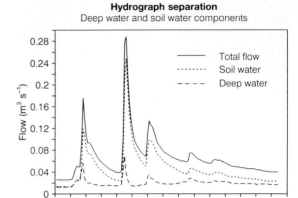

Fig. 6.1. Storm hydrograph separation using acid neutralizing capacity (ANC) conservative tracing techniques for the Afon Hafren, Plynlimon mid-Wales (after Neal *et al.* 1990*a*).

results in an increase of pH (Reuss and Johnson 1986).

Thus, in general, during baseflow conditions water is supplied from the lower soil horizons and deeper groundwater; stormflow water is a mixture of soil water, deeper groundwater, and atmospheric inputs. Even for areas with thin soils and impermeable bedrock, the groundwater contribution can be larger than the soil and atmospheric inputs (Neal *et al.* 1990*a*; Robson and Neal 1990; Kleissen *et al.* 1990). This feature is illustrated in Fig. 6.1 where hydrograph separation has been applied using the conservative parameter acid neutralizing capacity (ANC) for the Hafren forest stream at Plynlimon in mid-Wales (data from Neal *et al.* 1990*a*). Comparisons of rainfall, soil, and stream water chemistry reveal major gaps in knowledge over the routes water takes to enter the stream and the nature of the chemical reactions during this transport (Neal *et al.* 1989; Christophersen and Neal 1990). Occasionally, small storm events, particularly after prolonged dry periods, can result in stream waters with increased alkalinities or decreased acidities. This can be attributed to the derivation of significant quantities of the water from the near stream bank areas (Harriman *et al.* 1990; Chappell *et al.* 1990; Muscutt *et al.* 1990). In some instances, saturated conditions can occur near the stream bank leading to the accumulation of waters with a similar composition to that of the groundwaters upslope. Redox reactions may also be important here and consumption of hydrogen ions will, in part, be related to sulphate reduction and bicarbonate generation (Fiebig *et al.* 1990; Vogt *et al.* 1990).

Lake water chemistry is usually considerably less variable than that of stream water due to the much longer water residence times involved. In lakes, chemical processes continue to modify the water chemistry and these are related to biological activity whereby redox processes assimilate sulphate and, in some cases, generate anoxic conditions near the lake bottom (Davison and Tipping 1984; Davison and Woof 1984; Hall and Jeffries 1984; Brock 1985). High daytime pH values can occur when photosynthetic activity reduces the partial pressure of CO_2 to very low levels.

The chemistry of soil solutions and shallow groundwaters

Sampling the soil solution

The growing interest in measuring and modelling the transport of pollutants through the soil and unsaturated zone has reawakened interest in the chemistry of soil solutions *sensu stricto*, i.e. the chemistry of water extracted from field-moist soils with minimal disturbance of pre-existing solid-solution equilibria. Soil chemists interested in the response of plants to specific nutrients (P and K) or toxic species (e.g. inorganic Al and B) have also been exploring the use of soil solution chemistry to explain their results sometimes with considerable success (Aiken and McCallum 1988).

The isolation and analysis of the soil solution has always presented a problem although this has been overcome to some extent by the increased sensitivity of modern analytical methods. Now just a few millilitres of solution are all that are required for a fairly comprehensive chemical analysis.

The two most popular methods of sampling the soil solution involve using either a porous cup sampler in the field or the centrifugation of samples of field-moist soil in the laboratory. There have been few studies which compare these two approaches. The centrifuge method can be carried out either by drainage of the soil water through a porous plate or by immiscible liquid displacement using a heavy organic liquid (Kinniburgh and Miles 1983).

The centrifuge methods also work well with aquifer core material obtained by conventional drilling techniques and so can be used to investigate the variation of interstitial water chemistry down to considerable depths. The problems encountered in obtaining deeper profiles largely reflect the nature of the formation. Problems tend to be encountered when crossing either particularly hard beds such as a band of large flints, or in soft unconsolidated sands where core loss tends to be high. Percussion drilling is preferable where feasible; rotary air-flush is necessary in harder formations.

The major practical difficulties encountered in extracting interstitial water are when the material is either too dry or when exposure to air leads to serious contamination problems, for example through oxidation of Fe^{2+} or by loss of gaseous CO_2. These problems are beginning to be tackled (Takkar *et al.* 1987). Disturbance of field-moist soils during sampling usually encourages nitrification and so displacement of the soil solution within a day of sampling is advisable. Estimating a reliable calcite saturation index for soil solutions and soil extracts is made more complicated by the presence of high concentrations of dissolved organic matter. This can complex Ca^{2+} and will contribute to the titratable alkalinity (Reddy *et al.* 1990).

Soil solutions versus shallow groundwaters

Even within the UK, the chemistry of soil solutions varies greatly and it is difficult to define any single characteristic that uniquely and reliably differentiates a soil solution from a deeper groundwater. The top few centimetres of soil frequently contain a relatively high concentration of many solutes reflecting both concentration by evaporation and the cycling of solutes in the root zone. Trace element concentrations tend to be higher in the root zone of actively growing crops than in corresponding fallow land (Linehan *et al.* 1989).

Probably the single most diagnostic feature of soil solutions is their high DOC concentration (Kinniburgh and Miles 1983). This is reflected in their characteristic yellow-brown colour; concentrations of 50 mg DOC l^{-1} are common in surface soils, and in forest soils TOC concentrations can exceed 200 mg l^{-1} (Van Praag and Weissen 1984). Campbell *et al.* (1989) found a median concentration of 52 mg l^{-1} DOC for a range of near neutral Oxfordshire topsoils (0–15 cm). In wetter parts of Britain, the DOC concentrations of soil solutions are lower, 2–13 mg l^{-1} at Plynlimon in Wales, for example (Reynolds *et al.* 1988). The DOC concentration falls off rapidly with depth.

As well as high concentrations of DOC in soil solutions, there are often significant enhancements of those solutes that are associated with organic matter such as P, B, and trace metals, particularly Cu, Fe, and Al. Phosphate-P concentrations are often about 1 mg l^{-1} in topsoil solutions but tend to drop to less than 0.1 mg l^{-1} in the deeper parts of the unsaturated zone. In groundwaters, phosphate-P concentrations are often less than 0.01 mg l^{-1}.

Forest and woodland soils

The soil solutions derived from forest soils tend to be somewhat different from surrounding soils for several reasons: there is often quite a large difference in inputs (e.g. the absence of fertilizers and lime and increased atmospheric deposition under forest) and outputs (differences in evaporation); the absence of cultivation; the efficient cycling of solutes in forests, and sometimes the very existence of a forest in a particular place may reflect some soil characteristic that made the site favoured for tree planting in the first place. The particularly high concentrations of DOC found in forest soils have already been mentioned but differences can also be found in more 'benign' solutes. For example,

1. There is evidence that groundwater draining broadleaf woodland may sometimes contain higher than expected concentrations of solutes such as chloride and sulphate. This seems to arise in two circumstances: in clearings within woodland where old trees have been lost by senescence or windfall, and close to the edges of woodland.

2. Profiles of chloride, sulphate, and total oxidiz-able nitrogen (TON) beneath a beech copse high on the Berkshire Downs (Upper Chalk) showed Cl concentrations that were generally high (Fig. 6.2). In profile LMO8, a broad peak was found between 3–8 m with a maximum Cl concentration of 930 mg l^{-1} (compared with an average concentration of approximately 30 mg l^{-1} Cl in throughfall). The SO_4 concentrations were also enhanced beneath the trees. There is a marked contrast between the two 'beneath tree' profiles and the two 'open site' profiles in all three solutes: the TON/Cl ratio is much lower beneath the trees than between them. There could well be an edge effect operating since these four profiles were all taken within 50 m of the edge of the wood and the site was in a generally exposed situation. However, without independent estimates of evaporation it is difficult to know to what extent the high concentrations are due to increased inputs or to increased evaporation.

Agricultural soils

The purpose of fertilizers is to change the concentration of solutes in the soil solution. This change can be by as much as two orders of magnitude for the major nutrients. The main impact of the regular cultivation of land and the input of N–P–K fertilizers is seen in the elevated concentrations of NO_3–N in the soil solution and underlying groundwater. Water draining unfertilized permanent grassland in the UK typically contains less than 2 mg NO_3–N l^{-1}. These concentrations are less than expected from the atmospheric deposition of nitrate and ammonium indicating the efficient cycling and utilization of N in undisturbed grasslands. Very low nitrate concentrations are also found in acidic heathland and bog soils, and sometimes (but not always) in woodland as demonstrated above. An extreme case is seen in the native tallgrass prairies of the mid-western USA where nitrate concentrations in soil solutions rarely exceed 0.03 mg NO_3–N l^{-1}. In contrast, concentrations greater than 50 mg NO_3–N l^{-1} are common beneath intensively managed agricultural land including grass leys (see also Chapter 10).

The impact of K and P fertilizers on groundwater is much less owing to their strong retention by soils. The greatest effect is likely to be seen in the incidental components of these fertilizers (e.g. Cl

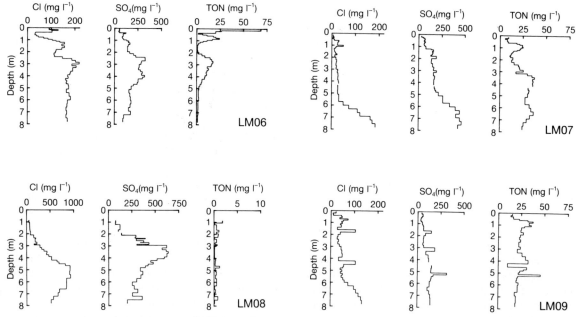

Fig. 6.2. Profiles of chloride, sulphate, and TON concentrations in the interstitial water of the Chalk's unsaturated zone beneath a beech copse high up on the Berkshire Downs near Lambourn. Profiles LM06 and LM08 were sampled directly beneath trees and LM07 and LM09 were sampled in clearings within the copse (probably where trees had once been).

and SO_4). Since the Cl concentration in rainfall normally averages 3–6 mg l^{-1} (more within 10 km of the coast) and the ratio of rainfall/(recharge+runoff) is normally less than 5, Cl concentrations in the subsoil and unsaturated zone much greater than 30 mg l^{-1} may indicate a fertilizer input. Similarly, sulphate concentrations greater than 50 mg l^{-1} are often indicative of a fertilizer input.

Seasonal variation in soil solution chemistry

It is to be expected that the soil solution chemistry will change in response to the various inputs and outputs of solutes and water and that this change will diminish with depth as the extent of mixing of soil water increases. In general, the highest soil solution concentrations are found in the summer-early autumn period (Campbell *et al.* 1989; Linehan *et al.* 1989) and the range is greater than would be expected from the change in soil moisture content alone.

It is clear from Fig. 6.2. that steady state profiles are not necessarily found beneath woodland; woodland represents an intrinsically rather variable land use both in space and time. However, where permanent, unfertilized grassland and heathland overlie aquifers, profiles of the interstitial water chemistry tend to lose much of their variation below about 1.5 m and provide better examples of steady state profiles.

The stable isotopes of oxygen and hydrogen can be useful tracers for the movement of water through the unsaturated zone since summer rainfall tends to be isotopically heavier than winter rainfall. Evidence from studies of stable isotopes in the Chalk's unsaturated zone has shown that in some, but not all, cases seasonal cycles are preserved with diminishing amplitude down to depths of 6 m (Darling and Bath 1988).

Shallow groundwater chemistry

The dominant geochemical process beneath the soil zone is the continued slow weathering of primary minerals with the release of base cations and the production of bicarbonate alkalinity. The development of this weathering front can be looked upon as a chromatographic process in which the mineral transformations gradually move from the surface downwards over timescales of decades and centuries. These mineral–water reactions are reflected in changes in the groundwater chemistry: there is often an increase in Si and HCO_3 with depth for example; in acid areas, there is usually a decrease in acidity; and in the unsaturated zone of the Chalk, a decrease in Mg with increasing depth.

Decalcification is perhaps the most important of these reactions not only in limestone environments, but also in situations where $CaCO_3$ cement is present in minor (< 1 per cent) but geochemically significant amounts as for example in the Triassic sandstones. This explains why most groundwaters from the Triassic sandstones have higher alkalinities and are close to calcite saturation even though the infiltration occurs through soils in which calcium carbonate is absent.

Acidic soils and groundwaters

Acid soils develop on parent materials where the rate of mineral weathering is insufficient to neutralize the inputs of acidity. Typically, their cation exchange capacity has a high saturation with H^+ and Al^{3+} and a low 'base' (Ca^{2+}, Mg^{2+}, Na^+, and K^+) saturation. The pH of the soil solutions from such acid soils is usually less than pH 5 and total Al concentrations are frequently greater than 1 mg l^{-1} and occasionally exceed 10 mg l^{-1}. The most acid soil solutions in the UK are probably found in the humus-iron podzols of southern England especially those under conifers. Here inputs of acidity in rainfall and dry deposition are high, rainfall is low and evaporation is high, and some of the sands on which these soils develop (especially the Lower Greensand and Bagshot Beds) are very low in feldspars and other readily weatherable minerals. A soil solution with a pH of 3.2 (E_a horizon) was found in a podzol soil situated under a mature stand of Scots Pine in Berkshire; H^+ was one of the principal ions (Kinniburgh 1986).

The soil solution from the uppermost horizons of this profile had a K/Na (molar) ratio of 1.3–1.7 reflecting the efficient cycling of K in soils in general and in forest soils in particular. It is unusual for deeper groundwaters to have a K/Na ratio greater than one. Lower down the profile the acidity and K/Na ratio decreased but the dissolved Al had increased. Broadly similar results have been found elsewhere (Mulder *et al.* 1987; Reynolds *et al.* 1988, 1989). Dissolved Ca and Mg concentrations in such soils are very low, normally less than 3 mg l^{-1}.

Other acid soils such as those developed on decalcified parts of the Triassic sandstones tend to be less acid (pH 4.0–4.5). Normally at a depth of 5 m most of the acid has been consumed by reaction with minerals and the pH has risen to pH 5 or more (Edmunds *et al.* 1989). Assuming that water moves downwards through the unsaturated zone at an average rate of about 1 m a^{-1}, then the timescale of these mineral-water reactions must be of the order of a few years.

Acid groundwaters are generally confined to areas with acid soils but even in areas with strongly acid soils, it is unusual to find groundwater sources (pumped boreholes or springs) with a pH of less than 5.5, an alkalinity of less than 1 meq l^{-1} and significant dissolved Al (Kinniburgh and Edmunds 1986). For example, even in an area well known for its acidic surface waters, Loch Fleet (Galloway), groundwater derived from a shallow (9.9 m) borehole in the underlying granite had a pH of 7.2, an alkalinity of 2.3 meq l^{-1} and total Al < 0.003 mg l^{-1} (Cook *et al.* 1987). This reflected the presence of small amounts of calcite in the granite.

A recent survey of UK groundwaters showed that only six UK aquifers—the Millstone Grit, Lower Greensand, Triassic sandstones, Old Red Sandstone (Moray), Tunbridge Wells Sand, and Carboniferous Limestone (one sample)—contain groundwaters with Al_T concentrations in excess of 0.003 mg l^{-1} and only the first two contained concentrations in excess of the EC maximum admissible concentration (0.02 mg l^{-1}) (Edmunds *et al.* 1989). By way of contrast, acidic surface waters frequently have Al_T concentrations of 0.2–0.3 mg l^{-1}. Therefore, unlike surface waters, acidic deep groundwaters are a rarity. They generally only arise where the acidity is generated internally by the oxidation of pyrite and other Fe^{2+}- or S^{2-}- containing minerals.

Surface water and groundwater acidification in the UK

Historical evidence

The first phase of soil acidification probably occurred several thousand years ago (up to 10 000 BP) with the development and subsequent loss of deciduous forest: birch (*Betula*), hazel (*Corylus*), and alder (*Alnus*) which dominated the higher open plateau areas; oak (*Quercus*) occupied the plateau fringes (Dimbleby 1952; Taylor 1974; Pennington 1984). The deforestation was extensive from the Neolithic and Bronze age times due to a deteriorating micro-climate of the sub Atlantic period (about 2700 BP). Local deforestation by man began before this period (5000–3000 BP). The net effect of this deforestation was the formation of moorland vegetation with the thin acidic soils characteristic of much of the uplands today. Natural peat deposits, accumulating in waterlogged areas where reducing conditions ensured limited breakdown of organic matter, also provided acidic conditions. Since the late nineteenth century, many parts of the uplands have been conifer afforested in a drive to improve the UK 'home produced' timber supplies (Calder and Newson 1979; Binns 1985). This led to further acidification of the soil with the development of 'thicker' organic rich upper soil horizons.

Much of the effort within acidification research has centred around the role of man-induced changes that have occurred over the past one hundred years, particularly the increases (and latterly decreases) in industrial acidic oxide emissions and the extensive conifer afforestation programmes.

Extent

From the early 1970s, there has been widespread concern over the acidification of surface waters and the possible impact of acidic atmospheric emissions on tree growth in Scandinavia, Western Europe, and North America. This followed the pioneering work in Scandinavia (SNSF 1981) where it was shown that many lakes were becoming more acidic with consequent deterioration in lake ecology. Heated debate followed: the SNSF (1981)

strongly argued that 'acid rain' was the major cause; others, in particular Rosenqvist (1978 1980 1987 1990), argued that conifer development of moorland was the major cause. The result was widespread pressure for reductions in acidic oxide emissions. The UK debate has taken on a greater significance since it has been realized that similar declines in aquatic ecology and tree growth are evident in the UK also (UKAWRG 1988). The distribution of acidic surface waters covers much of the upland UK (Fig. 6.3) with only NW Scotland north of the Great Glen escaping. There remains a controversial scientific and political debate about the precise causes and consequences of surface water acidification (e.g. Krug and Frink 1983) but there is now strong evidence to implicate acidic oxide emissions as a major causal factor.

Much of the British uplands (Scotland, Wales, the Lake District, the Pennines, south-west England, and parts of the Midlands) comprise bedrock with acid sensitive lithologies (Kinniburgh and Edmunds 1986; UKAWRG 1988; Harriman 1989; Hornung *et al.* 1990). This distribution corresponds closely with that of acid soils (Catt 1985; Wilson *et al.* 1989) and surface waters (Fig. 6.3). While streams draining these areas tend to be acidic, there are insufficient long-term records of stream water chemistry to indicate how the water quality has changed over long periods of time. Indirect evidence based on the diatom remains in lake sediments (Battarbee 1984, 1989; Battarbee *et al.* 1985), on stream biota assays (Gee and Radford 1982; Stoner and Gee 1985; Ormerod and Edwards 1987; Ormerod *et al.* 1987; UKAWRG 1988) and on modelling (Jenkins *et al.* 1990a; Whitehead *et al.* 1990) suggest that many of the streams are becoming more acid or have been acidified in the past.

The chemical hydrograph

In acid-sensitive areas, the major changes in stream water chemistry occur with changing flow. Stream baseflow is derived mainly from water draining the lower soil horizons and from fissures within the bedrock and tends to have significant alkalinity whereas stormflow is largely derived from subsurface flow through the uppermost acidic soil horizons with a relatively small deep ground-

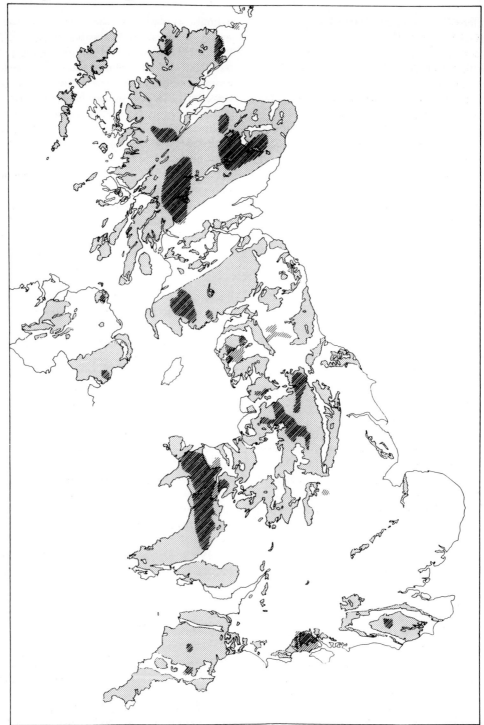

Fig. 6.3. The distribution of acid (pH less than 5.6) surface waters in the UK. The hatched areas represent the locations of known areas of acid waters while the stippled areas correspond to the areas where the underlying rock type is capable of giving rise to acidic waters (after UKAWRG 1988).

water component. Therefore peak acidity is often associated with stormflow or snowmelt events.

In the soil, the waters have a chemistry determined by the rainfall's chemical signal, modified by cation exchange and limited weathering reactions: components such as sulphate and chloride pass through the soil as 'mobile anions' carrying with them cations whose concentrations depend upon the composition of the ion exchange complex (the mobile anion concept, Reuss and Johnson 1986). Thus the stream chemistry during a storm event is determined by both the changes in the 'end member' composition of the soil solution and the relative contributions of ground and soil waters to the stream water. These concepts are incorporated in the 'the Birkenes model' which attempts to describe short-term changes in stream water chemistry (Christophersen *et al.* 1982, 1984; Whitehead and Neal 1987).

Time scale for acidification

On a longer time scale (decades), the composition of the soil water changes as the exchangeable cation complex changes composition. The principal effect of increasing the input of 'acidic deposition' is to deplete the soil's cation exchange store of base cations, gradually replacing it with hydrogen and aluminium. In the process, the soil solution becomes more acidic and gives rise to soil solutions high in dissolved Al. The rate at which the soils acidify depends not only on the size of the cation exchange complex and the rate of replenishment by weathering reactions, but also on the degree to which the soil and associated vegetation absorb the mobile anions and sulphate in particular. The greater the sorptive capacity of the soil for anions, the slower will be the rate of acidification since the loss of base cations requires the equivalent loss of an anion. In effect, for every mole of sulphate absorbed so too are two moles of hydrogen ion. These features are incorporated in the MAGIC model (Cosby *et al* 1985).

Effect of afforestation

Field data provides conflicting evidence over the dominance of a forest effect in surface water acidification. For example, diatom, geochemical, and microfloral reconstruction of lake sediments back to medieval times shows stream acidity to have increased dramatically for some acidic environments at the time of the 'black death' as agricultural practice declined and the land reverted to conifer vegetation (Rosenqvist 1987, 1990). Set against this, diatom evidence from Scottish lochs (Battarbee 1989) suggests that recent afforestation only exacerbates the acidification process in polluted areas: with low atmospheric acidic oxide depositions, no evidence for acidification is found. Furthermore, streams in relatively unpolluted parts of Scotland (Harriman 1989) and central Norway (Christophersen *et al.* 1990*a*) have a relatively low acidity and inorganic aluminium content.

Whatever the relative influences of conifer afforestation and atmospheric deposition, there is overwhelming evidence showing, on a regional scale, that conifer afforested areas are more acidic and give rise to surface waters with higher aluminium concentrations than their moorland counterparts (Harriman and Morrison 1982; Stoner and Gee 1985; Bull and Hall 1986; Ormerod *et al.* 1989). When a conifer forest is introduced onto acid moorland, the reactions occurring are even more complex than those described above: acid soils are being developed while base cations are accumulated by the biomass; soil microbiological reactions also change. For this reason, it is difficult to describe quantitatively the reactions occurring.

Qualitatively, the following processes all seem to conspire to produce more acidic and aluminium rich stream waters in acid-sensitive areas (Calder 1985; Miller 1989; Reynolds *et al.* 1988):

(i) 'base' cations are transferred from the soil to the biomass with the risk that weathering reactions may replace them in part by acidic cations such as Al^{3+};

(ii) an upper organic soil develops which is highly acidic;

(iii) the tree canopy has the ability to scavenge pollutants and salts from the atmosphere more efficiently than grasses;

(iv) the trees increase evapotranspiration thereby

increasing the concentration of any residual non-volatile acidity in the soil solution;

(v) artificial drainage channels promote greater near surface flow during storm events thereby effectively short-circuiting the deeper flow-paths where acid neutralization is greatest.

None of these processes necessarily leads to acidic waters but the presence of non-acidic streams in areas remote from industrial pollution suggests that (iii) above may be the determining factor in many acid-sensitive areas (Miller 1989).

Modelling the effects of acidification

Initial modelling work confirms that the deterioration in stream water quality over the past century is essentially caused by the input of atmospheric pollutants, the trees simply enhancing the effect (Neal *et al.* 1986). Simulating the effect of forest growth on moorland soils under pristine conditions showed an initial decrease in stream pH followed by an increase back to pristine moorland values (Neal *et al.* 1986). However, only factors (iii) and (iv) above were included in this simulation. Subsequent modelling development has allowed a more comprehensive assessment with the inclusion of a forest growth component. This indicates a more extensive and less reversible acidification (Jenkins *et al.* 1990*b*; Whitehead *et al.* 1990). Other recent work is aimed at improving the sophistication of the soil chemistry model (Tipping and Hurley 1988).

Reversal of acidification

While the wide extent of acidification in the British uplands is becoming clearer (UKAWRG 1988; Harriman 1989; Battarbee 1989; Jenkins *et al.* 1990*b*; Whitehead *et al.* 1990), the challenge now is to be able to predict accurately the degree to which reversibility can be achieved following proposed reductions in atmospheric acidic oxide emissions. To achieve this, models are required to make regional predictions. The extent to which stream acidification can be reversed is difficult to estimate but models such as MAGIC (Cosby *et al.*

1985) provide a valuable benchmark and enable the sensitivity to various emission strategies to be explored. The initial modelling exercises are showing that large scale reductions in acidic oxide depositions (> 60 per cent) are required to stabilize stream acidity or to provide modest improvments (Neal *et al.* 1988*b*; Jenkins *et al.* 1990*a,b*; Whitehead *et al.* 1990; Christophersen *et al.* 1990*b*). Improvements to revitalize fish stocks are likely to take decades to centuries.

Radioactivity in shallow groundwater

Man-made radioactivity

The testing of atomic weapons in the atmosphere during the 1950s and the early 1960s introduced large quantities of radioactive material into the natural environment. Debris released into the atmosphere has continued to be deposited on the earth's surface ever since (Cambray *et al.* 1987*b*). The peak of the fallout from the testing of weapons occurred in 1963 and after a fairly rapid drop over a four to five year period reduced only slowly to almost undetectable concentrations by the late 1980s. ^{90}Sr and ^{137}Cs were the most persistent (half-lives of 28 and 30 years, respectively) and potentially dangerous isotopes produced during this period. It is extremely difficult to measure the concentration of these radionuclides in the soil solution *per se* but regular monitoring for ^{90}Sr and ^{137}Cs in milk over the last 30 years has shown a peak in 1964 which was followed by a rapid decline and then a slower decline which continued into the 1980s (Government Statistical Service 1988).

Tritium (^3H) released by testing weapons has been monitored in groundwaters by Smith *et al.* (1970) and Foster and Smith-Carington (1980). Using the characteristic 1963–1964 peak in the unsaturated zone of the Chalk produced by tests of weapons, they estimated that the rate of movement of tritium through the soil and the unsaturated zone of the aquifer was approximately 1 m yr^{-1} (Fig 6.4). Tritium has provided a generally useful tracer for identifying recently introduced groundwater but since the half-life of tritium is only 12.4 yr, it is becoming increasingly difficult to use tritium in this way.

Early studies showed that tritium concentrations

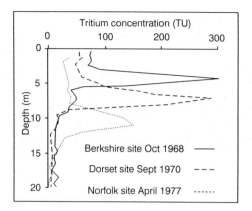

Fig. 6.4. Tritium profiles from the Chalk's unsaturated zone at sites in Berkshire, Dorset, and Norfolk. The peaks in the profiles correspond with the peak deposition of tritium during 1963–4 (after Foster and Smith-Carington 1980).

in stream water responded to rainfall. For example, tritium concentrations increased following a storm event on forest and moorland catchments at Plynlimon (Biggin 1971) (Fig. 6.5). Tritium concentrations were considerably higher in the forested catchment due to the enhanced drainage and rapid response of the system. In the moorland catchment, infiltration of rainfall through the soil and into the groundwater was much more significant leading to lower tritium concentrations in the stream and increased concentrations in both soil water and groundwater.

The consequences of the fallout of radioactive material were clearly demonstrated in 1986 when on April 26 the accident at the Chernobyl nuclear reactor occurred. Radioactive material was released into the lower atmosphere over a period of several days (Cambray *et al.* 1987*a*). A radioactive cloud of debris from Chernobyl moved rapidly across Europe. Only small quantities of Chernobyl fallout were deposited over most of the UK as dry deposition. However, in a few areas (Wales, Cumbria, south-west Scotland), the passage of the Chernobyl cloud coincided with isolated, very heavy, advective rainstorms so that the deposition occurred over a very short time period in these areas, e.g. 12 hours in Cumbria. The effect of this was to create rapidly changing radionuclide concentrations in most compartments within the eco-

system (Bennett *et al.* 1989). ^{137}Cs is normally strongly retained by micaceous minerals in soils but the low abundance of these minerals in the acid organic topsoils of the upland regions concerned and the efficient nutrient (especially of K) cycling mechanisms developed by the plants of these regions meant that Cs was retained by the vegetation longer than originally expected (Livens and Loveland 1988; Bell *et al.* 1988). Concentrations of ^{137}Cs in milk increased after the Chernobyl accident but those of ^{90}Sr did not (Government Statistical Service 1988).

The spatial distribution of ^{137}Cs across the UK following Chernobyl has been mapped by Horrill *et al.* (1989) (Fig. 6.6) and the movement and transport of this Cs through soils, vegetation, catchments, lakes, and fish have been extensively studied by scientists at the Institutes of Freshwater Ecology (Windermere) and Terrestrial Ecology (Merlewood). Models have also been developed by Jenkins *et al.* (1988) for predicting the movement and distribution of Cs through the natural environment. Detailed studies of the soil and soil solution chemistry of ^{137}Cs have recently been undertaken by Nisbet and Lambrechts (1989).

One of the by-products of the reprocessing of nuclear fuels is the production of ^{129}I some of which is discharged to the atmosphere. It has been estimated that by the year 2000 the total amount of ^{129}I generated by man will exceed its natural abundance by a factor of 5 (Haury and Schikarski

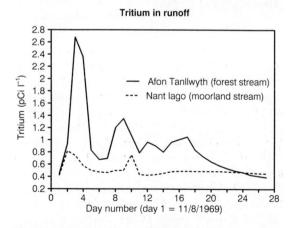

Fig. 6.5. Tritium concentration in stream waters draining acid moorland and spruce forest catchments at Plynlimon in mid-Wales (after Biggin 1971).

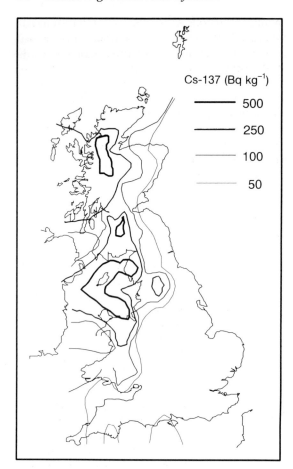

Fig. 6.6. Caesium-137 distribution across the UK following the Chernobyl nuclear accident (after Horrill *et al.* 1989).

1977). While other radioactive isotopes of iodine are produced by the nuclear industry, most are short-lived (^{125}I and ^{131}I have half-lives of 60 and 8 days, respectively) and only ^{129}I is present in the environment in significant amounts. Its half-life is extremely long (1.6×10^7 yr) and so any release to the environment will be essentially permanent. There is, therefore, a strong possibility that whatever mode of disposal is used, it will ultimately be dispersed globally just as the more persistent pesticides have become. Although the amounts of ^{129}I and ^{131}I normally produced are insufficient to constitute a major hazard, any major development of nuclear power or increased reprocessing of nuclear fuels may well also increase their dispersal.

The processes by which iodine enters the food chain and the hydrogeochemical processes regulating its distribution are extremely complex. For example iodine may be (i) cycled through vegetation; (ii) specifically adsorbed by soil minerals; (iii) immobilized by, and mineralized from, soil organic matter; and (iv) released back to the atmosphere as iodine vapour. These processes have been reviewed by Whitehead and Truesdale (1982).

Natural radioactivity

Natural sources of radioactivity provide approximately 90 per cent of the average effective human dose of ionizing radiation in the UK with radon and its daughters providing by far the largest dose and contributing to about 40 per cent of the total annual dose. Medical irradiation is by far the largest source of man-made radiation (Green *et al.* 1988). Whilst most public concern is related to accidents such as Chernobyl, it is also necessary to consider natural sources of radionuclides. Castle (1988) has summarized the impacts of such sources on groundwater supplies. ^{40}K is usually the most abundant radioactive isotope in groundwater (its activity is directly related to the K concentration) but it is of little concern because of its low specific activity. The radionuclides of most concern, and their half-lives, are radium (^{226}Ra; 1600 yr), uranium (^{234}U, ^{235}U, and ^{238}U; 2.5×10^5 yr, 7.1×10^8 yr, and 4.5×10^9 yr, respectively), radon (^{222}Rn, 3.8 d), ^{210}Pb (21 yr), and ^{210}Po (138 d). The concentration of a particular radionuclide in groundwater depends primarily on its concentration and rate of production in the surrounding rocks (see Chapter 15). For example, uranium exists naturally in the form of the three isotopes, ^{234}U, ^{235}U and ^{238}U, and although present at extremely low concentrations in groundwater, uranium and its daughter products (particularly ^{226}Ra) can reach significant concentrations in water supplies (Government Statistical Office 1988). ^{222}Rn gas dissolves readily in water and diffuses through fissures into the surrounding groundwater. Igneous rocks, particularly the granites found in south-west England, give rise to locally high concentrations of radon in groundwater, and ultimately, the water supply. These high concentrations are reflected in the build-up of radon gas in homes. Recent surveys of ^{222}Rn in homes (Institution of Environmental Health Officers 1988; Wrixon *et al.* 1988) have confirmed the high concentrations in south-west England with smaller concentrations in parts of the Pennines (associated with phosphatic limestones with high uranium contents) and in granites in Scotland. The much lower radon concentrations associated with the Scottish granites, compared with those of south-west England, has been attributed to their lower uranium content and to their less fractured and altered nature (Wrixon *et al.* 1988). Although the concentrations of ^{210}Pb and ^{210}Po in groundwater are usually extremely small, and drinking water does not normally provide a significant source of daily intake of these radionuclides, Salonen (1988) has shown that the presence of ^{210}Pb and ^{210}Po in groundwater in southern Finland could provide a significant contribution to the received radiation dose there. Users of water from deep wells were particularly at risk.

Some bottled mineral waters have high concentrations of natural radionuclides, particularly ^{222}Rn (Bettencourt *et al.* 1988), and could provide regular drinkers with significant doses.

Conclusions

Over the past decade major advances have been made in our understanding of the hydrogeochemical behaviour of shallow groundwater systems. This understanding has developed as the need to assess man's impact on the upland environment has become ever more important. The effects of atmospheric input of acidic oxides and radionuclides and of long-term land use changes such as conifer afforestation have received much attention and have attracted considerable research support. Despite this, the means of measuring, quantifying and modelling the flow of water and solutes from the atmosphere to the vegetation and then through the soil to the stream is still in its infancy. Even with the simplest of components (chloride), there is difficulty not only in obtaining a catchment input-output budget, but also in explaining the high degree of smoothing found between the chloride concentrations in rainfall and stream water. Similarly, while the importance of groundwater inputs to storm runoff has been established, the mechanisms by which water is transferred and the

chemical reactions within the soil, and the unsaturated and groundwater zones have not been well established. Further linking of surface water and groundwater studies is needed.

Acknowledgment

This chapter is published with the permission of the Directors of the Institute of Hydrology and the British Geological Survey (NERC).

References

Adamson, J. K. and Hornung, M. (1990). The effect of clearfelling a Sitka spruce (*Picea sitchensis*) plantation on solute concentrations in drainage waters. *J. Hydrol.*, **116**, 297–8.

Aiken, R. L. and McCallum, L. E. (1988). Boron toxicity in soil solutions. *Austr. J. Soil Res.*, **26**, 605–10.

Battarbee, R. W., (1984). Diatom analysis and the acidification of lakes. *Phil. Trans. Royal Soc. (London) B*, **305**, 451–77.

Battarbee, R. W. (1989). The acidification of Scottish Lochs. Patterns of surface water acidification in Scotland. In *Acidification in Scotland*, pp. 104–10. Scottish Development Department, Edinburgh.

Battarbee, R. W., Flower, R. J., Stevenson, A. C., and Rippey, B. (1985). Lake acidification in Galloway: a palaeoecological test of competing hypotheses. *Nature*, **314**, 350–2.

Bell, J. N. B., Minski, M. J., and Grogan, H. A. (1988). Plant uptake of radionuclides. *Soil Use Manag.*, **4**, 76–84.

Bennett, P. J. P., Leeks, G. J. L., and Cambray, R. S. (1989). Transport processes for Chernobyl-labelled sediments: preliminary evidence from upland mid-Wales. *Land Degrad. Rehabil.*, **11**, 39–50.

Bettencourt, A. O., Teixeira, M. M. G. R., Faisca, M. C., Vieira, I. A., and Ferrador, G. C. (1988). Natural radioactivity in Portuguese mineral waters. *Radiation Protect. Dosimetry*, **24**, 139–42.

Biggin, D. S. (1971). The use of natural tritium in hydrograph analysis. *Subsurface hydrology*, Report 26a, Institute of Hydrology, Wallingford.

Binns, W. O. (1985). Weather and woodlands. In *Weather and water*. Proc. Inst. Biol. Symp., Edinburgh, 1984, pp. 1–12. Scottish Development Department, Edinburgh.

Brock, T. D. (1985). A eutrophic lake: Lake Mendota, Wisconsin. *Ecol. Studies*, **55**, 1–308.

Bull, K. R. and Hall, J. R. (1986). Aluminium in the River Esk and Duddon, Cumbria, and their tributaries. *Environ Poll. Ser. B.*, **12**, 165–319.

Calder, I. C. (1985). Influence of woodlands on water quantity. In *Weather and water*. Proc. Inst. Biol. Symp., Edinburgh, 1984, pp. 13–31. Scottish Development Department, Edinburgh.

Calder, I. C. and Newson, M. D. (1979). Land use and upland water resources in Britain—a strategic look. *Water Resources Bull.*, **15**, 1628–39.

Cambray, R. S., Cawse, P.A., Garland, J. A., Gibson, J. A. B., Johnson, P., Lewis, G. N. J., Newton, D., Salmon, L., and Wade, B. O. (1987*a*). Observations on radioactivity from the Chernobyl accident. *Nuclear Energy*, **26**, 77–101.

Cambray, R. S., Playford, K., Lewis, G. N. J., and Burton, P. J. (1987*b*). Radioactive fallout in air and rain—results for 1985 and 1986. *AERE-R12872*. HMSO, London.

Campbell, D. J., Kinniburgh, D. G., and Beckett, P. H. T. (1989). The soil solution chemistry of some Oxfordshire soils: temporal and spatial variability. *J. Soil Sci.*, **40**, 321–39.

Cape, J. N., Fowler, D., Kinniard, J. W., Nicholson, I. A., and Paterson, I. S. (1987). Modification of rainfall chemistry by a forest canopy. In *Pollutant transport and fate in ecosystems*, British Ecological Society Special Publication Number 6 (ed. P. J. Coughtrey, M. H. Martin, and M. Unsworth), pp. 155–69. Blackwells, Oxord.

Castle, R. G. (1988). Radioactivity in water supplies. *J. Instn Water Engrs Mgmt* **2**, 275–84.

Catt, J. A. (1985). Natural soil acidity. *Soil Use Manag.*, **1**, 8–10.

Chappell, N. A., Ternan, J. L., Williams, A. G., and Reynolds, B. (1990). Preliminary analysis of water and solute movement beneath a conifer hillslope in mid-Wales. *J. Hydrol.*, **116**, 201–16.

Christophersen, N. and Neal, C. (1990). Linking hydrological, geochemical, and soil chemical processes on the catchment scale: an interplay between modelling and field work. *Water Resour. Res.*, **26**, 3077–86.

Christophersen, N., Seip, H. M., and Wright, R. F. (1982). A model for stream chemistry at Birkenes, a small forested catchment in southernmost Norway. *Water Resour. Res.*, **18**, 966–77.

Christophersen, N., Rustad, S., and Seip, H. M. (1984). Modelling streamwater chemistry with snowmelt. *Phil. Trans. Royal Soc. (London) B*, **305**, 427–39.

Christophersen, N., Vogt, R. D., Neal, C., Anderson, H. A., Ferrier, R. C., Miller, J. D., and Seip, H. M. (1990*a*). Controlling mechanisms for streamwater at the pristine Ingerbekken site in mid Norway. Some implications for acidification models. *Water Resour. Res.*, **26**, 59–67.

Christophersen, N., Neal, C., and Mulder, J. (1990*b*). Reversal of acidification at the Birkenes catchment southern Norway: predictions based on potential ANC changes. *J. Hydrol.*, **116**, 77–84.

Cook, J. M., Edmunds, W. M., and Robins, N. S. (1987). Groundwater contribution to Lock Fleet, Galloway. *Hydrogeol. Report 87/4*. British Geological Survey, Wallingford.

Cosby, B. J., Hornburger, G. M., Galloway, J. N., and Wright, R. F. (1985). Freshwater acidification from atmospheric deposition of sulphuric acid: a quantitative model. *Environ. Sci. Tech.*, **19**, 1144–9.

Cryer, R. (1986). Atmospheric solute processes. In *Solute processes* (ed. S. T. Trudgill), pp. 15–84. Wiley, Chichester.

Darling, W. G. and Bath, A. H. (1988). A stable isotope study of recharge processes in the English Chalk. *J. Hydrol.*, **101**, 31–46.

Davison, W. and Tipping, E. (1984). Treading in Mortimer's footsteps: the geochemical cycling of iron and manganese in Esthwaite Water. *Freshwater Biol. Ass.*, **52**, 91–101.

Davison, W. and Woof, C. (1984). A study of the cycling of manganese and other elements in a seasonally anoxic lake, Rotherne Mere, UK. *Water Res.*, **18**, 727–34.

Derwent, D. G., Hopper, S., and Metcalf, S. E. (1989). Computer modelling studies of the origin of acidity deposited in Scotland. In *Acidification in Scotland*, pp. 28–39. Scottish Development Department, Edinburgh.

Dimbleby, G. W. (1952). Soil regeneration on the north-east Yorkshire moors. *J. Ecol.*, **40**, 331–41.

Eckhardt, F. E. W. (1985). Solubilization, transport and deposition of mineral cations by micro-organisms—efficient rock weathering agents. In *The chemistry of weathering* (ed. J. I. Drever), NATO ASI series C: Mathematical and Physical Sciences, Vol. 149, pp. 161–74. Reidel, Dordrecht.

Edmunds, W. M., Cook, J. M., Kinniburgh, D. G., Miles, D. L., and Trafford, J. M. (1989). Trace element occurrence in British groundwaters. *Research Report SD/89/3*. British Geological Survey, Wallingford.

Ferrier, R. C., Jenkins, A., Miller, J. D., Walker, T. A. B., and Anderson, H. A. (1990). Assessment of wet deposition mechanisms in an upland Scottish catchment. *J. Hydrol.*, **116**, 251–66.

Fiebig, D. M., Lock, M. A., and Neal, C. (1990). Soil water in the riparian zone as a source of carbon for a headwater stream. *J. Hydrol.*, **116**, 217–38.

Foster, S. S. D.. and Smith-Carington, A. (1980). The interpretation of tritium in the Chalk unsaturated zone. *J. Hydrol.*, **46**, 343–64.

Fowler, D. and Irwin, D. (1989). The pollution climate of Scotland. In *Acidification in Scotland*, pp. 10–17. Scottish Development Department, Edinburgh.

Gee, A. S. and Radford, P. (1982). The regulation of stock characteristics of Atlantic salmon population. *Fisheries Res.*, **1**, 105–16.

Georgii, H. W. (ed.) (1988). *Mechanisms and effects of pollutant-transfer into forests*, pp. 1–343. Kluwer Academic Press, Dordrecht.

Government Statistical Service (1988). *Radioactivity*. Department of the Environment Statistical Bulletin (88)3. HMSO, London.

Green, B. M. R., Lomas, P. R., and O'Riordan, M. C. (1988). Action on radon in UK homes. *Radiation Protect. Dosimetry*, **24**, 541–5.

Hall, G. H. and Jeffries, C. (1984). The contribution of nitrification in the water column and profundal sediments to the total oxygen deficit of the hypoliminum of a mesotrophic lake (Grasmere, English Lake District). *Microb. Biol.*, **10**, 37–46.

Harriman, R. (1989). Patterns of surface water acidification in Scotland. In *Acidification in Scotland*, pp. 72–9. Scottish Development Department, Edinburgh.

Harriman, R. and Morrison, B. R. S. (1982). Ecology of streams draining forested and non forested catchments in an area of central Scotland subject to acid deposition. *Hydrobiologia*, **88**, 252–63.

Harriman, R., Gillespie, E., King, D., Watt, A. W., Christie, A. E. G., Cowan, A. A., and Edwards, T. (1990). Short-term ionic responses as an indicator of hydrochemical processes in the Allt a Mharcaidh catchment, Western Cairngorms, Scotland. *J. Hydrol.*, **116**, 267–86.

Haury, G. and Schikarski, W. (1977). Radioactive inputs into the environment: comparisons of natural and man-made inventories. In *Global chemical cycles and their alterations by man* (ed. W. Stumm), pp. 165–88. Proceedings Dahlem Conference, Berlin.

Hornung, M., Le-Grice, S., Brown, N., and Norris, D. (1990). The role of geology and soils in controlling surface water acidity in Wales. In *Acidification in Wales* (ed. R. W. Edwards, J. H. Stoner, and A. S. Gee), pp. 55–66. Kluwer Academic Press, Dordrecht.

Horrill, A. D., Lowe, P. W., and Howson, G. (1989). *Survey of caesium for the UK*. Report No. DoE/RW/88.101. Dept of the Environment, London.

Institution of Environmental Health Officers (1988). *Radon. Report of the I.E.H.O. survey on radon in homes 1987/8*. Institution of Environmental Health Officers, London.

Jenkins, A., Whitehead, P. G., and Hunt, J. (1988). *Modelling caesium transport—a review*. Institute of Hydrology, Wallingford.

Jenkins, A., Whitehead, P. G., Musgrove, T. J., and Cosby, B. J. (1990*a*). A regional model of acidification in Wales. *J. Hydrol.*, **116**, 403–16.

Jenkins, A., Cosby, B. J., Miller, J. D., Ferrier, R. C., and Walker, T. A. B. (1990*b*). Modelling stream acidification in afforested catchments: long term reconstruction at two sites in Central Scotland. *J. Hydrol.*, **120**, 163–81.

Kinniburgh, D. G. (1986). Towards more detailed methods for quantifying the acid susceptibility of rocks and soils. *J. Geol. Soc. (London).*, **143**, 679–90.

Kinniburgh, D. G. and Edmunds, W. M. (1986). The susceptibility of UK groundwaters to acid deposition. *Hydrogeol. Report 86/3*. British Geological Survey, Wallingford.

Kinniburgh, D. G. and Miles, D. L. (1983). Extraction and chemical analysis of interstitial water from soils and rocks. *Env. Sci. Tech.*, **17**, 362–8.

Kleissen, F. M., Wheater, H. S., Beck, M. B., and Harriman, R. (1990). Conservative mixing of water sources: analysis of the behaviour of the Allt a Mharcaidh catchment. *J. Hydrol.*, **116**, 365–74.

Krug, E. C. and Frink, C. R. (1983). Acid rain on acid soil: a new perspective. *Science*, **221**, 520–5.

Krumbein, W. K. and Dyer, B. D. (1985). This planet is alive—weathering and biology, a multi facetted problem. In *The chemistry of weathering* (ed. J. I. Drever). NATO ASI series C: Mathematical and Physical Sciences, Vol. 149, pp. 143–60. Reidel, Dordrecht.

Langan, S. (1986). *Atmospheric deposition, afforestation and water quality at Loch Dee, S.W. Scotland*. PhD. thesis, University of St Andrews.

Linehan, D. J., Mitchell, M. C., and Sinclair, A. H. (1989). Seasonal changes in Cu, Mn, Zn, and Co concentrations in soil solution in the root zone of barley (*Hordeum vulgarea* L.). *J. Soil Sci.*, **40**, 103–15.

Livens, F. R. and Loveland, P. J. (1988). The influence of soil properties on the environmental mobility of caesium in Cumbria. *Soil Use Manag.*, **4**, 69–75.

MacMahon, R. and Neal, C. (1990). Aluminium disequilibrium solubility controls in Scottish acidic catchments. *Hydrol. Sci. Bull.*, **35**, 21–8.

Mayer, R. and Ulrich, B. (1974). Conclusions on the filtering actions of forests from ecosystem analysis. *Oecologia Planetarum*, **9**, 157–68.

Miller, H. (1989). Forests and acidification. In *Acidi-*

fication in Scotland, pp. 52–7. Scottish Development Department, Edinburgh.

Mulder, J., Van Grinsven, J. J. M., and Van Breemen, N. (1987). Impacts of acid deposition on woodland soils in the Netherlands: III. Aluminium chemistry. *Soil Sci. Soc. Am. J.*, **51**, 1640–6.

Mulder, J., Breemen, N., and van Eijck, H. C. (1989). Depletion of soil aluminium by acid deposition and implications for acid neutralization. *Nature*, **337**, 247–9.

Muscutt, A. D., Wheater, H. S., and Reynolds, B. (1990). Stormflow hydrochemistry of a small Welsh upland stream. *J. Hydrol.*, **116**, 239–50.

Neal, C., Whitehead, P. G., Neale, R., and Cosby, B. J. (1986). Modelling the effects of acidic deposition and conifer afforestation on stream acidity in the British uplands. *J. Hydrol.*, **86**, 15–26.

Neal, C., Christophersen, N., Neale, R., Smith, C. J., Whitehead, P. G., and Reynolds, B. (1988*a*). Chloride in precipitation and streamwater for the upland catchment of River Severn, mid-Wales; some consequences for hydrochemical models. *Hydrol. Proc.*, **2**, 155–65.

Neal, C., Whitehead, P. G., and Jenkins, A. (1988*b*). Are present UK SO$_2$ emission declines sufficient to reverse the long-term stream acidification in the British uplands? *Nature*, **334**, 109–10.

Neal, C., Reynolds, B., Stevens, P., and Hornung, M. (1989). Hydrogeochemical controls for inorganic aluminium in acidic stream and soil waters at two upland catchments in Wales. *J. Hydrol.*, **106**, 155–75.

Neal, C., Smith, C. J., Walls, J., Billingham, P., Hill, S., and Neal, M. (1990*a*). Hydrogeochemical variations in Hafren forest stream waters, mid-Wales. *J. Hydrol.*, **116**, 185–200.

Neal, C., Mulder, J., Christophersen, N., Neal, M., Waters, D., Ferrier, R. C., Harriman, R., and MacMahon, R. (1990*b*). Limitations in the understanding of ion-exchange and solubility controls for acidic Welsh, Scottish and Norwegian sites. *J. Hydrol.*, **116**, 11–24.

Nisbet, A. F. and Lambrechts, J. F. (1989). The dynamics of radionuclide behaviour in soil solution with special reference to the application of countermeasures. *CEC Conference on the Transfer of Radionuclides in Natural and Semi-Natural Environments*. Udine, Italy.

Ormerod, S. J. and Edwards, R. W. (1987). The ordination and classification of macroinvertebrate assemblages in the catchment of the River Wye in relationship to environmental factors. *Freshwater Biol.*, **17**, 341–56.

Ormerod, S. J., Wade, K. R., and Gee, A. S. (1987). The macrofloral assemblages in upland Welsh streams in relation to acidity and their importance to invertebrates. *Freshwater Biol.*, **18**, 545–57.

Ormerod, S. J., Donald, A. P., and Brown, S. (1989). The influence of plantation forest on the pH of upland streams: a re-examination. *Environ. Poll.*, **62**, 47–62.

Pennington, W. (1984). Long-term natural acidification of upland sites in Cumbria: evidence from post-glacial lake sediments. *Rep. Freshwater Biol. Assoc.*, **52**, 28–46.

Reddy, K. J., Lindsay, W. L., Workman, S. M., and Drever, J. I. (1990). Measurement of calcite ion activity products in soils. *Soil Sci. Soc. Am. J.*, **54**, 67–71.

Reuss, J. O. and Johnson, D. W. (1986). *Acid deposition and the acidification of soils and waters*, Ecological Studies Series, Vol. 59. Springer-Verlag, New York.

Reynolds, B. and Pomeroy, A. B. (1988). Hydrochemistry of chloride in an upland catchment in mid-Wales. *J. Hydrol.*, **99**, 19–32.

Reynolds, B., Neal, C., Hornung, M., and Stevens, P. (1988). Impact of afforestation on the soil solution chemistry of stagnopodzols in mid-Wales catchments. *Water Air and Soil Poll.*, **38**, 167–85.

Reynolds, B., Hornung, M., and Hughes, S. (1989). Chemistry of streams draining grassland and forest catchments at Plynlimon, mid-Wales. *Hydrol. Sci. Bull.*, **34**, 667–86.

Robson, A. and Neal, C. (1990). Hydrograph separation using chemical techniques: an application to catchments in Wales. *J. Hydrol.*, **116**, 345–64.

Rosenqvist, I. Th. (1978). Acid precipitation and other possible sources of acidification of rivers and lakes. *Sci. Total Env.*, **10**, 39–49.

Rosenqvist, I. Th. (1980). Influence of forest vegetation and agriculture on the acidity of fresh water. In *Advances in environmental science and engineering* (ed. J. R. Pfafflin and E. N. Ziegler), Vol. 3, pp. 56–79. Gordon and Breach, London.

Rosenqvist, I. Th. (1987). Acidity of surface waters in Norway. In *Acidification and water pathways*, pp. 223–35. Norwegian National Committee for Hydrology, Bolkesjo, 4–5 May, 1987.

Rosenqvist, I. Th. (1990). From rain to lake: pathways and chemical change. *J. Hydrol.*, **116**, 3–10.

Salonen, L. (1988). Natural radionuclides in ground water in Finland. *Radiation Protect. Dosimetry*, **24**, 163–6.

Sklash, M. and Farvolden, R. N. (1979). The role of groundwater in storm runoff. *J. Hydrol.*, **43**, 45–65.

Smith, D. B., Wearn, P. L., Richards, H. J., and Rowe, P. C. (1970). Water movement in the unsaturated zone of high and low permeability strata by measuring natural tritium. In *Proc. IAEA symposium on isotope hydrology*, pp. 73–87. IAEA, Vienna.

SNSF. (1981). *Acid precipitation—effects on forest and fish*. Final report of the SNSF—project 1972–1980 (ed. L. N. Overrein, H.M. Seip, and A. Tollan), pp. 1–175. Norwegian Council for Scientific and Industrial Research, Oslo.

Stoner, J. H. and Gee, A. S. (1985). Effects of forestry on water quality and fish in Welsh rivers and lakes. *J. Instn Water Engrs Sci.*, **39**, 27–45.

Swank, W. T. (1986). Biological control of solute losses from forest ecosystems. In *Solute processes* (ed. S. T. Trudgill), pp. 85–140. Wiley, Chichester.

Takkar, P. N., Bernhard, U., and Meiwes, K. J. (1987). Method for estimation of $CO_2(aq)$ plus $H_2CO_3°$, HCO_3^- and pH in soil solutions collected under field conditions. *Z. Pflanzen-ernahr. Bodenk.*, **150**, 319–26.

Taylor, J. A. (1974). Organic soils in Wales. In *Soils in Wales* (ed. W. A. Adamson), Report 15, 30–43. Welsh Soils Discussion Group.

Tipping, E. and Hurley, M. A. (1988). A model of solid-solution interactions in acidic organic soils, based on the complexation properties of humic substances. *J. Soil Sci.*, **39**, 505–19.

UKAWRG (1988). *United Kingdom acid waters review group, second report: acidity in United Kingdom fresh waters*, pp. 1–61. HMSO, London.

Van Praag, H. J. and Weissen, F. (1984). The intensity factor in acid forest soils: extraction and composition of the soil solution. *Pedologie*, **34**, 203–14.

Vogt, R. D., Andersen, D. O., Andersen, S., Christophersen, N., and Mulder, J. (1990). Streamwater, soilwater chemistry and flow paths at Birkenes during a dry wet hydrological cycle. *Trans. Royal Soc., London*: final conference of the Surface Water Acidification Programme, pp. 149–54.

Walling, D. E. and Webb, B. W. (1986). Solutes in river systems. In *Solute processes* (ed. S. T. Trudgill), pp. 251–328, Wiley.

Wellings, S. R. and Bell, J. P. (1980). Movement of water and nitrate in the unsaturated zone of Upper Chalk near Winchester, Hants., England. *J. Hydrol.*, **48**, 119–36.

Whitehead, D. C. and Truesdale, V. W. (1982). *Iodine: its movement in the environment with particular reference to soils and plants*, pp. 1–83. A report prepared for the National Radiological Protection Board. Grassland Research Institute, Hurley.

Whitehead, P. G. and Neal, C. (1987). Modelling the effect of acid deposition in upland Scotland. *Trans. Royal Soc. of Edinburgh*, **78**, 385–92.

Whitehead, P. G., Jenkins, A., and Cosby, B. J. (1990). Modelling long term trends in surface water acidification. *Trans. Royal Soc., London*: final conference of the Surface Water Acidification Programme, pp. 31–41.

Wild, A. (ed.) (1988). *Russell's soil conditions and plant growth*. Longman, Harlow.

Wilson, M. J., Lilly, A., and Nolan, A. J. (1989). Vulnerable soils and their distribution. In *Acidification in Scotland*, pp. 60–70. Scottish Development Department, Edinburgh.

Wrixon, A. D., Green, B. M. R., Lomas, P. R., Miles, J. C. H., Cliff, K. D., Francis, E. A., Driscoll, C. M. H., James, A. C., and O'Riordan, M. C. (1988). *Natural radiation exposure in UK dwellings*. Report NRPB-R190. National Radiological Protection Board, Chilton.

7. Data collection, storage, retrieval, and interpretation

G. P. Jones and F. C. Brassington

Introduction

The main developments in groundwater data collection and processing which have taken place over the past ten years have been made possible by the increased use of computers and electronic instrumentation. Rapid and significant improvements in microcomputers and their widespread availability have greatly increased the capability for handling large quantities of data in all branches of science, and hydrogeology is no exception. The hydrogeologist's ability to assemble large collections of reliable field data in a computer-compatible form has been made possible by the introduction of solid-state data loggers. These may be used to store water level data measured by pressure transducers, as well as water quality information monitored using conductivity meters, pH meters, and similar instruments. Such instrumentation has improved the quality of available data sets and raised the standard of hydrogeological investigations by allowing a greater use of statistical and analytical techniques. It has also had a considerable impact on the standard of presentation of data in graphical form. Perhaps the greatest advantage of these systems, however, is the saving of time by the direct recording of data in a form suitable for computer processing thereby avoiding laborious graph-plotting by hand.

The introduction of these new instruments has been slow and there is a long way to go before they totally replace the more traditional methods. This is partly because of the large capital investment required. Improvements in electronic engineering and the use of computers have also greatly enhanced the quality of geophysical data which has increased the value of geophysics in hydrogeological investigations.

During the 1980s, better information on aquifer properties and well-performance were made available by advances in pumping test techniques. Data loggers and pressure transducers have been introduced for field measurement as well as computer-based methods for data analysis and interpretation. There have also been advances in other field techniques such as the use of inflatable borehole packers for detailed field studies of hydraulic conductivity, variations with depth, and the use of slug-tests for the evaluation of low permeability rocks. In addition, there have been developments in methods for the analysis of step pumping test data.

This chapter reviews these advances in data handling and processing in British hydrogeology, together with a survey of current practices.

Data collection

Groundwater data collection is essentially a field activity, often involving the use of specialist equipment. Laboratory measurements are also necessary in the case of groundwater chemistry and some aspects of aquifer properties.

Groundwater levels

Groundwater levels are measured on a regional scale in observation well networks or, more locally, as part of a pumping test or other special investigations. The traditional ways of measuring groundwater levels are electronic tape water level probes (dippers) which are used for instantaneous measurements, or float-operated chart recorders which are used for a continuous record. These instruments have been in use for many years and are very familiar to practising hydrogeologists.

Observation well network

The foundation of the observation well network for monitoring groundwater levels in England and Wales was laid by the former 29 river authorities and the Water Resources Board during the period 1965–74. By the time that the water industry was re-organized in 1974, 1393 observation wells were known to be in use, of which 212 were fitted with automatic groundwater level recorders. 55 per cent of these were located in the Chalk and Upper Greensand aquifer, and 24 per cent in the Permo-Triassic sandstones (Anon. 1985). Responsibility for maintaining a national record of groundwater levels fell to the British Geological Survey (BGS) (then the Institute of Geological Sciences) in 1981, when a review was carried out with the aim of selecting between 200–300 sites to be used for periodical assessments of the national groundwater situation (Monkhouse and Murti 1981). At the request of the Department of the Environment in 1983, the archive was drastically reduced to fewer than 200 representative wells (Rodda and Monkhouse 1985). One observation well in Scotland and three in Northern Ireland were added in 1985 to form the Register of Selected Groundwater Observation Wells (Anon. 1989). This consists of some 175 sites which are occasionally changed when there are difficulties in continuing the field measurements.

Data logger systems

Electronic water level monitoring equipment has been developed specifically for monitoring groundwater levels. These systems use a submersible pressure transducer which is hung down the borehole on a cable at a sufficient depth below the water surface to ensure that it will not be exposed by changes in water level. The pressure transducer consists of a semi-conductor electrical strain gauge which is attached to a diaphragm and housed in a stainless-steel cylinder some 15–20 mm in diameter. Fluctuations in water level cause a variation in pressure which flexes the diaphragm, thereby moving the strain gauge. These movements change its electrical resistance which is amplified electronically to a variation in voltage.

It is possible for a transducer to be connected to a chart recorder to produce a paper record of groundwater level changes. The usual arrangement, however, is for the transducer to be connected to a solid state data logger. This device accepts the analogue signal from the transducer, converts it to a digital form and stores it in RAM (random accessible memory). Data stored in this form are computer-compatible and are collected either by down-loading the data from the logger into a portable hand-held computer, or by bringing the logger into the office for down-loading onto a computer. Removing the data logger from site would usually be normal practice only where the equipment has been installed for short-term monitoring such as a pumping test. Data loggers are installed at the well-head, either in a suitable weather-proof box or in a cylindrical case which is designed to be located within the borehole and secured to the borehole casing flange.

It is possible to pre-programme the data logger to collect readings at a variety of time intervals. This means that short-term readings (e.g. every 30 seconds) can be taken during pumping tests or to monitor changes in groundwater levels caused by tidal effects etc. or at much longer time intervals for regional groundwater level monitoring. Data loggers can usually store several thousand readings and are equipped with long-lasting lithium battery packs. Consequently, they can be left for periods as long as two years without being visited. In practice, more frequent visits are made to inspect the device and ensure that no breakdown has occurred.

Pressure transducers are rated according to the range of pressures over which they operate. Although they may vary from one manufacturer to another, generally each one has the same electrical output and level of accuracy. It is important, therefore, to match the transducer pressure range with the anticipated variation in water level to optimize the accuracy of the record. For example, with water level changes over a 50 m range, the measurement resolution will be typically 10 mm. The need to pre-select the suitable transducer means that the likely range of water levels must be known in advance. Furthermore, a selection of transducers will be needed, since those used for monitoring natural groundwater fluctuations are not likely to be suitable for use during pumping tests.

Pumping tests

The use of pressure transducers and data loggers for collecting water level data has greatly enhanced the ease with which pumping tests can be conducted in the field and the subsequent data analysed for aquifer properties. Transducers are installed both in observation boreholes and the pumping borehole, where it is prudent to place it within a dip-tube or similar pipe, to protect it from damage caused by turbulent flow around the pump. A dip-tube also ensures that the transducer can be lowered into the borehole and recovered at the end of the test without risk of snagging on the rising main flanges or electrical cables to the pump.

A data logger/pressure transducer system allows more frequent readings of water level to be taken than with a manual dipper. Computer analysis of all the data is possible and is made easy because the data are already in a form suitable for downloading onto the computer. Vines (1989) strongly recommends taking a large number of readings during step drawdown tests as this significantly improves the statistical confidence of the results and permits well-storage effects to be defined.

Optical shaft encoders

In other hydrometric installations, such as river gauging stations, electronic equipment for monitoring water level fluctuations includes optical shaft encoders. These devices are attached to a float via a small diameter steel cable which passes over a pulley wheel and is attached to a counter-weight in a similar fashion to a standard float-operated chart recorder. Water level movements rotate the wheel, causing a digital signal to be produced which can then be stored in a data logger. This type of equipment is generally more accurate than a pressure transducer but, being float-operated is more difficult to install in a borehole and is therefore only occasionally used for groundwater level measurements.

Telemetry

Once hydrometric field measurements have been converted to an electrical signal, it is possible for the data to be collected remotely via the public telephone system using special telemetry equipment. Occasionally, telemetry is carried out using radio signals. These systems are very costly, however, and it is most unusual for groundwater meas-

urements to be monitored by telemetry because changes in level usually take place slowly. Telemetry may be used where close monitoring is needed to control the operation of supply wells.

The past decade has seen a significant growth in the use of pressure transducer and data logger systems for monitoring groundwater levels, but the total number of such systems in use at the moment is small compared with the overall number of observation wells which are monitored. This is largely because of the high capital cost (*c.*£1,500 per borehole) and the low frequency of water level measurements which are normally required for regional monitoring purposes (at 1–3 month intervals).

Geophysical borehole logging

Borehole (or down-hole) geophysics has played an important role in hydrogeological investigations for many years. Tate *et al.* (1970) demonstrated how fissure-flow could be investigated using borehole logging techniques and Robinson (1974) reviewed the use of geophysical well logs in hydrogeological investigations. The value of these techniques in groundwater investigations and the importance of common standards has been recognized in the publication of a guide by the British Standards Institution (1988). These geophysical techniques are used to obtain information on:

(a) the geological formations through which the borehole is drilled;

(b) the presence, quantity, location, and the quality of the groundwater within the borehole;

(c) the dimensions, construction, and physical condition of the borehole.

Geophysical logging equipment consists essentially of four units: a down-hole instrument which is usually called a sonde; a cable and winch; power and control equipment; and data recording equipment. Each sonde contains the appropriate sensors which enable specific properties to be measured. The sensors have an output in the form of electronic signals in either analogue or digital form which are transmitted to the surface instruments via the cable and winch. The miniaturization of electronic components and other improvements in technology have enabled the development of more reliable and slimmer sondes.

The cable supports the sonde in the borehole and also conveys electrical power and electronic signals between the sonde instrumentation and the surface control equipment. The cable is wound on a winch which is used to raise or lower the sonde within the borehole. The cable is passed over a measuring pulley of known diameter which is connected to a recording device in order that an accurate record of the depth of the sonde can be made. The surface control equipment typically consists of two parts which provide power to the sonde and process the electronic signals from the sonde. Data is recorded in the form of an on-site graphical plot and where the signal is appropriate, data is stored on magnetic tapes.

Early borehole geophysical logging equipment used in hydrogeological investigations produced analogue signals. In this system, the measurement is affected by the length of the cable, and compensation must be made when interpreting the signals from the sonde. The data recording in an analogue system is typically in the form of multi-channel graphs plotted on-site. This produces long rolls of paper which require re-plotting or digitizing before they can be used easily.

During the 1980s, geophysical logging equipment which works on digital signals became available. Measurements taken within the sonde are converted into digital electronic signals which are not affected by the length of the cable. They are also in a form that is suitable for recording directly on magnetic tape and later down-loading into micro-computers. The data is computer processed on-site to produce a graphical output and to allow the operator to decide whether adequate field readings have been taken. All subsequent processing is carried out on an office-based computer using the raw data recorded on magnetic tape (Walters and Lloyd 1985). Standard output allows a full suite of logs for a particular borehole to be plotted on a single sheet of paper for ease of interpretation. Figure 7.1(a) shows an example of such a summary. Where required, a single parameter could be plotted for a number of boreholes to allow correlations to be made (Fig. 7.1.(b)). This is usually carried out by eye, simply by curve-matching logs of similar type using similar scales. Alternatively, correlation between boreholes may be achieved by a statistical comparison of the digitized logs from each location. These methods of processing geophysical data represent a vast improvement over earlier techniques where all site measurements were recorded graphically on long rolls of paper. The graphs could be digitized for computer data processing, but this was rare and graphs were usually laboriously re-plotted by hand.

Groundwater chemistry

Growing concern over the quality of groundwaters in the United Kingdom and the associated studies of hydrogeochemical processes has resulted in a many-fold increase in groundwater sampling and analysis, with the consequent accumulation of a large volume and wide range of information. The specific parameters requiring determination in any project will depend on the aims of the study which are of such diverse scope that no fixed rules can apply.

Hydrochemical work over the past decade has emphasized the variability of groundwater in both its natural state and particularly under the influence of man's activities. With aquifers coming under increasing stress in the UK from extraction and contamination, groundwater monitoring systems are an essential requirement of successful management. Costs of operating such systems are not inconsiderable, and should be reviewed regularly to ensure optimum efficiency. The principle of conceptualizing the hydrogeological conditions and likely hydrochemical controls affecting groundwater in the aquifer, allows the development of alternative hypotheses. These regional and or local-scale regimes must be taken into consideration in the planning of the sampling programme.

Wilkinson and Edworthy (1981) have emphasized that the aim of such a programme 'is to draw a picture of spacial distribution of each groundwater quality parameter and how that distribution changes in time'. Regardless of the purpose of the exercise, prior thought needs to be given to the selection of sample locations, type of source, method of sampling, and choice of analytical method(s).

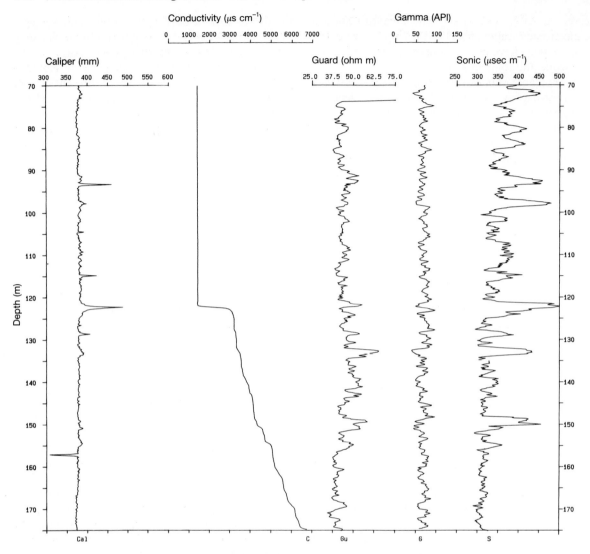

Fig. 7.1(a). Representative geophysical logs.

Sample locations

The location and number of points for sampling depends on the level of understanding of the 3D subsurface flow system, and especially on the availability of sources suitable for sampling purposes. Despite the numerous sources available in those parts of the UK where groundwater is of some importance, it is rare that existing conditions are adequate for satisfactory monitoring, and new dedicated boreholes are frequently necessary. The optimum number of points and their frequency of sampling are sometimes amenable to statistical treatment (Ward 1979), though the procedure is not customarily followed in the UK.

Types of source

Unsaturated zone

Although groundwater may be the main objective, the significance of the unsaturated voids in the vadose zone in initiating hydrochemical changes is

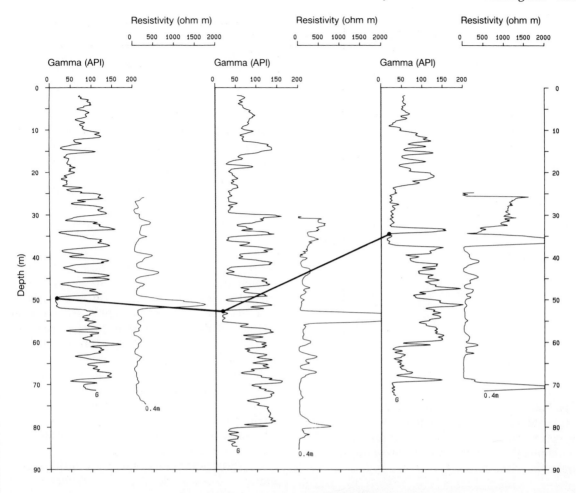

Fig. 7.1(b). Comparative geophysical logs.

well attested. With a thickness ranging up to tens of metres, it is an important physical and chemical buffer providing considerable storage as well as serving as an early warning of downward contamination.

Saturated zone

In this zone, the voids may be intergranular or of fracture origin but they are fully saturated. Depending upon whether the aquifer is confined or unconfined, homogenous or layered, the sources to be sampled will be either springs or wells (free-flowing or pumped), and their flow simple or complicated by mixing.

Methods of sampling

Water samples in the unsaturated zone are customarily taken by suction methods, but the pore-water extraction system described below is equally effective for unsaturated as for saturated strata.

In the saturated zone, traditional methods of taking depth or pumped samples are still in common use, though the former is open to criticism in being representative of the borehole rather than the aquifer. Because of the general lack of control over pumped water, the composition of which may be affected by depth of borehole and/ or pump, disposition of linings, pumping regime, and local hydrogeology, there is some difference

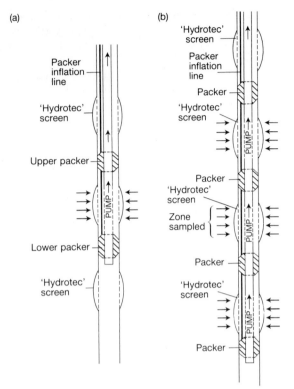

Fig. 7.2. Lightweight packer installations for sampling discrete zones (after Stuart 1984). (a) Isolation of one zone for sampling within a borehole; (b) 4-packer system for sampling central zone of three.

of opinion regarding the effectiveness of samples obtained in this manner. Parker *et al.* (1983) described their inadequacies with particular reference to vertical groundwater quality stratification, and Marsh and Lloyd (1980) demonstrated the modifications caused by the effects of metal casing on water from lined sections of boreholes.

The newly flushed, aquifer-derived waters from pumped wells can be used to study regional variations if precautions are taken and their limitations are fully understood, despite their composite chemistry. A borehole should be pumped long enough to remove four times the volume of the water column to ensure that representative aquifer water is being sampled (Clark 1988).

The complementary use of fluid-property logging such as conductivity, fluid flow, and temperature, allows the identification of inflow zones so

that sequential depth sampling can be effectively implemented. More desirable in this regard is the use in open-holes of inflatable packers (Fig. 7.2) to provide isolated samples from discrete sections of the borehole; these are usually temporary installations (Stuart 1984). Kay and Holmes (1983) describe the operation of a sophisticated wireline double-packer system, which combines hydraulic testing with representative sampling of water for chemical analysis and dissolved gas extraction.

Minimizing water movement within the borehole improves representative sampling of water in aquifers, and can best be attained by the installation of *in situ* sampling devices. A feature of British hydrogeological conditions, to which Clark and Baxter (1989) drew attention, is the depth to water (> 10 m) in the main aquifers, necessitating submersible pumps or deep location of samplers. The WRC-designed *in situ* sampler (Fig. 7.3) can be set at a number of different depths separated by bentonite seals, with the water drive actuated by compressed nitrogen. Similarly permanent in construction are the multiple-port samplers originating from North America (Westbay and Waterloo systems) that allow piezometric measurements and groundwater sampling to be made at different depths via a purpose-built string of devices. To date, such installations in the UK have been used primarily for geotechnical purposes.

Interstitial water extraction by centrifuging of saturated cores as developed by Edmunds and Bath (1976) has become well established, and is considered to be a most effective though necessarily expensive method. Current procedures are described in Lloyd and Heathcote (1985) and Gale and Robins (1989), who outline means of overcoming the difficulties associated with core sampling and pore-water contamination by drilling fluids.

One feature common to most pollution studies but exacerbated in those involving organic micropollutants, is the likelihood of cross-contamination from the sampling equipment. A planned, careful approach to the taking and handling of samples, together with the use of inert materials in equipment will minimize, if not prevent, anomalous results.

A groundwater sample will usually contain particulate, colloidal, biological, and dissolved fractions, so that some form of filtration is usually necessary (Edmunds 1986). Such filtration should

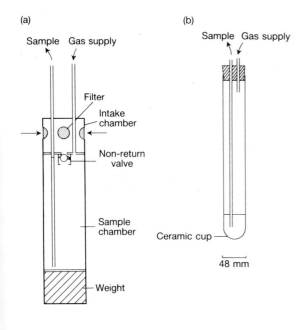

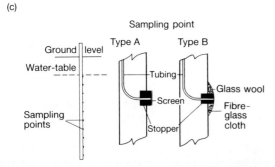

Fig. 7.3. Permanent sampling installations. (a) W.R.C. *in situ* sampler; (b) Soil moisture sampler; (c) Schematic diagram of the multilevel sampling device (after Pickens *et al.* 1978).

chemical analysis can represent as closely as possible the *in situ* aquifer water, preservation is required to stabilize the sample. This commonly involves the addition of strong ultra-pure acid to lower the pH to less than pH2 which will prevent the precipitation of metals. Methods of preservation for different determinands are given by Gale and Robins (1989) as well as recommendations for on-site treatment, transport and storage (see Fig. 7.4).

Methods of analysis

It is claimed that analytical accuracy is the most important and neglected factor after correct sampling. The hydrogeologist needs to be concerned with accuracy since most modern instruments are

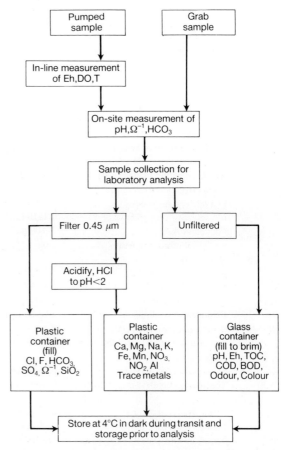

Fig. 7.4. Flow chart for the optimal collection of groundwater samples for the analysis of common determinands (after Gale and Robins 1989).

be carried out in the field, before preservation, preferably using an in-line filtration unit with 0.45 μm pore-size papers. The choice of filter medium will depend on the determinands being investigated. If organic solvents are likely to be encountered then the commonly used cellulose acetate papers should be replaced by glass-fibre or PTFE filter media (Gale and Robins 1989).

Some species, such as those involving carbonate and redox reactions, will undergo changes on reaching atmospheric conditions. In order that the capable of giving high analytical precision. Recommended techniques for the analysis of elements in groundwater are given in Table 7.1 after Cook and Miles (1980). It is clear that certain parameters must be measured in the field, particularly those which define the carbonate system and oxidation regime. Thus, pH, Eh, temperature, dissolved oxygen, and electrical conductivity should normally be measured at the well-head. The remaining major elements, and such minor and trace elements as are needed will then be determined in the laboratory by the most appropriate method. A check on the ionic balance will give some indication of the accuracy of the analysis which would be expected to be better than 2.5 per cent (Edmunds 1981). Quality control measures, considered by Hunt and Wilson (1986), will ensure that data is collected with a known degree of accuracy.

Data storage and retrieval

The introduction of data-logger systems for collecting hydrogeological field data has meant that

Table 7.1 Recommended analytical techniques (after Cook and Miles 1980)

Element	Analytical method	Analytical working range	Approximate detection limit	Minimum sample volume (ml)	Typical concentration in groundwaters	Drinking water EC limits
Al	Complex with 8-hydroxyquinoline, extract in field into MIBK* Analysis by flame AA† ($N_2O–C_2H_2$)	5–50 μg l^{-1}	2 μg l^{-1}	500	1–60 μg l^{-1}	50 μg l^{-1} (G)
Ba	Flame AA ($N_2O—C_2H_2$)	0.1–1.0 mg l^{-1}	0.1 mg l^{-1}	2	0.01–2.0 mg l^{-1}	100 μg l^{-1} (G)
HCO_3	Potentiometric filtration with 0.01 M H_2SO_4	100–400 mg l^{-1}	10 mg l^{-1}	25	50–400 mg l^{-1}	–
Br	Neutron activation analysis	0.5–100 mg l^{-1}	0.1 mg l^{-1}	1	50–500 μg l^{-1}	–
Ca	Flame AA (Air–C_2H_2)	5–40 mg l^{-1}	<1 mg l^{-1}	2	5–500 mg l^{-1}	100 mg l^{-1}
Cl	$AgNO_3$ titration using Ag/AgS ion selective electrode as monitor	5–800 mg l^{-1}	1 mg l^{-1}	10	10–200 mg l^{-1}	25 mg l^{-1} (G)
Cu	Flameless AA–direct	1–50 μg l^{-1}	1 μg l^{-1}	1	1–100 μg l^{-1}	3 mg l^{-1} (M)
F	Ion selective electrode	0.1–10 mg l^{-1}	0.05 mg l^{-1}	10	0.1–5 mg l^{-1}	1.5 mg l^{-1} (M)
Fe (total)	Flameless AA	2–100 μg l^{-1}	3 μg l^{-1}	1	10–20 000 μg l^{-1}	50 μg l^{-1} (G)
Fe^{2-}	Spectrophotometric with 2,2'-bipyridyl	0.08–1.0 mg l^{-1}	0.01 mg l^{-1}	25	0.01–20 mg l^{-1}	50 μg l^{-1} (G)
Li	Flame AA (Air–C_2H_2)	20–200 μg l^{-1}	5 μg l^{-1}	2	2–60 μg l^{-1}	–
Mg	Flame AA (Air–C_2H_2)	0.5–6.0 mg l^{-1}	0.05 mg l^{-1}	2	1–200 mg l^{-1}	30 mg l^{-1} (G)
Mn	Flameless AA–direct	1–50 μg l^{-1}	0.5 μg l^{-1}	1	1–50 μg l^{-1}	20 μg l^{-1} (G)
Ni	Flameless AA–direct	10–100 μg l^{-1}	5 μg l^{-1}	1	1–50 μg l^{-1}	50 μg l^{-1} (M)
NO_3-N	Spectrophotometric (UV)	0.1–1.0 mg l^{-1}	0.01 mg l^{-1}	1	1–100 mg l^{-1}	5.6 mg l^{-1} (G)
HPO_4	Spectrophotometric-molybdenum blue	0.06–3.0 mg l^{-1}	8 μg l^{-1}	40	0.01–0.5 mg l^{-1}	0.35 mg l^{-1} (G)
K	Flame AA (Air–C_2H_2)	0.5–10 mg l^{-1}	0.1 mg l^{-1}	2	0.2–30 mg l^{-1}	10 mg l^{-1} (G)
SiO_2	Spectrophotometric-molybdenum blue	0.5–4.0 mg l^{-1}	0.05 mg l^{-1}	25	1–10 mg l^{-1}	–
Na	Flame AA (Air–C_2H_2)	0.25–400 mg l^{-1}	0.1 mg l^{-1}	2	5–1000 mg l^{-1}	20 mg l^{-1} (G)
Sr	Flame AA (Air–C_2H_2)	0.05–1 mg l^{-1}	0.02 mg l^{-1}	2	0.05–10 mg l^{-1}	–
SO_4	Titration with $Ba(HClO_4)_2$ using thorin as indicator	10–100 mg l^{-1}	10 mg l^{-1}	70	10–500 mg l^{-1}	25 mg l^{-1} (G)

* MIBK methyl isobutyl ketone.
† AA–atomic adsorption

(G) = EC guide limit (1976)
(M) = EC maximum permissible limit (1976)

microcomputers have been used increasingly for the processing and validation of raw data and for data storage. There are a number of proprietary software systems available which can be used on personal computers (PC). Companies which market data-logging systems usually provide suitable software to process and store data collected from the field, once it has been down-loaded from the data logger. These computer systems allow groundwater level data to be stored for each borehole where measurements are taken. Usually, information regarding the depth, diameter, location, and aquifer are included to identify each borehole. Water level information can be provided from the computer, either as a graph, or in tabular form. The operator is able to select the scales for the graphs to meet the needs of a particular project. Figure 7.5 shows an example of the graphical presentation of groundwater level data. Where the appropriate software is available, it is possible to plot maps of groundwater levels for a given time, or to correlate information between various boreholes using standard statistical packages. The main advantage in the use of computers is that large volumes of data can be stored and easily accessed. When a record is accessed from a computer database, it is in a form where it can be processed further to aid interpretation.

The British Geological Survey (BGS) manages a number of databases for groundwater information. The National Well Archive contains details of a large number of water wells in the UK, but it is a paper-based system. The groundwater level archive is computer-based and is used to produce the data published in the Hydrological Data United Kingdom Yearbook series (e.g. Anon. 1989). BGS is currently constructing a new computer-based archive for groundwater quality information.

Water quality data for groundwaters in the United Kingdom are being and will continue to be collected at a rate which requires care as to the means of data storage. The sheer volume and diversity of data makes a national depository impractical, except for representative data that may be part of a national monitoring scheme, including water levels as well as water quality. The ten regions of the National Rivers Authority (NRA) provide the most realistic basis for data storage, with some older and specific research-orientated data kept on paper files, but with most of the data kept on computer archive. The advan-

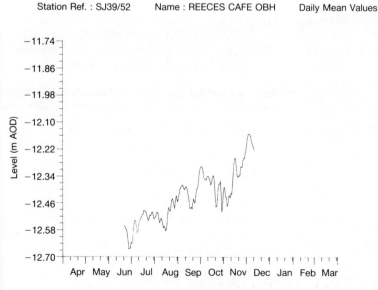

Fig. 7.5. Graphical presentation of groundwater level data.

tages of the latter system are essentially comparable to those described for the handling of water level data, particularly in a form amenable to data processing and presentation for further interpretation.

Measurements of aquifer properties

Pumping tests remain the most usual field technique to determine the hydraulic properties of aquifers. They are also used to evaluate the yield of wells and to assess the impact of a new abstraction on the local groundwater conditions. A British Standard code of practice was published (1983) in order to raise the standards of these tests, particularly the measurement of the pump discharge and the frequency and accuracy of water level measurements.

Pumping test analysis

Most of the graphical methods for pumping test analysis have now been produced as commercial software packages. Field data can often be downloaded from data loggers. The high quality graphics which are available on most PCs mean that it is now possible to curve-fit on the computer screen.

As part of an exercise to reduce both operating costs and the consumption of energy, a number of regional water authorities undertook programmes of standard borehole performance tests, based on stepped pumping procedures. Skinner (1988) described such a testing programme undertaken by Severn Trent Water on their 250 borehole pumping stations. The main objectives of this programme were to establish reliable yield–drawdown relationships for each borehole which will enable a direct cost comparison to be carried out between individual sources and to establish a 'baseline' record of borehole performance for comparison with future tests so that changes in borehole performance can be assessed.

The stepped pumping test method consists of pumping a borehole from rest in a series of four or more increasing steps. Each step is for a duration of 60–100 minutes, with a recovery test being carried out at the end. Conversely, where it is difficult to rest a production borehole, the test consists of reducing pumping in a series of steps.

There are a number of methods available to analyse step-drawdown tests, and the most important one has been summarized by Clark (1977, 1979). The methods are essentially graphically based and sometimes difficult to interpret. Indeed, it is not unusual to obtain different results from the different methods. Attempts to explain the complexity of stepped pumping tests have been made by Rushton and Rathod (1988) who examined the effect of the development of a seepage face within the pumping well, zones of lower or higher hydraulic conductivity within the aquifer, and the change from confined to unconfined state in the vicinity of the well, using a radial and a vertical flow numerical flow model developed by Rushton and Redshaw (1979).

A Fortran program for analysing step-drawdown tests has recently been presented by Vines (1989). This is based on the method developed by Eden and Hazel (1973) and has several advantages over graphical methods, offering a more flexible approach to the planning and analysis of such tests, including:

(1) the ability to cope with pump failures without restarting the test;

(2) the easy analysis of a step test starting from a steady pumping condition;

(3) the easy treatment of unequal time periods;

(4) the production of statistical parameters which give an idea of the 'goodness-of-fit';

(5) the ability to process pumping test data rapidly and thereby complete an extensive testing programme in a relatively short period.

Borehole packer tests

The use of inflatable borehole packers as a field technique in groundwater studies was pioneered in Britain by Price *et al.* (1977, 1982) using double-packer injection tests. Brassington and Walthall (1985) describe the use of borehole packer tests involving both injection of water and out-flow pumping tests on a sandstone aquifer in north-west

England. They showed that tests involving pumping from between the packers produces more reliable results than injection tests which are often subject to error except when very low injection pressures are used. In-flow testing can block aquifer pores or even cause fractures when high pressures are used. Out-flow testing has a further advantage in that water samples are obtained from each test zone, thereby permitting vertical changes in groundwater chemistry to be studied. The field equipment and methodology are described in detail by Price and Williams (1989).

The tests described by Price *et al.* and Brassington and Walthall relate to sandstone and chalk aquifers where a significant part of the overall transmissivity is due to the presence of fissures. Packer testing enables small vertical intervals of the aquifer to be tested in detail to quantify the role of fissures in controlling groundwater flow. The results obtained by Brassington and Walthall (1985) in a Permo-Triassic sandstone show that some 75–80 per cent of the overall transmissivity is caused by fissures. They also compared the field values of hydraulic conductivity with laboratory measurements made by Campbell (1982) using core samples taken from the test boreholes. Similar values were obtained from each technique, which gave confidence both to the field methods and to deductions made about the significance of fissures on transmissivity values.

The reliability of packer tests for estimating the hydraulic conductivity of aquifers was investigated by Bliss and Rushton (1984). A mathematical model of the radial and vertical flow from a borehole in a typical aquifer was devised and used to investigate the nature of flow in the aquifer caused by the injection of water between two packers. They showed that the formulae used by Barker (1981) and Hvorslev (1951) provide adequate estimates for zones of both high and low hydraulic conductivity. A number of other aspects were investigated from which it was found that packer tests only disturb the groundwater flow at a distance up to about 10 m from the borehole. Consequently, an equilibrium condition is achieved within a few minutes; though for regions having very low hydraulic conductivities, equilibrium may not be reached for about an hour. The investigation showed that fissures collect most of their inflow from the first 10 m and therefore there is

little difference between the inflow to a borehole from a fissure which is 10 m long and one which is much more extensive.

Slug tests

The need to make field measurements of the hydraulic properties of poorly-permeable rocks has increased markedly over the past fifteen years or so. This has been the result of investigation programmes into radioactive waste disposal, toxic waste disposal, and the development of geothermal energy resources (Mather *et al.* 1982; Black 1985). A significant part of the UK programme to investigate suitable sites for radioactive waste disposal involved a large number of slug and pulse tests, based around a wire-line straddle packer system. A slug test involves the rapid, almost instantaneous removal of a known volume of water from a test borehole, using a bailer or similar device, and the monitoring of subsequent water level changes (Black 1978). Early results produced inconsistencies between some field measurements and expected values based on laboratory measurements especially for specific storage values. Consequently, a new method of analysis was devised by Barker and Black (1983) which overcame the problems of earlier methods and produced sensible results.

A survey of current practices

The majority of routine hydrogeological work in England and Wales is carried out by the National Rivers Authority (NRA). The NRA was established in September 1989 and inherited the responsibilities for groundwater management and protection from the former regional water authorities which were described by Hardcastle and Hunter Blair (1981). To assist with this review, the ten NRA regions kindly provided information on their current practices with regard to groundwater data collection and management which they had continued from the former regional water authorities.

Hydrogeological fieldwork in Scotland and Northern Ireland is carried out on a lesser scale than in England and Wales as groundwater forms only a small part of water supplies. Most of the work is carried out by the British Geological Survey in both countries, usually on behalf of local authorities. In Scotland, several River Purification Boards make hydrogeological assessments of waste disposal sites and monitor to ensure that groundwater is not polluted. Some local groundwater information is also available from the Water Departments in the Regional Councils and the Scottish Development Department. In Northern Ireland, limited groundwater records are kept by the Department of the Environment (NI).

In England and Wales, major groundwater research projects have been carried out by the regional water authorities, the British Geological Survey, and the Water Research Centre (WRC). The latter two organizations are the two major national bodies involved in such work, although important groundwater studies have also been carried out by the Institute of Hydrology. Several university departments and some hydrogeological and engineering consultants have completed groundwater research projects, often in conjunction with a Regional Water Authority or on contract to a government department.

Groundwater levels

The NRA operates a total of some 5750 observation boreholes to monitor groundwater levels in the major and some minor aquifers throughout England and Wales. The majority are read at monthly or more frequent intervals, with most of the rest being read once every three months. Over 90 per cent of the groundwater level measurements are taken using a dipper, and a dipper is used for a check reading in the majority of other instances. Where a constant record is required, a float-operated chart recorder is usually used. Pressure transducer and data logger systems are used in 70 per cent of the regions, but on only about 2 per cent of the observation boreholes. It was reported that generally the number of pressure transducer/data logger systems installed is being increased.

The majority of groundwater level data is stored on computers, generally using commercially available software packages, although there is also some purpose-written software. Both mainframe and personal computers are used, with a trend to

change to PCs. Besides archiving data from the observation boreholes, water level information is also stored from a selected number of pumping boreholes and non-NRA observation boreholes.

Aquifer properties

Over the past decade, more than 2500 pumping tests have been carried out which have been analysed for transmissivity values and 1500 analysed for storage values. These data on aquifer properties are mostly recorded only in project reports. There are also a large number of pumping tests carried out which are not suitable for analysis to determine the hydraulic properties. Most of these tests are carried out to assess the impact of new abstractions prior to an application for an abstraction licence.

Borehole packers have been used by half the regions during the past ten years as part of special investigations into aquifer properties or groundwater quality. Laboratory measurements of permeability have generally not been made, and only one region has the in-house capability of making such measurements.

Geophysical borehole logging

Three-quarters of the NRA regions have borehole geophysical logging equipment, although only half operate digital computer-based equipment. All those regions with the equipment use geophysics as part of routine groundwater work. Borehole geophysics is a routine part of investigations carried out by BGS, WRC, and several universities, and each organization keeps its own records.

Data handling and processing

Computer systems are in widespread use for storing groundwater level and chemistry information. Computer systems are used for the routine analysis of pumping tests, using either software versions of graphical methods, or radial flow models. Most regions have computer models of groundwater flow in the principal aquifers which were developed as part of major groundwater investigations. These models are not used on a routine basis except in the single example of controlling saline intrusion into the Chalk at and near Grimsby.

Groundwater sampling

There is wide variation in sampling practice between the ten different regions. Separate monitoring networks, sampling public and private sources, are maintained by half, but private sources are excluded by the remainder. It is usual for the public sources to be sampled by the operations staff and private sources by pollution control staff with the majority, but not all, of the samples being of raw water. There has been no consistency to date in the manner of sampling from multiple well sources, the frequency of sampling, and the range of parameters analysed. The amount of historical data kept by the regions varies but most data sets go back to 1974. The majority of regions store data on a mainframe computer archive, with down-loading to microcomputers for data manipulation. Table 7.2 summarizes the main features of the groundwater sampling practices as inherited by the NRA. Work is currently in hand to develop a common national sampling programme which will cover sampling frequency, sampling technique, types of sources, and the analysis of the sample.

Future trends

It is likely that the developments in data collection, handling, and interpretation seen during the 1980s will continue into the 1990s. The majority has been made possible by the continuing advances in electronic engineering, computers, and software development. These improvements are likely to be in three main areas: an overall improvement in the accuracy of field measurements, the integration of different types of data, and the development of more sophisticated mathematical models.

Accuracy of measurements

There have been such great steps forward in the way in which groundwater levels have been measured during the 1980s, that it is difficult to imagine that any further significant improvements in accuracy are possible. The overall quality of groundwater level measurements is likely to be improved by a more widespread use of pressure transducer systems. Improvements in instrumentation used for water quality sensing and a more widespread use in groundwater applications are also likely. A better understanding of groundwater quality will also be brought about by improved methods of analysis, particularly in determining low concentrations of complex organic substances.

The trend to collect more field data in a computer compatible form is likely to mean that portable 'lap-top' PCs will be used increasingly in the field. This is especially likely for pumping tests.

Data management

The main trends established towards the end of the 1980s in data management were the development of integrated computer databases which are capable of handling large volumes of different types of information. Previously, it was uncommon for quantity (river flow, groundwater levels, and pumping rates) data to be stored in the same computer archive system as water quality data. For a wide range of environmental management applications, these types of data need to be available with geographical (or spatial) information (i.e. details of the location of pumping wells, observation boreholes, landfill sites, etc.), with data accessed on the basis of location. This means that new integrated databases are likely to be part of a geographical information system (GIS).

Groundwater modelling

Groundwater models have been used for many years to aid the understanding of groundwater systems and to make predictions for future changes. The majority of groundwater models predict groundwater levels in response to variations in recharge and groundwater abstraction. It is likely that there will be an increase in the routine use of groundwater models for small scale projects using largely standard commercial software packages on PCs. Models are available to assess groundwater pollution problems (Bear and Verruijt 1987) but

Table 7.2 Summary of groundwater sampling practices inherited by the NRA

Area	Separate groundwater monitoring network/archive	Public and/or private sources	Approximate percentage of raw to treated in public samples	Approximate number of public supply sources sampled		Approximate number of private sources sampled		Frequency of sampling	Range of analysis
				Boreholes	Springs	Boreholes	Springs		
Anglian	Yes	All public some private	Varies between Divisions	150	–	350	130	Variable	3 standard suites plus specific ones Basic Ca, Mg, Na, K, Cl, SO$_4$, alkalinity, TON, pH, temp, EC, TOC, NH$_3$, pH (field), temp, (field) Ferruginous Basic plus DO, PO$_4$, Mn, Fe Extended Ferruginous + BOD, TDS, T hardness, F, SiO$_2$, Li, B, As, Cd, Hg, Se, Cr, Cu, Pb, Ni, Zn, V, EC (field)
North-west	Yes	All public some private	95:5	240	20	20 (+ 400 observation boreholes)	20	Private source variable Public 12+ p.a.	Standard suite Ca, Mg, Na, K, Cl, SO$_4$, alkalinity, pH, EC, hardness, NO$_3$, NO$_2$, NH$_3$, CO$_3$, bicarbonate, metals
Northumbrian	Yes	All public some private	100:0	30	–	40 (+ 40 observation boreholes)	–	1 p.a.	Standard suite Ca, Mg, Na, K, Cl, SO$_4$, alkalinity, TON, pH, EC, T hardness, Fe
Severn-Trent	Yes	All public some private	95:5	155	9+	164 (plus past data on 89 disused boreholes)	–	2 p.a. except for new sites & variable quality sources 4 p.a.	2 Standard suites plus other specific analyses, if required. Basic Ca, Mg, Na, K, Cl, SO$_4$, alkalinity, TON, pH, EC, T hardness, temp, Basic plus DO, TOC, F, PO$_4$, Fe, Mn, Ba, Cd, Cr, Cu, Pb, Ni, Zn
Southern	Yes	Some public some private	Varied	100	–	40	–	2 p.a. (not all parameters on all sites)	Standard suite Ca, Mg, Na, K, Cl, SO$_4$, alkalinity, pH, EC, T hardness, F, NH$_3$, Fe, Mn, B

South-west	No	50:50	28	12	–	4 p.a.	*4 p.a.* Ca, Mg, Na, K, Cl, SO$_4$, alkalinity, pH, temp, EC, T hardness, Fe, Mn, Al *1 p.a.* pesticides
Thames	Yes	25:75	310	10	–	Varies according to EC Drinking Water Directive (at least 1 p.a.)	*4 p.a.* Ca, Mg, Na, K, Cl, SO$_4$, alkalinity, TON, pH, temp, EC, T hardness, NH$_3$, F, TOC, Sr, DO *1 p.a.* pesticides, heavy metals, Ba, Mn, B, Al
Welsh	No	20:80	14	70	–	12 p.a.	*12 p.a.* EC, temp, pH, NH$_3$ *4 p.a.* TON, Cl, Al, DO *2 p.a.* SO$_4$, Fe, Mn *1 p.a.* pesticides, heavy metals, F, Ba, B
Wessex	No	Varied	Varied	–	–	26 p.a.	*26 p.a.* Cl, TON, pH, EC, temp, NH$_3$, Fe *4 p.a.* Ca, Mg, Na, K, SO$_4$, alkalinity, F, TOC *1 p.a.* pesticides, PAH, Hg
Yorkshire	No	30*:70	65	51	11	Varies on size and use	Basic analysis on raw waters includes pH, alkalinity, EC, Al, TON, Fe, Mn, NH$_3$. Treated station output analysis (30% single borehole sources)* includes Ca, Mg, Na, K, Cl, SO$_4$, alkalinity, TON, EC, T hardness, metals

these models are not yet in common use. It is likely that increasing processing power of computers will enable more complex groundwater models to be developed for both groundwater quantity and quality investigations. The development of the graphical display capabilities of computers is likely to mean that by the end of the next decade groundwater models will be available which display fluctuations in levels and the dispersion of groundwater pollution in graphical form on the computer screen.

Acknowledgements

The authors gratefully acknowledge the willing assistance of colleagues employed by the National Rivers Authority and other organizations who provided basic information for this chapter. They are also grateful to the NRA for permission to publish summary information on groundwater sampling practices.

References

Anon. (1985). *Hydrological data United Kingdom 1981 yearbook*. Natural Environment Research Council, Institute of Hydrology.

Anon. (1989). *Hydrological data United Kingdom 1988 yearbook*. Natural Environment Research Council, Institute of Hydrology.

Barker, J. A. (1981). A formula for estimating fissure transmissivities from steady state injection test data. *J. Hydrol.*, **5**, 337–46.

Barker, J. A. and Black, J. H. (1983). Slug tests in fissured aquifers. *Water Resour. Res.*, **19**, 1558–64.

Bear, J. and Verruijt, A. (1987). *Modelling groundwater flow and pollution*. Reidel, Dordrecht.

Black, J. H. (1978). The use of the slug test in groundwater investigations. *Wat. Serv.*, **82**, 174–8.

Black, J. H. (1985). The interpretation of slug tests in fissured rocks. *Q. J. Eng. Geol.*, **18**, 161–71.

Bliss, J. C. and Rushton, K. R. (1984). The reliability of packer tests for estimating the hydraulic conductivity of aquifers. *Q. J. Eng. Geol.*, **17**, 81–91.

Brassington, F. C. and Walthal, S. (1985). Field techniques using borehole packers in hydrogeological investigations. *Q. J. Eng. Geol.*, **18**, 181–93.

British Standards Institution, (1983). *British Standard code of practice for test pumping water wells*, BS6316; 1483.

British Standards Institution, (1988). *British Standard guide for geophysical logging of boreholes for hydrogeological purposes*.

Campbell, J. E. (1982). *Permeability characteristics of the Permo-Triassic sandstones of the Lower Mersey basin*. Unpublished MSc thesis, University of Birmingham.

Clark, L. (1977). The analysis and planning of step drawdown tests. *Q. J. Eng. Geol.*, **10**, 125–44.

Clark, L. (1979). The analysis and planning of step drawdown tests: a clarification. *Q. J. Eng. Geol.*, **12**, 124.

Clark, L. (1988). *The field guide to water wells and boreholes*. Geological Society of London Professional Handbook. Open University Press, Milton Keynes.

Clark, L. and Baxter, K. M. (1989). Groundwater sampling techniques for organic micropollutants: UK experience. *Q. J. Eng. Geol.*, **22**, 159–68.

Cook, J. M. and Miles, D. L. (1980). Methods for the chemical analysis of groundwater. *Hydrogeology Report*, No. 80/5. British Geological Survey, Keyworth.

Eden, R. N. and Hazel, C. P. (1973). Computer and graphical analysis of variable discharge pumping tests of wells. *Civil Eng. Trans. Inst. Eng. Australia*, **15**, 5–10.

Edmunds, W. M. (1981). Hydrochemical investigation. In *Case-studies in groundwater resources evaluation* (ed. J. W. Lloyd), pp. 87–112. Oxford Science Publishers, Oxford.

Edmunds, W. M. (1986). Groundwater chemistry. In *Groundwater occurrence, development and protection* (ed. T. W. Brandon), pp. 49–107. Inst. Water Engrs. Sci., London.

Edmunds, W. M. and Bath, A. H. (1976). Centrifuge extraction and chemical analysis of interstitial waters. *Env. sci. tech.*, **10**, 467–72.

Gale, I. N. and Robins, N. S. (1989). The sampling and monitoring of groundwater quality. *Hydrogeology Report*, No. 89/37. British Geological Survey, Keyworth.

Hardcastle, B. and Hunter Blair, A. (1981). The legislative framework for aquifer management. In *A survey of British hydrogeology 1980*, Royal Society, London.

Hunt, D. T. E. and Wilson, A. L. (1986). *The chemical analysis of water: general principles and techniques* (2nd edn). The Royal Society of Chemistry, London.

Hvorslev, M. J. (1951). Time lag and soil permeability in groundwater observations. *US Army Corps.*, *Vicksburg Eng. Waterways Exp. Stn. Misc. Bulletin No.36*.

Kay, R. L. F.. and Holmes, D. C. (1983). Combined

water sampling and hydraulic testing in poorly permeable rocks. In *Methods and instrumentation for the investigation of groundwater systems*, pp. 546–55. Proc. Int. Symp., Noordwijkerhout (May 1983). TNO, The Netherlands.

Lloyd, J. W. and Heathcote, J. A. (1985). *Natural inorganic hydrochemistry in relation to groundwater*. Clarendon Press, Oxford.

Marsh, J. M. and Lloyd, J. W. (1980). Details of hydrochemical variations in flowing wells. *Ground Water*, **18**, 366–73.

Mather, J. D., Chapman, N. A., Black, J. H., and Lintern, B. C. (1982). The geological disposal of high-level radioactive waste—a review of the IGS research programme. *Nuclear Energy*, **2**, 167–73.

Monkhouse, R. A. and Murti, P. K. (1981). *The rationalisation of groundwater observation well networks in England and Wales*. Report No. WED/81/1. Institute of Geological Sciences, London.

Parker, J. M., Perkins, M. A., and Foster, S. S. D. (1983). Groundwater quality stratification—its relevance to sampling strategy. *Methods and instrumentation for the investigation of groundwater systems*, pp. 43–54. Proc. Int. Symp., Noordwijkerhout (May 1983), TNO, The Netherlands.

Pickens, J. F., Cherry, J. A., Grisak, G. E., Merritt, W. F., and Risto, B. A. (1978). A multilevel device for groundwater sampling and piezometric monitoring. *Ground Water*, **16**, 322–7.

Price, M. and Williams, A. T. (1989). *Using a pumped double-packer system in groundwater studies.*, Technical report WD/89/55, Hydrogeology Series. British Geological Survey, Keyworth.

Price, M., Robertson, A., and Foster, S. S. D. (1977). Chalk permeability—a study of vertical variation using water injection tests and borehole logging. *Wat. Serv.*, **81**, 603–10.

Price, M., Morris, B., and Robertson, A. (1982). A study of intergranular and fissure permeability in Chalk and Permian aquifers, using double packer injection testing. *J. Hydrol.*, **54**, 401–23.

Robinson, V. K. (1974). Low cost geophysical well logs for hydrogeological investigations. *Q. J. Eng. Geol.*, **7**, 207–15.

Rodda, J. and Monkhouse, R. A. (1985). The national archive of river flows and groundwater levels for the UK. *J. Instn. Water Engrs.*, **39**, 358–62.

Rushton, K. R. and Rathod, K. S. (1988). Causes of non-linear step pumping test responses. *Q. J. Eng. Geol.*, **21**, 147–58.

Rushton, K. R. and Redshaw, S. C. (1979). *Seepage and groundwater flow*, Wiley. Chichester.

Skinner, A. C. (1988). Practical experiences of borehole performance evaluation. *J. Instn. Water Env. Mgmt.*, **2**, 332–40.

Stuart, A. (1984). *Borehole sampling techniques in groundwater pollution studies*. Technical Report FLPU 84/15. British Geological Survey, Keyworth.

Tate, T. K., Robertson, A. S. and Gray, D. A. (1970). The hydrogeological investigation of fissure-flow by borehole logging techniques. *Q. J. Eng. Geol.*, **2**, 195–215.

Vines, K. J. (1989). Ednhaz: A program for analysing step drawdown tests. *Computers and Geoscience*, **15**, 965–78.

Ward, R. C. (1979). Statistical evaluation of sampling frequencies in monitoring networks. *J. Water Poll. Con. Fed.*, **51**, 2292–300.

Walters, M. and Lloyd, J. W. (1985). The use of a microcomputer for recording and analysis of borehole logging data in hydrogeological investigations. *Q. J. Eng. Geol.*, **18**, 381–9.

Wilkinson, W. B. and Edworthy, K. J. (1981). Groundwater quality monitoring systems—money wasted? *Quality of groundwater. Proc. Int. Symp.* Noordwijkerhout, The Netherlands. Studies in Environmental Science, **17**, 629–42. Elsevier.

8. Landfill disposal of wastes

G. M. Williams, C. P. Young, and H. D. Robinson

Introduction

Investigating landfill sites and tracing pollution plumes has prompted hydrogeologists to focus their attention on the fundamental mechanisms affecting the flow of water and the behaviour of solutes in the subsurface. In Britain, a major research project on the behaviour of hazardous wastes in landfill sites commenced in 1974 with a three year programme (Department of the Environment 1978) and similar programmes have continued throughout the 1980s. In contrast to the original programme which set out to identify the mechanisms controlling the behaviour of various hazardous wastes and the appropriate hydrogeological environments in which they could be placed, work in the last decade has been concerned more with the movement of leachate and the production of gas from domestic wastes. Particular attention has been given to the role of microbes in degrading organic components of leachate in both the saturated and unsaturated zones, and studies have attempted to isolate and qualify microbial populations.

Driven by EC Water Quality Directives, water authorities have been more vigilant in their efforts to avoid groundwater pollution and have forced a swing away from 'dilute and disperse' sites, where detailed hydrogeological surveys are required, towards engineered containment with natural or artificial barriers. In a few cases, artificial fine-grained materials or 'permeable liners' have been installed to increase the opportunity for attenuation reactions beneath the landfill and to control the release of leachates to formations which otherwise would be considered unsuitable for waste disposal activities.

With the consequent need to deal with leachate, research has been carried out on a variety of leachate treatment options, including aerobic and anaerobic biological methods.

In spite of an ongoing research programme there have been several serious explosions in the last ten years as a result of migrating landfill gas. Partly to control the spread of landfill gas and to recover an energy resource, the technology for methane recovery and its utilization has now been successfully developed in many situations.

Since hydrogeological studies have focused on leachate migration through aquifers rather than on ensuring containment, this chapter is divided into studies on the unsaturated zone, the saturated zone, and problems of methane migration.

The unsaturated zone

Introduction

During the 1970s, investigations of the impact of landfills on groundwater identified the unsaturated zone as an area of particular interest, because of its potential to allow physical, chemical, and biological processes to attenuate leachate (Department of the Environment 1978; Royal Society of London 1981). The work begun in the 1970s included both field studies at existing landfills and laboratory and lysimeter based experiments and whilst some evidence of attenuation by processes other than dilution of leachate with clean pore-waters was apparent, some of the more optimistic statements on the effectiveness of unsaturated zones were beginning to be questioned by the close of that decade. This gave impetus to studies designed to elucidate the processes operating in the unsaturated zone and their relationships to the various lithologies found in British aquifers. The work took two principal directions, which are reviewed here: detailed field investigations on the Sherwood Sandstone and the Chalk, and studies of the potential for enhancing the attenuating capacity of unsaturated zones by modifying their mineralogical and physical properties.

Field studies

Sherwood Sandstone

During the late 1970s, hydrogeologists of the then Severn Trent Water Authority initiated detailed field studies of a series of dilute and disperse landfills based on the Sherwood Sandstone in Nottinghamshire. The investigations included sequential profiling of leachate components in the unsaturated zone on two dates (1978 and 1981) in order to assess changes in the vertical distribution of leachate-derived contaminants. Analysis of the results (Harris and Parry 1982; Harris and Lowe 1984) suggested that attenuation of organic compounds was minimal and that significant contamination of the aquifer was possible. In order to clarify the situation further, the Water Research Centre, funded by the Department of the Environment's landfill research programme, has continued the study since 1985 with the co-operation of Severn Trent Water and, since 1989, with the National Rivers Authority Severn Trent Region.

Particular effort has been directed at Burntstump landfill, located about midway between Nottingham and Mansfield. The site receives controlled wastes and commenced with the infilling of two shallow valleys (about 5 m deep) in 1964. In 1978 the site was extended into an adjacent sandstone quarry, with disposal to a depth of up to 15 m. The infill is composed predominantly of municipal solid wastes with up to 30 per cent non-notifiable industrial and commercial wastes. The Sherwood Sandstone at Burntstump comprises red, cross-bedded, medium to coarse grained sandstones with thin intercalations and lenses of red and green siltstones and mudstones. The primary intergranular porosity is relatively high but significant secondary permeability is developed by fissures. The water-table is about 55 m below the landfill surface.

Including the early work (Harris and Parry 1982), four sequential profiles (1978, 1981, 1985, and 1987) have been obtained through the landfill and into the underlying Sherwood Sandstone. Evidence of the downward migration of the geochemically conservative chloride front over the nine year period is apparent (Fig. 8.1), with a similar

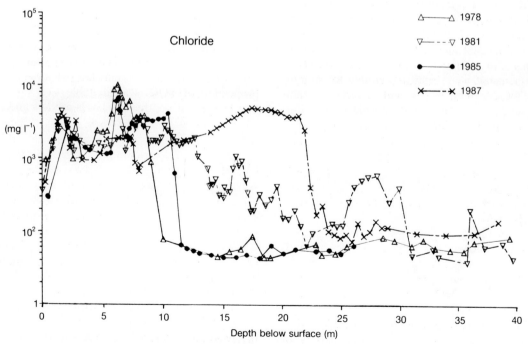

Fig. 8.1. Changes in chloride profiles, Burntstump, 1978–1987.

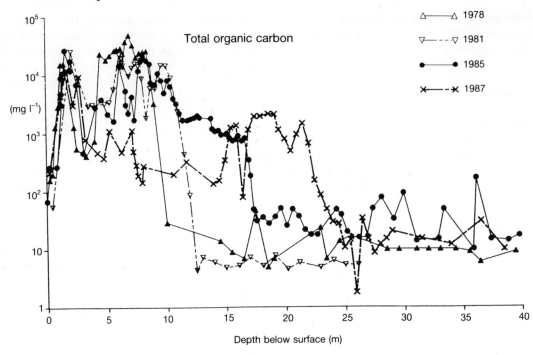

Fig. 8.2. Changes in total organic carbon profiles, Burntstump, 1978–1987

movement indicated by total organic carbon (TOC) for the organic components (Fig. 8.2). The 1985 profile passed through a zone of gravelly, fractured sandstone from 11–29 m below ground level, giving rise to dispersed migration fronts. The ammonia profiles also exhibited a progressive downward movement but were retarded relative to chloride and TOC. A point of particular interest was the recognition of a zone of reduced pH values (Fig.

8.3) that migrated downwards consistently with the TOC, and in which the lower edge of the low pH zone corresponded closely with the leading edge of the TOC front. Comparison of the TOC profiles with measurements of total volatile acids (TVA) in pore-waters (Blakey and Towler 1988) has shown that within the zone of reduced pH a very high proportion of TOC is attributable to TVA, but that below that depth TVA levels are low and pH

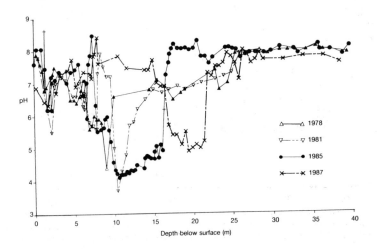

Fig. 8.3. Changes in pH profiles, Burntstump, 1978–1987.

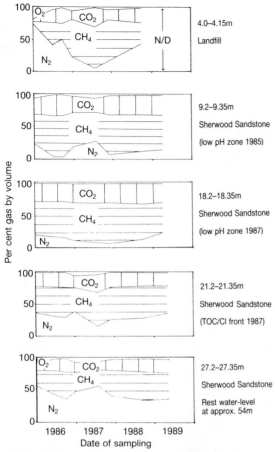

Fig. 8.4. Penetration of land fill gas beneath Burntstump and its change in composition with time.

values increase rapidly to neutral. Overall, the combination of the low buffering capacity of the sandstone, due to its limited carbonate content, the high concentrations of organic acids, and the dissolution of carbon dioxide generated within the wastes depress the pH of the interstitial water at the leachate front.

The indigenous microbial population in the aquifer penetrates to depths of up to 50 m, and there is a correlation between increased populations and horizons where anaerobic degradation has been demonstrated by pore-water analyses and methane/carbon dioxide production, and between the zone of low pH and a lack of identifiable populations (Blakey and Towler 1988). A more complete description of the study is given in Chapter 11.

In addition to obtaining vertical pore-water profiles, gas probes were emplaced at various depths in two boreholes. Routine gas monitoring (Fig. 8.4) has shown progressive invasion of the unsaturated zone by landfill gas, moving in front of the dissolved leachate components.

The sequential profiles (Figs. 8.1 and 8.2) indicated that the rate of vertical leachate migration has varied with time. It is inferred from water balance estimates, using the model proposed by Blakey and Craft (1986), that between 1978 and 1981 infiltration was being stored in the wastes and that it was not until between 1981 and 1985 that leachate generation commenced in direct response to infiltration (Blakey and Towler 1988). Plotting the depths of the chloride, TOC, and pH fronts

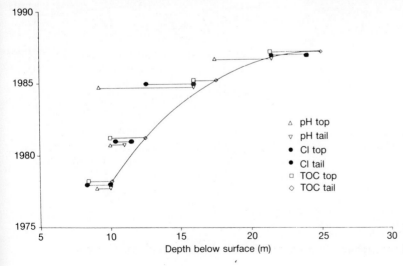

Fig. 8.5. Migration of chloride, TOC, and low pH fronts, Burntstump, 1978–1987.

over the period of monitoring (Fig. 8.5) reveals that the rate of movement has increased from 0.8 m yr^{-1} in 1978–81, to 1.3 m yr^{-1} (1981–1985) and 3.7 m yr^{-1} (1985–1987).

Overall the results of the Burntstump study indicate that significant penetration of organic and other components of leachate can take place through the unsaturated zones of aquifers with low buffering capacities, due to the persistence of conditions not conducive to microbial degradation.

Chalk

Possibly the longest continuous record of site investigation of chalk based landfills in the United Kingdom is that for sites at Ingham, Suffolk, about midway between Bury St Edmunds and Thetford. The area is underlain by the lower part of the Upper Chalk. The upper surface of the Chalk has been subjected to periglacial weathering including solution hollowing, and is overlain by glacial sands and gravels capped by a chalk-rich till. Two land-

fills are present (Fig. 8.6), both occupying gravel workings that extended in places to the underlying Chalk. The earlier landfill (Folly site) began operation in 1968 with the deposit of sawdust, latex, egg packing wastes, builders' rubble, and paper wastes. Between 1970 and 1972 aqueous industrial wastes, including oils and sludges were discharged into lagoons excavated in the solid wastes. Filling ceased in 1973 and the surface was covered with a clay cap in 1974. The site has subsequently been restored to arable farmland. The second landfill (Culford Road site) began to be used in the late 1970s, taking principally municipal solid wastes, with filling commencing in the north-west and progressing south-eastwards across the site. The filling of the site is now substantially completed.

The initial phase of investigation at the Folly site began in 1974 with an auger survey to determine the dimensions of the completed landfill. At the time of the first major summary of work (Stiff *et al.* 1978), 14 exploratory/observation boreholes had been constructed in and around the landfill; subsequently a further six boreholes have been

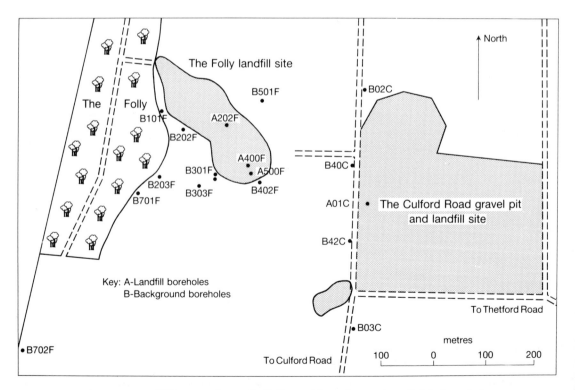

Fig. 8.6. Landfills at Ingham, Suffolk, and locations of available boreholes.

completed. Active study of the Culford Road site began in 1980, with six boreholes drilled between 1980 and 1984.

As at Burntstump, sequential profiling has been carried out to examine the vertical flux of contaminants through the unsaturated zone. The sequence of chloride and TOC profiles from 1975 to 1986 showed an apparent rapid initial movement of the pollution front (about 1.6 m yr^{-1}), followed by a prolonged period, lasting until 1986, during which no significant downwards migration occurred; at the end of this period increased concentrations were found above the water table. Tritium profiling surveys (in 1977 and 1981) in the adjacent farmland, immediately north of the Folly landfill,

had indicated an average vertical migration rate of 0.6 m yr^{-1} in the Chalk. Interpretation of the erratic behaviour of the chloride front suggests that the initial, rapid movement (1975–77), represented delayed drainage from the waste mass, which had previously been subject to substantial liquid waste inputs. It is also noted that although reinstatement of the surface commenced in 1974, it was not until about 1977 that complete conversion to a useful agricultural soil was achieved. The subsequent stagnation of the chloride front until 1986, is related to the effectiveness of the completed capping and evapotranspiration losses from crops, which limited infiltration to very low values. The increase in concentration in 1986 marked an increase in direct infiltration through the wastes with a consequent rise in concentrations in the unsaturated zone. Confirmation of a persistent change in dynamics was found in routine monitor-

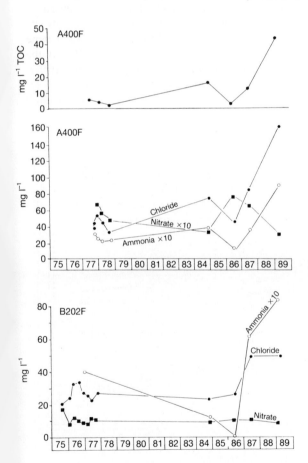

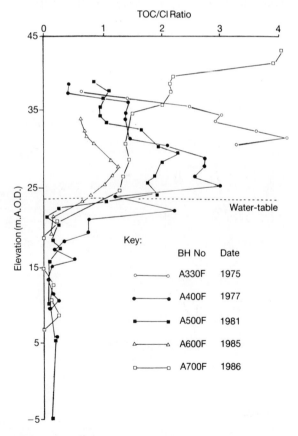

Fig. 8.7. Changes in time of leachate derived contaminants in groundwater beneath the landfill (in borehole A400F) and down gradient (in borehole B202F) at Ingham.

Fig. 8.8. Changes in TOC:chloride ratios, Ingham, 1975–1986.

ing surveys of groundwater, in which there has been consistent increases in leachate-derived contaminants, especially in borehole A400F, sampling groundwater beneath the centre of the landfill, and borehole B202F, sited to the south-west of the fill, in the direction of groundwater flow (Fig. 8.7).

During the eight year period of apparent stagnation in the unsaturated zone, a persistent decrease in the TOC:Cl ratio occurred (Fig. 8.8) indicating continuous removal of organic carbon. Early surveys of organic compounds at the site (Stiff *et al.* 1978) showed the presence of phenols, volatile fatty acids, mineral oils, and halogenated solvents. The readily degradable fatty acids were not detected in the 1981 survey, nor were phenols after the 1984 survey, but total halogenated solvents remained present at levels of up to 50 μg l^{-1} beneath the landfill during a special packer pumping test on borehole A400F in 1987, and at similar levels during a survey in March 1989 (Fig. 8.9).

Both the Folly and Culford Road landfills have been sites for the investigation of the effectiveness of indigenous microbial populations in attenuating leachate components (Blakey and Towler 1988). The high buffering capacity of the Chalk was found to be conducive to microbial metabolism, in contrast with the Sherwood Sandstone, providing a possible explanation for both the decrease in TOC:Cl ratio with time and the disappearance of readily degradable organic compounds in the groundwater.

Engineered unsaturated zones

In addition to the investigations of the attenuating process in the unsaturated zones of the principal aquifers, attention had been focused on the development of techniques to enhance the attenuating capacity of strata immediately underlying landfills (Robinson and Lucas 1984, 1985; Robinson 1989).

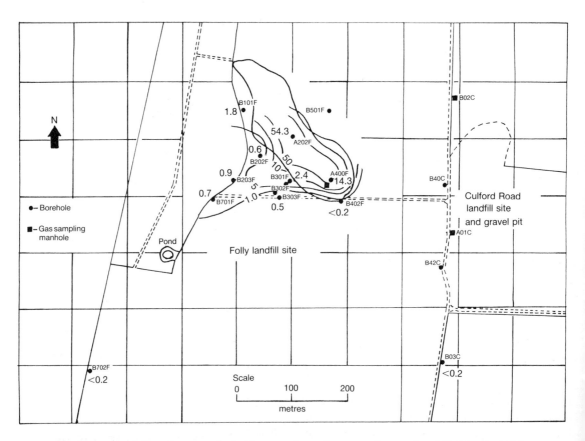

Fig. 8.9. Areal distribution of trichloromethane in groundwater (μg l^{-1}), March 1989.

Lysimeter studies of the attenuation of leachate components during the 1970s (Department of the Environment 1978; Newman and Ross 1985) had shown that enhanced levels of attenuation were present in fine to medium grained, granular strata of complex mineralogy, when compared with coarser materials and with those of limited mineralogical diversity. The principal mechanisms, identified in the case of strata with a high attenuating capacity, were the dominance of primary intergranular flow providing prolonged contact between fluids and solids and high surface areas of solids available for exchange and sorption reactions and, potentially, colonization by active microbial populations. In addition, strata of mixed mineralogy are likely to possess a substantial neutralizing capacity, which would both limit the mobility of many metals and prevent the development of low pH conditions which inhibit the action of certain micro-organisms in degrading organic contaminants.

Stangate East quarry, in Kent, has been excavated into the Lower Greensand formation, which at that location comprises alternations of calcareous silty sands (hassock) and limestone (rag), with the former lithology rejected as waste material.

Site investigations in 1979 showed that the interbedded sandstone and limestone sequence formed a local aquifer, with a water-table less than 10 m below the quarry base and groundwater discharging to the River Bourne within 100 m of the site. Characterization of the physical and chemical properties of the waste sands and silts had shown that a mixture of 80 per cent hassock and 20 per cent silt would provide a material with substantial neutralizing and cation exchange capacities, and a permeability of about 10^{-9} ms^{-1}. Design of the attenuation blanket by Aspinwall and Co. (Robinson and Lucas 1984) required a minimum thickness of 6 m above the maximum water-table, composed of at least 4 m of *in situ* strata or compacted hassock overlain by a 2 m layer of the 80:20 mixture. As a result of funding under the Department of the Environment's Landfill Research Programme, instrumentation of the engineered zone was possible, comprising vertically nested arrays of vacuum lysimeters, to obtain direct samples from both the unsaturated and saturated zones; gas samplers, gypsum blocks, and thermocouple psychrometers to measure soil water

pressures; pneumatic piezometers for the measurement of changes in hydraulic head in the saturated zone; and electrical conductivity sensors. Disposal of wastes, at a rate of about 1300 t d^{-1} commenced in September 1981. Regular monitoring has shown that a strong leachate (COD 4000–> 50 000 mg l^{-1}, ammonia 450–1900 mgN l^{-1}, chloride 1400–3000 mg l^{-1}) accumulated at the base of the wastes which migrated into the engineered zone in the vicinity of each of the four instrument clusters. Measurements of pore gas composition (Fig. 8.10) demonstrated that anaerobic conditions were rapidly established within the engineered layer and penetrated into the underlying strata. Monitoring of organic leachate components in the engineered layer (Fig. 8.11) indicated that, following the initial invasion, active microbial populations became established in the anaerobic conditions, providing

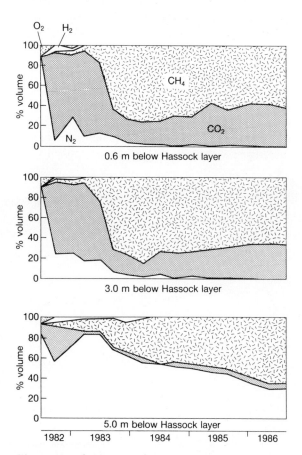

Fig. 8.10. Changes with time of gas phase, Stangate East quarry.

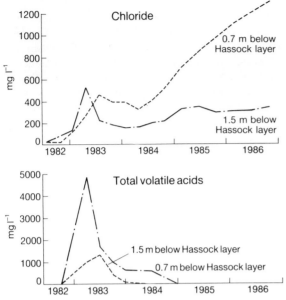

Fig. 8.11. Changes with time in leachate component concentrations beneath Stangate East quarry.

effective degradation of volatile acids. During the first six years of operation, movement of ammonia through the engineered zone is reported to have been minimal, because of cation exchange (Robinson 1989), with no detectable contamination of the underlying water table by leachate components.

The saturated zone

The Villa Farm research site

To provide quantitative data on leachate attenuation reactions for future landfill assessment modelling, the British Geological Survey has been studying the mechanisms controlling waste migration from the Villa Farm lagoons (Williams *et al.* 1984). The site, near Coventry, fell into disuse in 1980 having received a wide variety of industrial wastes for over thirty years. These include oil/water mixtures, effluent treatment sludges containing heavy metals, acids and alkalis, organic solvents, and paint wastes, all of which were deposited directly into lagoons in hydraulic continuity with a shallow lacustrine sand aquifer. An extensive monitoring network has been established around the site to determine the distribution of inorganic and organic components in the groundwater and changes in concentration with time (Figs 8.12 and 8.13).

Although not a solid waste landfill, the inorganic reactions that are observed in the groundwater are identical with those reported elsewhere for domestic waste sites (Baedecker and Back 1979). The organic compounds detected in the groundwater (Table 8.1) are similar to those found in domestic leachate and are probably similar to those

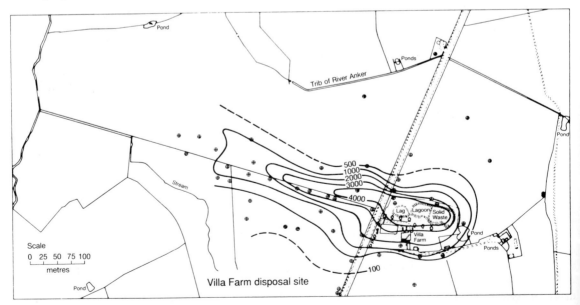

Fig. 8.12. Distribution of chloride (in mg l^{-1}) in groundwater around the Villa Farm lagoons

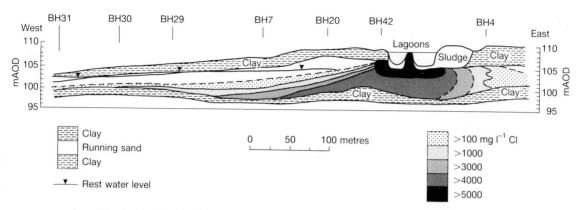

Fig. 8.13. Vertical cross section along the Villa Farm pollution plume.

Table 8.1   Organic compounds identified in a borehole in the pollution plume at Villa Farm

Aerobic zone only	Both aerobic and anaerobic zones	Anaerobic zone only
2,2,3,3-tetramethylbutane	4-methyl-2-pentanone	3-methylcyclohexanol
2,3,3-trimethylpentane	2-methyl cyclopentanone	4-methylcyclohexanol
2,3-dimethylhexane	2-methyl-2,4-pentanediol	2-methylphenol
methylbenzene [toluene]	2,4-dimethylcyclopentanone	octanoic acid
2,2,4-trimethylhexane	4-methylphenol	1-methoxy-4-methyl-benzene
1,2,3-trimethylbenzene	3,5,5-trimethylhexanoic acid	3-methylbenzoic acid
tributyl ester of phosphoric acid	4-(1,1-dimethylethyl)benzoic acid	methyl ester of 2-ethylheptanoic
2,2,4-trimethylpentane-1,3-diol	n-n-butylbenzenesulphonamide	acid
isobutyrate (isomer I)	tetradecanoic acid	1-(2-methoxy-1-methyl-ethoxy)-2-
	pentadecanoic acid	propanol
	hexadecanoic acid	1,7,7-trimethylbicyclo-
	1-ethoxy-1-methoxyethane	[2.2.1]heptan-2-one
	2-methyloctanoic acid	
	5-methyl-3-oxohexanoic acid	
	3,7-dimethyl-6-octenoic acid	
	2-methyl-2-propylhexanoic acid	
	dodecanoic acid	
	tetradecanoic acid	
	hexadecanoic acid	
	9-octadecanoic acid	
	16-methylheptadecanoic acid	
	2,2,4,4-tetramethylpentanoic acid	
	3,3-dimethyl-2-(1-methylethyl)-	
	butanoic acid	
	1-propyl-cyclohexanecarboxylic	
	acid	
	5-(1,1-dimethylethyl)-1,3-	
	benzodioxol-2-one	

Concentrations range from < 1 μg l^{-1} to 10 mg l^{-1}
Many unidentified organic compounds were also found.

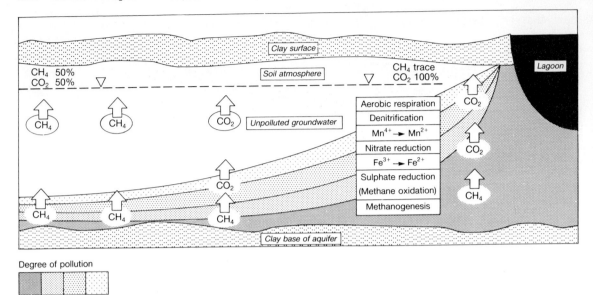

Fig. 8.14. Schematic diagram showing the redox zonation at Villa Farm.

around a co-disposal site after the labile short chain carboxylic acids have degraded. Villa Farm, therefore, provides an ideal site, in a relatively simple hydrogeological setting, in which to study the fate of important pollutants at reasonably high concentrations.

Chloride is a conservative species which can be used to identify the extent of contaminant migration. A geochemical zonation based on redox reactions has been recognized in the transition from the oxidizing conditions in the natural uncontaminated groundwater to the heavily polluted and highly reducing zone near the lagoons and at the

base of the aquifer (Fig. 8.14). These zones follow the theoretical sequence of redox reactions predicted from thermodynamic considerations in a closed organically polluted system (Stumm and Morgan 1981). Heavy metals are attenuated near the lagoon as carbonates and sulphides in the zone where sulphate is reduced to sulphide, but organic components of the waste have been detected in excess of 300 m from the lagoons. The measurement of total organic carbon (TOC) suggests little gross change in the organic content of the pollution plume, but more detailed characterization of the organics by gas chromatography/mass spectroscopy

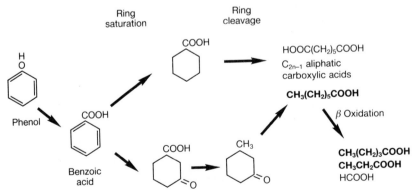

Fig. 8.15. Breakdown pathways for phenol at Villa Farm (ring compounds and those in bold type were identified by GC/MS).

shows that biotransformations occur. Changes in organic composition with increasing distance from the lagoons have been used to infer a breakdown pathway for aromatic compounds, and the predominance of benzoic acid derivatives, which are not known to have been discharged into the lagoon, is explained by synthesis from the primary disposal of phenol (Fig. 8.15).

The highly reducing conditions at the base of the aquifer initially suggested that methanogenesis takes place in this zone. However, a soil gas survey in the unsaturated zone immediately above the water-table revealed predominantly carbon dioxide and only traces of methane. This seemed widespread except at the leading edge of the plume where the zone of contamination is relatively thin and sulphate is depleted. Here, methane is found in the soil atmosphere at concentrations up to 55 per cent by volume.

It was originally considered that this distribution was consistent with the gradual breakdown of aromatic compounds as they migrated through the aquifer, ultimately to compounds (acetate, formate etc.) which can be utilized by methanogens. However, it is now clear that methane is being produced in the highly reducing conditions throughout the plume below the zone of sulphate reduction, but at the same time the methane appears to be consumed during oxidation to carbon dioxide in anaerobic conditions in the overlying zone of sulphate reduction. Where sulphate reduction is limited, such as at the leading edge of the plume, methane is probably able to diffuse upwards without oxidation to give the high concentrations found. However, since carbon dioxide is the product of all the redox reactions which use organic carbon as the electron donor, its presence in the soil atmosphere is a useful indicator of organically contaminated groundwater below.

In modelling reactions within the plume, particularly the fate of the dominant organic contaminants (phenol and benzoic acid), three approaches have been adopted:

(a) Small-scale laboratory experiments

Sediment cores removed from the field without exposure to the atmosphere have been incubated aerobically and anaerobically in the laboratory, under simulated field conditions. Respiration measurements, using Warburg apparatus, are used to infer biodegradation rates while breakdown pathways are being sought by spiking cores with ^{14}C labelled phenol and benzoic acid and determining active intermediate and end products (Dhillon *et al.* 1989).

(b) Intermediate-scale field tracer experiments

Small-scale radial-injection tracer tests are being used to determine relative migration and biodegradation rates of organics under different redox conditions, by comparing breakthrough curves with those of a conservative tracer. This work has been undertaken in an uncontaminated part of the aquifer where analysis of the organic tracer relative to the normal background organic composition poses no problems. The migration of phenol and chloride, introduced simultaneously as a slug into a recharge well, has been monitored in a series of multilevel sample tubes approximately 0.6 m

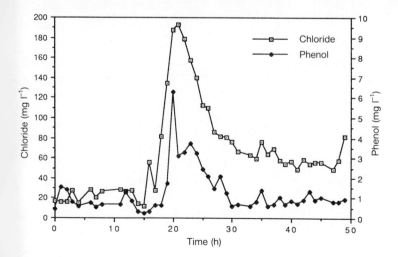

Fig. 8.16. Chloride and phenol breakthrough curves

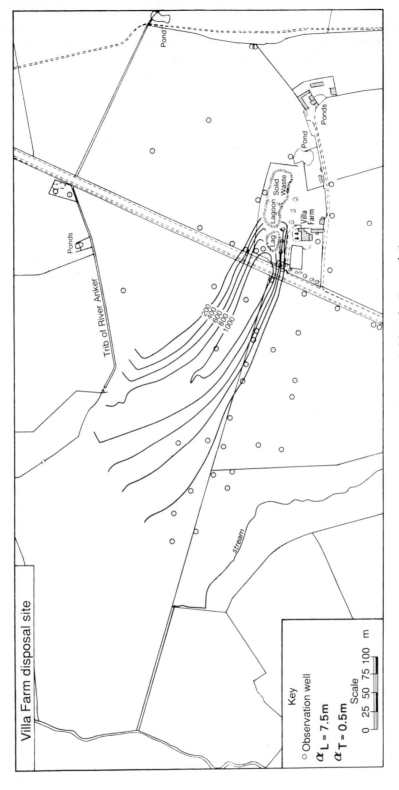

Fig. 8.17. Simulated distribution of chloride (in mg l^{-1}).

away. Results confirm expectations that phenol undergoes little degradation or sorption onto the sediment under these flow conditions (Fig. 8.16). The next stage of the work will involve the introduction of specific compounds into different redox zones of the plume under natural hydraulic gradients.

(c) Large-scale modelling studies

A two dimensional areal model has been developed to describe groundwater flow around the lagoons. This has been accomplished using field data (i.e. elevation of the top and base of the aquifer and water levels) which have been subjected to Universal Kriging as a means of interpolating across the finite element grid used, and for obtaining an estimate of the degree of uncertainty in the data (Mackay *et al.* 1986). Injection tests with packers were used to determine depth profiles of hydraulic conductivity in specially completed boreholes. Although these values were used in the model, the hydraulic conductivity of the aquifer was adjusted iteratively to fit the head distribution measured. This required a correlation between aquifer thickness and average hydraulic conductivity, which was supported by data on particle size distribution and which showed the aquifer becoming increasingly silty as it thinned out and coarser where it thickened.

This model has been extended to include conservative mass transport in which longitudinal and transverse dispersivities are inferred by calibration against the assumed steady state distribution of chloride (Fig. 8.17). Coupling reactive mass transport mechanisms into this model, initially for simple mechanisms such as biodegradation (first order decay) and reversible sorption (simple retardation) allow these mechanisms to be quantified, by fitting values to parameters to describe the observed steady-state distribution of total organic carbon and phenol, which compare favourably with laboratory measurements.

Drigg tracer experiments

Drigg, in Cumbria, is the site of the waste disposal facility in the UK for the shallow burial of low level radioactive waste and has, until recently, been operated in essentially a similar fashion to a

landfill. In part of the site outside the influence of the disposal operations the British Geological Survey has constructed a small-scale array of boreholes in a shallow confined sand aquifer to study the migration of cations and conservative species (Williams *et al.* 1985, 1991). By imposing groundwater flow conditions, the migration of radioactive species, such as ^{85}Sr, ^{131}I, ^{58}Co, ^3H, have been observed. The work has concentrated on two aspects. Firstly, the hydraulic characterization of the aquifer to try to predict the rate of flow of conservative species injected at a given rate; and secondly, the consideration, in the laboratory, of the chemical interaction of reactive species in the groundwater and the minerals of the aquifer, to try to predict retardation effects during migration, relative to a conservative tracer. The ability to attain these objectives of predicting solute transport was then assessed by comparing predicted with observed results in the field test.

The field test involved releasing ^{85}Sr and the conservative tracer ^{131}I, into the radially divergent flow field around a recharge well in a confined glacial sand aquifer (Fig. 8.18). Tracer breakthrough has been monitored in three multi-level sampling installations in tests involving two imposed rates of groundwater flow. At average flows of up to 5 md^{-1}, the retardation of ^{85}Sr at a given sampling point was in general lower than that observed with an average flow rate of up to 0.5 md^{-1}, suggesting a kinetic control on sorption. The retardations at lower flows were in relatively good agreement with laboratory values derived from batch sorption experiments lasting several days (Fig. 8.19).

Longitudinal dispersivity (α_L) has been determined by comparing the shape of the breakthrough curves for ^{131}I with dimensionless type curves resulting from an exponentially decaying source, based on the 1–D analytical solution of the advection-dispersion equation. At the high flow rate, α_L for ^{85}Sr was larger than that for ^{131}I, but this discrepancy reduced at the lower flow rate where convergence to the ^{85}Sr values was observed. A possible explanation for this phenomenon is that at the high flow the opportunity for ^{131}I diffusion into relatively static pore-water is limited whereas with solutes moving at a lower rate there is a greater opportunity for diffusion into 'dead end pores'.

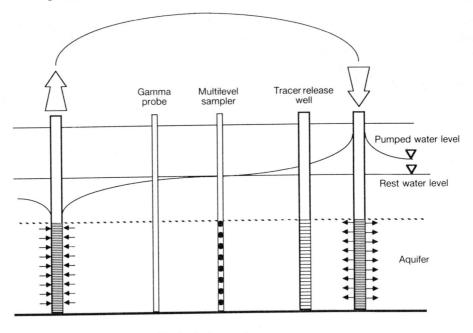

Recirculation system

Fig. 8.18. Schematic of the tracer test at Drigg.

The mechanism for [85]Sr interaction on core material has been studied by sequential selective leaching. In a core consisting predominantly of sand grade material, approximately 80 per cent of the [85]Sr was present in ion exchange sites (which agreed well with the measured cation exchange capacity of approximately 1 meq/100 g sediment), with 12 per cent in the carbonate phase. The remaining 8 per cent was recovered easily by extraction with distilled water. The amount of [85]Sr

recovered in the field test was estimated at 5–30 per cent with an average of around 15 per cent. The 'loss' of [85]Sr is probably partially due to isotopic dilution with stable Sr naturally present in the aquifer but may result from precipitation reactions due to non-equilibrium between flowing water and the aquifer minerals.

Although the geochemical interactions and retardation of [85]Sr have been reasonably well predicted in the heterogeneous system studied, initial

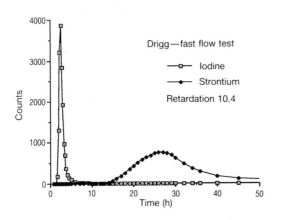

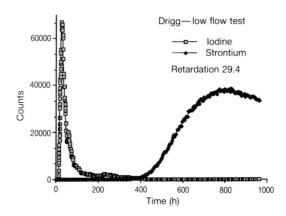

Fig. 8.19. Breakthrough curves for [85]Sr and [131]I.

estimates of groundwater flow velocities based on conventional hydraulic conductivity measurements averaged over the interval of the well screens, have proved inadequate to describe the flow of tracers. Small scale measurement of aquifer parameters have, therefore, been necessary to balance the scale of hydraulic testing to that of tracer monitoring.

This work has been extended to a study of the effects of dissolved fulvic and humic material in complexing and increasing the mobility of cobalt and nickel, which has direct application to the behaviour of metals in landfill leachate plumes (Warwick *et al.* 1991; Williams *et al.* 1991).

Karstic aquifers

Extensive karstic aquifers of the types developed in Mediterranean countries are absent in Britain, but localized karstic conditions are present, especially in the Carboniferous Limestone of the southern Pennines, the Mendip Hills, and in South Wales. Site investigations of landfills sited in potentially karstic areas in South Wales had been included in work begun in the 1970s (Department of the Environment 1978). Particular attention was paid to a landfill covering 9 ha at Tythegston, near Bridgend, which had received municipal solid wastes and some industrial wastes since 1968 (Young and Connor, 1978). Monitoring at the site showed that the landfill acted as a local recharge centre, with significant vertical movement of leachate beneath the landfill, prior to lateral dispersion in the groundwater. The problems of tracing or predicting solute movement through fissure flow,

or karst dominated aquifers was highlighted by the experiences at Tythegston and direct tracing techniques were employed in subsequent studies.

The tracing technique used employed fluorescent dye tracers because of the potential for the use of passive sampling, in the form of granular activated carbon bags placed at strategic points along possible flow paths (Aldous and Smart 1988), thereby extending the areas that could be economically monitored during each test. The successful use of the technique has been described by Clark (1984) for a test between sink holes in Cas Troggy Brook, near Caerwent, and the Great Spring in the Severn railway tunnel. The Great Spring has a mean daily discharge of 55 000 m^3 and provides both the principal groundwater outflow from the 44 km^2 Caerwent basin and an important water supply. Initial appearance of tracer at the spring occurred 42 days after injection, with the peak concentrations occurring after 130 days. Modelling of the smooth breakthrough curve (Fig. 8.20) provided an estimate of longitudinal dispersivity of 720 m indicating that the flow took place via a plexus of intersecting fissures, rather than through enlarged conduits.

Landfill gas

The explosion at Loscoe, Derbyshire

At 6.30 am on 24 March 1986 the bungalow at 51 Clarke Avenue, Loscoe, Derbyshire was completely destroyed by a methane gas explosion. The three occupants, though badly injured, were lucky to escape with their lives. Although natural gas

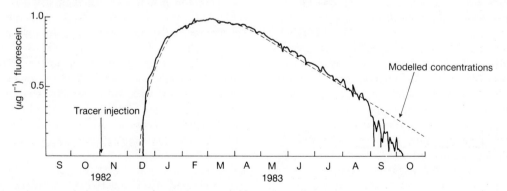

Fig. 8.20. Breakthrough curve, dye tracer study, Cas Troggy Brook to Great Spring, South Wales.

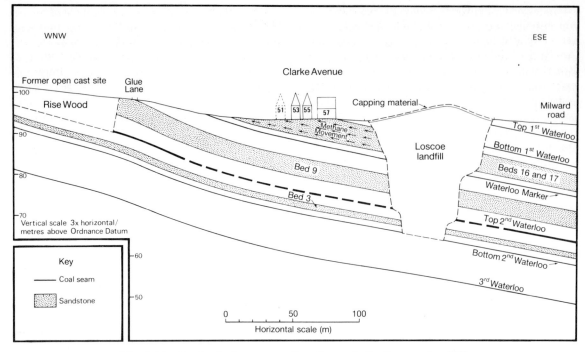

Fig. 8.21. Geological cross section through the Loscoe landfill.

was supplied to the bungalow, gas samples taken from the wreckage soon after the explosion were found to be generally similar to landfill gas which is typically composed of 60 per cent methane and 40 per cent carbon dioxide. Attention was directed to a landfill whose boundary lay 70 m from the bungalow and boreholes were drilled to extract and flare off landfill gas to prevent its lateral migration. A gas survey showed dangerous concentrations of methane in several adjacent houses and at least two families were evacuated. A detailed geological appraisal was made using reliable data collected by the Geological Survey while re-mapping the area in the mid-1960s. This revealed a fractured and gas permeable sandstone horizon as the pathway from the landfill. The area was also underlain by several shallow coal seams, and the ground was covered by a thin layer of impermeable clay or 'head'. Gas in the sandstone was expected to escape where this clay was absent or where it was breached by excavation for foundations or through laying underground services (Fig. 8.21).

Shortly before the explosion, the landfill had been capped with a layer of clay. At the time of the explosion a very deep depression was passing over the region and it is probable that this created a pressure gradient from the landfill helping the gas to escape in large volumes into the cavity below the suspended floor of the bungalow. Ignition was probably by the pilot flame of a gas central heating boiler.

One may think there is nothing unusual about this case history. However, during the gas survey of the area, it became apparent that for up to three years before the explosion there had been several reports of distressed vegetation in gardens. Typically the soil became warm, then dried out and crumbled, and the lawns died. The problem reappeared even after returfing. In one garden 90 m from the landfill, the occupier dug into a patch of lawn in an attempt to discover why it had died. An unpleasant smell 'like a sewer' was detected, along with white mould, heat and 'a rumbling noise like a burst pipe'. A hole was dug 0.5 m deep to expose muddy water bubbling with gas. The gas contained 50 per cent of the lower explosive limit (LEL) of methane but no carbon monoxide was recorded. It was assumed that the gas originated from underlying coal workings or from a burning coal seam, so the occupier contacted the National Coal Board (now British Coal) who installed a stand-pipe with flame trap to allow the soil gas to vent harmlessly to the atmosphere. Analyses of the issuing gas suggested to the Coal Board that it was unlike coal gas which has little carbon dioxide, but was probably derived from rotting vegetable matter. At that time British Coal was unaware of the existence of the landfill but considered that it had acted responsibly by making the situation safe.

Unfortunately, the full significance of the distressed vegetation was not realized until after the explosion when boreholes were drilled to determine the cause of the heating. In one borehole there was a gradual decrease in soil temperature with depth from 20.7°C at depths of 0–0.5 m below the surface to 18°C at 2.27 m below ground level (bgl). This was accompanied by an increase in methane composition from 2 per cent at the surface to 33.4 per cent at 2.27 m bgl. Gas samples taken at 1.65 m bgl in sandstone contained 29.6 per cent nitrogen suggesting that the methane could have been mixed with air. Coal was not intercepted.

Adjacent to the stand pipe installed by British Coal, which at the time of sampling no longer exhibited the same degree of heating and distress to the vegetation as previously recorded, a lower temperature profile was found with only a limited amount of carbon dioxide and methane in the soil atmosphere. However, gas similar in composition to that in the landfill (58 per cent methane and 39 per cent carbon dioxide) was encountered at 3.0 m bgl in a sandstone horizon. Again, no coal was intercepted.

If a burning coal seam were present the temperature would have been expected to increase with depth towards the coal seam. However, the reduction in temperature with depth and the absence of a coal seam was consistent with the theory that methane migrating to the surface was being oxidized, possibly by bacteria, in the soil zone. The gas below the zone of oxidation was, and would be expected to be, similar to landfill gas.

Had the significance of the distressed vegetation been accurately assessed, an evaluation of the geology would have allowed the area at risk to be identified and a gas survey would have revealed the existence of methane in several basements. The explosion could have been avoided (Williams and Aitkenhead, 1991).

The reasons for the spate of landfill gas problems through the 1980s may be the increasing content of putrescible material in present day waste, and the change to the rapid disposal of waste with high density compaction in the mid-1970s. Improved engineering works to reduce leachate generation, such as capping the site with an impermeable cover, may in fact prevent gas escaping through the surface of the landfill and thereby encourage its lateral migration. Recently, the government has issued guidelines to the waste disposal industry (Her Majesty's Inspectorate of Pollution 1989) in an attempt to avoid similar occurrences by appropriate site selection based on a knowledge of the geology, and by prior site engineering to seal off migration routes and allow controlled gas extraction, dispersal or utilization.

Effect of research on landfill practice

In 1979 it seemed that changes in landfill practice were taking place specifically because of economic constraints, in particular the closure of small local refuse tips during the 1970s and the concentration of resources into fewer but much larger landfills operated with high density compactors (Williams 1980). During the 1980s, the legacy of this approach has been manifest in an increased number of methane gas problems. Clay caps which were often specified to reduce leachate production, prevented gases diffusing upwards and encouraged lateral migration beyond the landfill boundary. Only recently have waste disposal authorities been alerted to the threat of landfill gas and urged to undertake surveys and remedial action to prevent methane migration.

While gas migration is relatively rapid, the generation and migration of landfill leachate is very slow. When waste is deposited with high density compactors, especially in deep sites, the delay before significant leachate is generated may be several, if not tens of years. The completed landfill of ten years ago may be just starting to produce leachate and the impact on groundwater quality may be just beginning to be seen now. Unfortunately, while landfills are getting larger, the level of site investigation often remains low, and the hydrogeology may not be known in sufficient detail to locate boreholes correctly so as to detect contamination. The need for effective, scientifically directed site investigations and, equally importantly, the need for both regular collection, interpretation and dissemination of data derived from the monitoring of landfills and their effects on groundwater systems remains as urgent now as it did ten years ago.

References

Aldous, P. J. and Smart, P. L. (1988). Tracing groundwater movement in abandoned and mined aquifers using fluorescent dyes. *Ground Water*, **26**, 172–8.

Baedecker, M. J. and Back, W. (1979). Hydrogeological processes and chemical reactions at a landfill. *Ground Water*, **17**, 429–37.

Blakey, N. C. and Craft, D. G. (1986). Infiltration and absorption of water by domestic wastes in landfills—leachate volume changes with time. *Proc. Harwell Symp. Landfill Water Management*, pp. 5–18. Harwell, UK.

Blakey, N. C. and Towler, P. A. (1988). The effect of unsaturated/saturated zone properties upon the hydrogeochemical and microbiological processes involved in the migration and attenuation of landfill leachate components. *Water Sci. Tech.*, **20**, 119–28.

Clark, L. (1984). Groundwater development of the Chepstow Block: a study of the impact of domestic waste disposal on a karstic limestone aquifer in Gwent, South Wales. *Proc. Int. Assoc. Hydrogeol. Groundwater Symp. Quebec, Canada*. II, pp. 300–9. Atomic Energy of Canada Ltd, Pinawa, Manitoba.

Department of the Environment (1978). *Co-operative programme of research on the behaviour of hazardous wastes on landfill sites*. Final report of the Policy Review Committee, HMSO London.

Dhillon, H. S., Ross, C. A. M., Williams, G. M. and Dart, R. K. (1989). *Aromatic degradation in contaminated cores from the Villa Farm disposal site*. Fluid Processes Research Group, Technical Report No. WE/89/20. British Geological Survey, Keyworth.

Harris, R. C. and Parry, E. .L. (1982). Investigations into domestic refuse leachate attenuation in the unsaturated zone of Triassic sandstones. In *Effects of waste disposal on groundwater and surface water*, pp. 147–55. Publ N° 139, Int. Assoc. Hydrol. Sci.

Harris, R. C. and Lowe, D. R. (1984). Changes in the organic fraction of leachate from two domestic refuse sites on the Sherwood Sandstone, Nottinghamshire. *Q. J. Eng. Geol.*, **117**, 57–69.

Her Majesty's Inspectorate of Pollution (1989). *The control of landfill gas*, Waste Management Paper No. 27. HMSO London.

Mackay, R., Porter, J. D., Williams, G. M., Ross, C. A. M., and Noy, D. J. (1986). *Modelling mass transport in the saturated zone—a case study*. Proc. Conf. water quality modelling in the inland natural environment. 10–13 June, 1986, Bournemouth, England, 259–76. BHRA., Cranfield, Beds.

Newman, J. R. and Ross, C. A. M. (1985). *Mineralogical and geochemical controls on heavy metal pollution in monolith lysimeters*. Report FLPU 85–5. British Geological Survey, Keyworth.

Robinson, H. D. (1989). Unsaturated zone attenuation of leachate. In *Sanitary landfilling: process, technology and environmental impact* (ed. T. H. Christensen, R., Cossu, and R. Stegmann), pp. 453–64. Academic Press, London.

Robinson, H. D. and Lucas, J. L. (1984). Leachate attenutation in the unsaturated zone beneath landfills: instrumentation and monitoring of a site in Southern England. *Water Sci. Tech*, **17**, 477–92.

Robinson, H.D. and Lucas, J. L. (1985). Attenuation of leachate in a designed, engineered and instrumented unsaturated zone beneath a domestic waste landfill. *Water Poll. Res. J. Canada*, **20**, 76–91.

Royal Society of London (1981). A survey of British hydrogeology 1980. The Royal Society, London.

Stiff, M. J., Young, C. P., Barber, C., Naylor, J. A., and Maris, P. J. (1978). *Investigation of a landfill site at Ingham, near Bury St Edmunds, Suffolk*. WRL Tech. Note No. 22, Department of the Environment, London.

Stumm, W., and Morgan, J. J. (1981). *Aquatic chemistry*. Wiley Interscience, New York.

Warwick, P., Hall, A., Shaw, P., Higgo, J. J. W., Williams, G. M., Smith, B., Haigh, D. G., and Noy, D. J. (1991). The influence of organics in field migration experiments:. Part 2—Radionuclide speciation and mobility studies. *Radiochimica Acta*, **52/53**, 465–71.

Williams, G. M. (1980). Control of Pollution Act 1974: Aspects of changes in landfill practice. *J. Instn. Water Engs. Sci.*, **34**, 153–60.

Williams, G. M. and Aitkenhead, N. (1991). Lessons from Loscoe—The uncontrolled migration of landfill gas. *Q. J. Eng. Geol.*, 24, 191–207.

Williams, G. M., Ross, C. A. M., Stuart, A., Hitchman, S. P., and Alexander, L. S. (1984). Controls on contaminant migration at the Villa Farm lagoons. *Q. J. Eng. Geol.*, **17**, 39–55.

Williams, G. M., Alexander, L. S., Hitchman, S. P., Hooker, P. J., Noy, D. J., Ross, C. A. M., Stuart, A., and West, J. M. (1985). *In situ radionuclide migration studies in a shallow sand aquifer (Parts 1 and 2)*. Report, Fluid Processes Research Group, FLPU 85–7 and FLPU 85–10. British Geological Survey, Keyworth.

Williams, G. M, Higgo, J. J. W., Sen, M. A., Falck, W. E., Noy, D. J., Wealthall, G. P. and Warwick, P. (1991). The influence of organics in field migration experiments: Part 1—*in situ* tracer tests and preliminary modelling. *Radiochimica Acta*, **52/53**, 457–63.

Young, C. P. and Connor, K. J. (1978). *Investigation of a site at Tythegston, Glamorgan*. Programme of research on the behaviour of hazardous wastes in landfill sites: WLR Tech Note No. 29. Department of the Environment, London.

9. Urban and industrial groundwater pollution

J. W. Lloyd, G. M. Williams, S. S. D. Foster, R. P. Ashley, and A. R. Lawrence

Introduction

Since the start of the industrial revolution in the eighteenth century the continual disposal and spillage of wastes containing potential groundwater contaminants has taken place in and adjacent to urban areas in the UK. Much of the early disposal was associated with mining and foundry wastes which were essentially of inorganic origin although phenolic wastes associated with coking undoubtedly occurred. In parallel with industrialization, the disposal of domestic wastes became concentrated, because of urbanization, adding to the inorganic pollution potential and introducing pathogenic contaminants to the ground.

During the Victorian era pollution controls were initiated chiefly with respect to domestic effluent so that in many of the large urban areas the first sewer systems date from this period. Unfortunately, at this time, little constraint was exercised upon industrial waste disposal which became more complex in form as the variety of manufacturing industries, producing metals, glass, leather, munitions, etc., grew and used increasing quantities of additives in their processes.

The advent of the First World War gave great impetus to industry with the allied increase in waste disposal. At this stage petroleum began to become important and the disposal of hydrocarbons and associated products added to the chemical complexity of the industrial wastes.

From the First World War to the end of the Second World War much of the traditional manufacturing industry in the UK expanded and with it the use of organic compounds. Both inorganic and organic wastes increased in concert. Between the wars the petrochemical industries commenced operations, developing rapidly following the Second World War at which time the pharmaceutical industrial expansion also occurred. With the appearance of these latter industries a new dimension was added to the complexity of organic wastes.

Prior to the 1970s it is difficult to find quantitative data for the wastes in an industrialized urban area but an indication may be given from later data as shown in Fig. 9.1, which identifies the present-day wastes that need to be disposed of under controlled conditions. Unfortunately, until the 1970s, many of the waste types listed in Fig.9.1 were disposed of at convenient uncontrolled locations within the urban area or on the industrial site where the waste was produced.

On the domestic scene the situation has been historically much better than for industry with programmes for the continual extension of the urban sewerage systems. However, as much of the original sewerage infrastructure was Victorian, sewers have degraded and leaks have become an increasing problem such that major renovation of the systems has proved necessary.

Domestic wastes have changed in character becoming more complex and toxic over the years but have largely been disposed of under reasonably controlled conditions away from urban areas so that post-Second World War domestic waste does not pose a major urban hazard to groundwater in the UK.

Within the urban environment potential groundwater contamination can also ensue from non-wastes. The application of sodium chloride to roads for de-icing is an example and a common

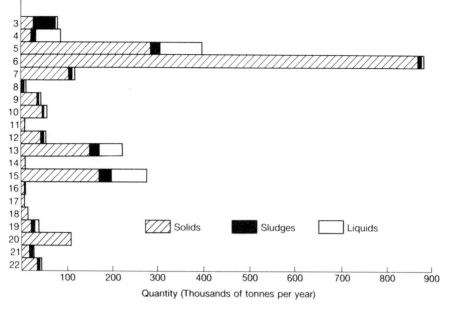

Source: West Midlands County Council

3	Food, drink, and tobacco	13	Motor vehicles
4	Chemicals, coal, and petroleum products	14	Other vehicles
5	Iron and steel, and steel tubes	15	Metal goods (not elsewhere specified)
6	Iron casting etc.	16	Textiles
7	Non-ferrous materials	17	Leather, leather goods and fur
8	Metal working machine tools	18	Clothing and footwear
9	Industrial plant and steelwork	19	Bricks, pottery, glass, cement, etc.
10	Other mechanical and marine engineering	20	Timber, furniture, etc.
11	Instrument engineering	21	Paper, printing and publishing
12	Electrical engineering	22	Other manufacturing

Fig. 9.1. An example of wastes generated by industry.

feature leading to direct contamination. It is limited to some extent by the storm-water systems which normally convey contaminants in runoff to rivers. Depending upon the local hydrogeological controls the contaminants may then be transmitted to the groundwater. Of less prominence in the potential non-waste contaminant field is the use of herbicides and pesticides, and the presence of chlorinated compounds (trihalomethanes) in leakages from mains supplies.

In any urban area the surface water–groundwater relationships are of importance as rivers are frequently used for the disposal of industrial effluent and partially treated domestic effluent and these may enter the groundwater environment under particular circumstances. Conversely, con-

tamination from a hydrocarbon spillage, for example, can frequently enter a river course from groundwater.

Under modern legislation, contamination from wastes such as those discussed above, has been effectively curtailed, although contamination from non-wastes continues and inevitably some contamination from accidental spillage is unavoidable. However, because of the very slow transmission times of contaminants in the unsaturated zone, the historical uncontrolled disposal of wastes may still provide a serious potential for groundwater pollution. In some urban areas it seems probable that the contaminants held in the unsaturated zone are being more rapidly enveloped in the groundwaters than would normally be the case because of the rise in groundwater levels resulting from the reduction in abstraction (Chapter 4).

The legacy of the contaminant potential may be attributed to the historical practices summarized below which can give rise to differing forms of contamination and concentration distributions in groundwaters:

1. Chimney fall-out that created contaminated land, primarily through metals, and also acid rain.

2. Surface disposal that produced a variety of contaminants in the shallow unsaturated zone which are often reactivated due to land redevelopment.

3. Disposal pits chiefly used for the disposal of fluids such as hydrocarbons, solvents, and sludge wastes. The pits were constructed either above or to below the groundwater level.

4. Disposal in major excavations such as surface mines or quarries; largely backfilled with industrial solid wastes, and in some older situations domestic waste, but which contain many diverse forms of waste.

5. Disposal in disused wells or mine-shafts of which some of the latter are still licensed. Such disposal is normally directly into the saturated zone. Commonly liquid wastes are disposed of in this manner, introducing contamination to significant depths and sometimes beneath a confining cover.

To the above may be added the historical and ongoing contamination from:

1. Small-scale leaks of sewage resulting in diffuse nitrate distributions.

2. Contamination by manifold chloride compounds producing diffuse chloride distributions.

3. Accidental spillages, chiefly of fluids.

The overall potential groundwater contamination scenario is depicted in Fig. 9.2. Potentially the situation may seem to be devastating but the

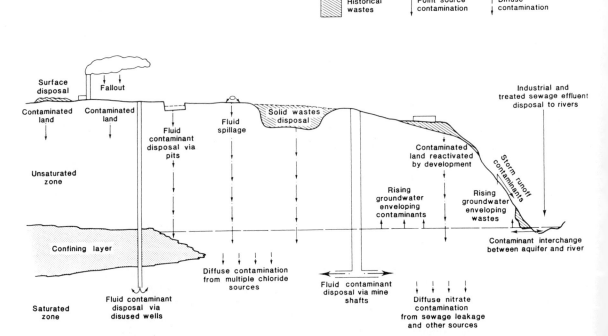

Fig. 9.2. The concepts of possible urban groundwater pollution.

buffering capacity of much of the unsaturated ground and the slow travel times have resulted in the attentuation of many of the contaminants before they enter the saturated zone, where they are further attenuated, dispersed, and diluted. However, the ability of the ground to attenuate contaminants depends upon the nature of the matrix and the hydraulic conductivity. Fractured or fissured media are likely to transmit contaminants more rapidly to the saturated zone than are materials with dominant intergranular hydraulic conductivity.

The variety in the contaminant legacy coupled with the complexity of hydrogeological conditions under urban areas provides considerable opportunity for study. However, research into urban groundwater pollution has been fairly limited in the UK for two reasons.

1. Public water supply sources are mainly located away from urban areas, partly in response to the identification of historical sources of contamination.

2. While the presence of inorganic pollutants has been recognized for a long time, significant organic pollution has only recently been identified as organic quality standards have been imposed, and improved field and laboratory techniques have allowed low concentrations to be measured.

The research fields reviewed below are considered under inorganic and organic pollution.

General inorganic pollution

Inorganic pollution has been identified under all of the major urban areas in the UK where groundwater conditions are unconfined or close to being unconfined. The dominant contaminants are chlorides and nitrates which tend to have higher diffuse background concentrations than occur in equivalent non-urban groundwaters. Other than those two ions, no general pollution patterns have been recognized and most research has only concentrated upon identifying inorganic contaminants in site specific circumstances. On Fig. 9.3 the range of inorganic pollution found in the Triassic sandstones under Birmingham is shown. The groundwaters are not used for public supply.

Heavy metal pollution

Heavy metal pollution of surface water has occurred as a result of a number of ancient industrial activities associated with ore processing and smelting, and from industrial effluent discharges (Forstner and Wittmann 1983). In contrast, the incidence of groundwater pollution by heavy metals is much lower and reflects the attenuation that occurs as a result of geochemical interaction between metals and rock-forming minerals. Even in the heavily industrialized region of the Midlands, where there is evidence in the Triassic sandstones of groundwater pollution by halogenated volatile organic compounds, the incidence of heavy metal pollution is minimal (Edmunds *et al.*

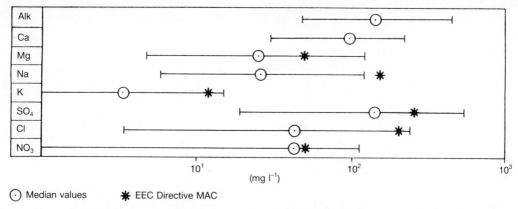

Fig. 9.3. Summary of major ion concentrations in groundwaters beneath Birmingham, largely related to urban pollution (after Ford 1989).

1989). This may be because the groundwater samples analysed have been in an oxidized state and that under these conditions metals are mainly present as free ions which are relatively reactive and easily sorbed or exchanged by clays, oxides, and silica minerals. Only in extreme conditions, for instance where acidic metal effluents have been discharged into quartzitic sandstones with a low ion exchange capacity, has widespread heavy metal pollution been observed. In one case, where acids and heavy metal sludges were deposited in a quarry in Coal Measures sandstones, heavy metals were found in pore-waters with a pH of 3.0, over 300 m downdip in the sandstone, and with Fe at a concentration of 2600 mg l^{-1}, Mn at 80 mg l^{-1}, Cu at 4.5 mg l^{-1}, Ni at 3.0 mg l^{-1} and Zn at 14 mg l^{-1} (Harrison *et al.* 1981).

The discharge of acidic heavy metal effluents into disused mine workings such as those which exist in the worked-out Warwickshire coalfield, has probably taken place since the cessation of mining (at least 100 years ago) and is also a potential cause of heavy metal pollution. The mine-water itself often has little buffering capacity and may be acidic because of the oxidation of sulphide minerals in shales (Dowse and Selby 1975). Conduit-flow through open mine-workings takes place with little opportunity for geochemical interaction between contaminants and the host rock, and the possibility of polluted mine-water discharges occurring is obviously of concern. Discharge to mines as a means of effluent disposal has, however, been largely phased out during the 1980s as a result of the licensing requirements brought in by the Control of Pollution Act 1974 (Williams 1981).

Research into the migration of heavy metals has been carried out during the 1980s mainly in controlled laboratory and *in situ* experiments. The movement of heavy metals in the unsaturated zone has been studied in detail using both large and small-scale lysimeters composed of a number of significant aquifer formations that occur in the UK, including the Lower Greensand, Lower Chalk, Plateau Gravels, and Triassic sandstones. The impetus for this work has been landfill research and the studies have focused on the migration of heavy metals in a synthetic landfill leachate containing high concentrations of volatile fatty acids (acetic propionic and butyric) at pH 5 (Table 9.1). The leachate was irrigated over the aerobic surface

of the lysimeters and, while not simulating true sub-landfill conditions, which are strongly reducing, these experimental conditions form a useful analogue for other industrial discharge scenarios. Results of sampling breakthrough curves at various depths in the lysimeters, and from selective extraction of geochemical phases following destructive sampling have shown that these formations have a considerable capacity for retaining metals. Even after irrigation for 1600 days, Pb and Cu were never detected at a depth of 400 mm in Lower Greensand, while Cr, Zn, and Cd were only present to depths of 620 mm. Ni was most mobile and broke through at 1080 mm depth after about 1550 days. Analysis of sediment cores showed that the total amounts of heavy metals retained was up to 75 meq kg^{-1} which was greater than the cation exchange capacity of the sand. Sequential extraction of increasingly stable mineralogical phases (Ross 1985) showed the mechanisms for attenuation are a combination of adsorption and precipitation. The order of retention was inversely related to the mobility of the metals in small laboratory columns, but directly related to the stability of the geochemical phases in which the metals were retained. For instance Pb and Cu were associated with sulphide and hydroxide phases (e.g. for metal sulphides and hydroxides, the solubility product $K_{sp} = 10^{-24}–10^{-42}$), while Ni and Cd were associated with the labile exchangeable and carbonate fractions (for metal carbonates $K_{sp} = 10^{-10}–10^{-14}$). The most obvious change in the mineralogy of the Lower Greensand was the loss of calcite from the surface of the lysimeter due to reaction with the

Table 9.1 Composition of heavy metals in a synthetic landfill leachate

	(mg l^{-1})		(mg l^{-1})
Na	1200	Cd	100
K	900	Cr(III)	100
Mg	100	Ni	100
Ca	500	Pb	100
Fe(II)	100	Phenol	20
Li	20	Acetic Acid	4800
Hg(II)	10	Propionic Acid	2700
Zn	100	N-Butyric	1500

Adjusted to pH 5 with ammonia

acidic leachate. This buffering, giving a neutral pH, is reported to be an important attenuation mechanism, especially in the surface layers of the lysimeter, which allows precipitation of the heavy metals as carbonates, oxides, and hydroxides. At lower levels, where heavy metal concentrations are not controlled by solubility, adsorption and ion exchange are the principal retention mechanisms.

In contrast to the Lower Greensand, the Triassic sandstones, which underlie major cities such as Birmingham, Nottingham, and Liverpool, are composed predominantly of quartz with iron oxide coatings and contain little calcite (only up to 10 per cent) with which to buffer the leachate. Thus heavy metals are not as well attenuated. However, where the pore-water contains low concentrations of heavy metals, the metals are almost exclusively associated with the clay- and silt-sized fractions suggesting adsorption and ion exchange mechanisms predominate. The research has suggested that the ideal formation for the attenuation of heavy metals is one which has a high carbonate content for good buffering capacity, a significant clay content for cation exchange, and a high proportion of amorphous oxides and oxyhydroxides to provide adsorption sites (Newman and Ross 1985).

While these mechanisms occur in the essentially aerobic conditions of the lysimeters, which simulate natural unsaturated formations, heavy metal attenuation in the saturated zone where redox conditions may be different may take place by different mechanisms. Studies of a Quaternary lacustrine sand aquifer at Villa Farm, near Coventry, has revealed that Ni migration in the saturated zone is significant but at the same time other heavy metals have been significantly attenuated, principally through the formation of sulphides and carbonates. In this case, Ni mobility is considered to result from its complexation with relatively high molecular weight organic compounds. (Warwick *et al.* 1991).

Organic pollution

In Table 9.2 the possible organic contaminants associated with the urban environment are listed. Many of these have been identified in minor concentrations in urban groundwaters in the UK but the chief concern has centred upon industrial solvents and petroleum hydrocarbons to which most of the research has been directed.

Pollution by chlorinated solvents

Chlorinated solvents are widely used in industrial processes and in domestic life and possess various physicochemical characteristics which make them especially insidious contaminants of groundwater (Lawrence and Foster 1987). The maximum recommended concentrations in drinking water are very low (Table 9.3), which implies that even a modest spill of a few litres in volume could potentially contaminate many millions of litres of groundwater. In general, these chemicals are significantly denser and less viscous than water and also have relatively low solubility in water (Table 9.3), which means that rapid and deep penetration of the immiscible phase into aquifers can be anticipated (Schwille 1981). The compounds are known to be toxic to bacteria and thus to be relatively resistant to biodegradation, especially at high concentrations. All such chemicals are highly volatile (Table 9.3), which, although of limited relevance to their transport in aquifers, causes substantial problems when attempting to sample them and to monitor their distribution in subsurface environments.

The soil zone is believed to have the capacity to retain such compounds by adsorption and to promote their elimination by volatilization and/or generally slow biodegradation, but groundwater pollution risks arise where the rate of discharge is such as to exceed this capacity or the discharge mechanism is such that the soil zone is bypassed.

Extent of the pollution problem

The casual disposal, leakage, or spillage of industrial solvents is known to have caused serious pollution of some 10–15 major groundwater sources used for public water supply, leading to their abandonment or to blending or treatment requirements. It is believed that the industrial activities involved were variously automobile and aircraft manufacture or maintenance, metal working and processing, chemical and pharmaceutical manufacture, and other activities including dry cleaning.

In addition it is suspected that a very much

Table 9.2 Sources of organic contaminants found in urban groundwaters

Chemical class	Sources	Examples
Aliphatic and aromatic hydrocarbons (including benzenes, phenols, and petroleum hydrocarbons)	Petrochemical industry wastes Heavy/fine chemicals industry wastes Industrial solvent wastes Plastics, resins, synthetic fibres, rubbers, and paints production Coke oven and coal gasification plant effluents Urban run-off Disposal of oil and lubricating wastes	Benzene Toluene Isooctane Hexadecane Phenol
Polynuclear aromatic hydrocarbons	Urban run-off Petrochemical industry wastes Various high temperature pyrolytic processes Bitumen production Electrolytic aluminium smelting Coal-tar coated distribution pipes	Anthracene Pyrene
Halogenated aliphatic and aromatic hydrocarbons	Disinfection of water and waste water Heavy/fine chemicals industry waste Industrial solvent wastes and dry cleaning wastes Plastics, resins, synthetic fibres, rubbers, and paints production Heat-transfer agents Aerosol propellants Fumigants	Trichloroethylene Trichloroethane Para-dichlorobenzene
Polychlorinated biphenyls	Capacitor and transformer manufacture Disposal of hydraulic fluids and lubricants Waste carbonless copy paper recycling Heat transfer fluids Investment casting industries PCB production	Pentachlorobiphenyls
Phthalate esters	Plastics, resins, synthetic fibres, rubbers, and paints production Heavy/fine chemicals industry wastes Synthetic polymer distribution pipes	

larger number of private borehole sources have also been affected, especially those located on, or adjacent to, industrial sites where solvents are widely used. A recent survey of groundwater supplies in the Birmingham conurbation (Lloyd *et al.* 1988; Rivett *et al.* 1990), in which 59 sites were sampled, some repeatedly, revealed widespread heavy contamination by trichloroethylene (TCE), and significant problems with trichloroethane (TCA) and tetrachloroethylene (PCE) (Table 9.4). Metal manufacturing and processing and mechanical engineering predominates in the area and

Table 9.3   Physicochemical properties of some common chlorinated solvents (after Schwille 1988)

Chemical compound	Chemical formula	Abbreviation	Density (g cm^{-3})	Kinematic viscosity (mm^2 s^{-1})	Solubility at 25°c (mg l^{-1})	Drinking water guideline concentration (μg l^{-1}) WHO	EC
Tetrachloroethylene (Perchloroethylene)	C$_2$Cl$_4$	PCE	1.6	0.5	160	10	1
Trichloroethylene	C$_2$HCl$_3$	TCE	1.5	0.4	1100	30	1
1,1,1-Trichloroethane (Methyl chloroform)	C$_2$H$_3$Cl$_3$	TCA	1.4	0.6	720	NS	1
Carbon tetrachloride (Tetrachloromethane)	CCl$_4$	CTC(PCM)	1.6	0.6	785	3	1
Trichloromethane (Chloroform)	CHCl$_3$	TCM	1.5	0.4	8 200	30	1
Dichloromethane (Methylene chloride)	CH$_2$Cl$_2$	DCM	1.3	0.3	20 000	NS	1

NS = not specified

almost half of the boreholes sampled in the survey were located on the sites of such industries (Fig. 9.4).

Behaviour of the immiscible phase

The subsurface transport of the immiscible phase is governed by completely different controls from those determining groundwater flow. The high density and low viscosity of chlorinated solvents promotes deep penetration into aquifers. The theoretical distribution of the immiscible phase following a major spill on an intergranular aquifer is shown (Fig. 9.5), illustrating the tendency for the accumulation of this phase at the base of the aquifer.

The situation in the British Chalk, a complex fissured microporous formation, has been studied in detail at a site on the Lower Chalk of eastern England with a shallow groundwater table (Chilton *et al.* 1990). A considerable volume, probably many thousands of litres of PCE, has been spilled or discharged, although the precise timing, exact location, and total quantity involved are uncertain. The subsurface distribution of PCE was determined by drilling a number of cored boreholes, using the dry-percussion technique, close to the likely location of the major spill, and recovering the miscible (dissolved) phase in pore-waters by using the pentane extraction method with GC–ECD analysis (Lawrence *et al.* 1990) and by sampling

Table 9.4   Summary of chlorinated solvents in groundwater in Birmingham

Industrial solvents	TCE	TCA	PCE	TCM	CTC
Proportion of boreholes exceeding (%)					
1 μg l^{-1}	62	22	9	17	2
10 μg l^{-1}	43	13	4	0	0
100 μg l^{-1}	30	5	2	0	0
Maximum concentration detected (μg l^{-1})	5500	780	460	5	1

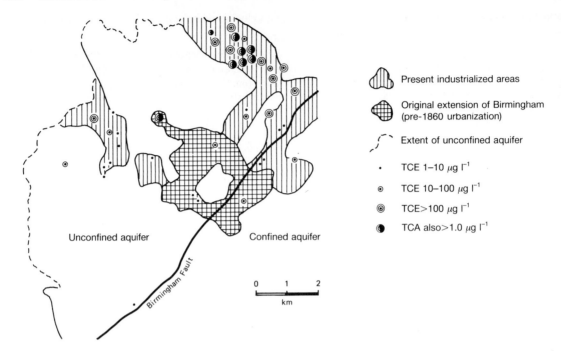

Fig. 9.4. Distribution of maximum TCE and TCA concentrations in a survey of groundwater supplies in the greater Birmingham area during 1986–88.

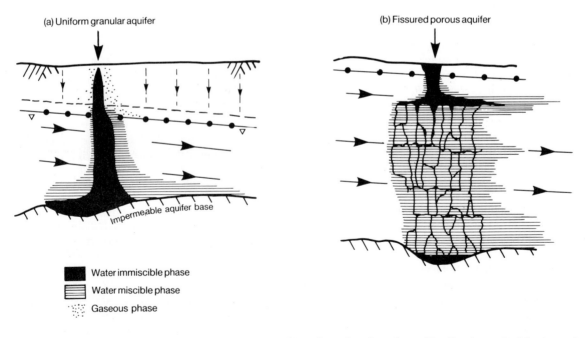

Fig. 9.5. Schematic sections showing theoretical/predicted subsurface distribution of chlorinated solvent following a major spillage in (a) uniform granular and (b) fissured porous aquifers.

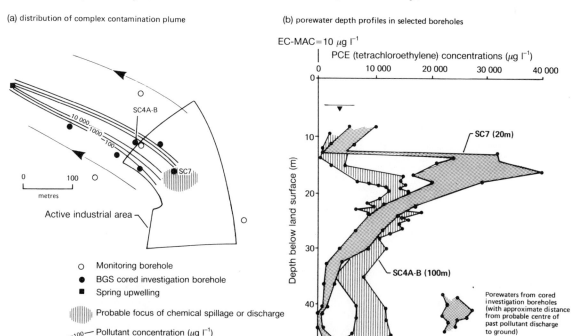

Fig. 9.6. PCE contamination of the Chalk at a site for leather processing (a) distribution of complex contamination plume (b) porewater depth profiles in selected boreholes.

the mobile fissure water. Very high dissolved solvent concentrations were found to depths of 50 m (Fig. 9.6), despite strong natural upward groundwater flow in the area. Solvent could only have migrated to such depths, against the upward hydraulic gradient if it were initially in the immiscible phase.

The downward migration of this phase appears to have been governed by variations in the lithology of the Chalk. Rapid penetration, with relatively limited opportunity for dissolution, appears to have occurred in the most permeable horizons. Downward migration was arrested and the solvent spread laterally at horizons of lower permeability, such as putty chalk and the Chalk Marl. More prolonged contact between immiscible solvent and groundwater at these levels permitted higher solvent concentrations to develop and diffuse into the chalk matrix.

With concentrations in the range 5000–40 000 μg l^{-1} (compared to a solubility limit of about

150 000 μg l^{-1}) still in the pore-water, downward solvent migration was almost certainly via the closely-spaced primary fracture system in the Lower Chalk (Fig. 9.5). The immiscible phase appears to have acted as a source of dissolved solvent for a long period of time. It is practically impossible to estimate the rate of dissolution because of the complex geometry of the interface between the solvent and the flowing groundwater.

Transport of the miscible phase

Once in the miscible phase, contaminant migration away from the site of the pollution is governed by the direction, paths, and velocity of groundwater flow. In a uniform granular aquifer these can be readily estimated but in a fissured aquifer a knowledge of the fissure geometry is required. The latter is difficult or impossible to determine in practice and thus statistical Monte Carlo techniques, coupled with a simplified model of fissured porous media (Barker and Foster 1981; Barker 1982),

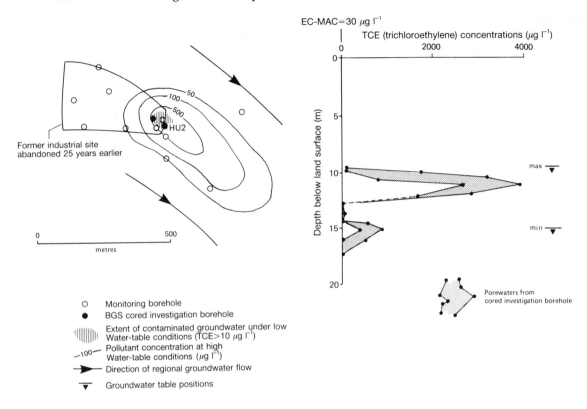

Fig. 9.7. TCE contamination of the Chalk at a site of a long-abandoned aircraft factory.

have been used to make predictions of first arrival and peak concentration probabilities (Price *et al.* 1989). The strong dependence of plume shape upon fissure development can be illustrated from the same site, where a narrow plume of high concentration has developed over considerable distances from the site (Fig. 9.6).

The chlorinated solvents are known to be highly resistant to biodegradation, except in very dilute concentrations, and are not subject to physical breakdown in the absence of ultraviolet light. Their persistence in subsurface environments has been clearly demonstrated by a parallel investigation at a site on the Upper Chalk in southern England. High levels of pore-water contamination (Fig. 9.7) and intermittent fissure water contamination by trichloroethylene (TCE) were detected, despite the fact that industrial activity (aircraft manufacture) had ceased at the site some twenty five years previously.

Pollution by hydrocarbons

The peculiar problems posed by the entry of mineral oils into urban groundwater systems stem from their behaviour as immiscible liquids. Arising as a pollutant spilled at or near the ground surface, the tendency for oil is to migrate downwards through the soil, subsoil, and unsaturated zone of the aquifer until, upon reaching the water-table of an unconfined aquifer, the oil spreads laterally to form a 'layer' floating on the water by virtue of its lower density.

Research into oil pollution of groundwater in the UK has focused to a large extent on the requirement to be able to predict the direction, extent, and rate of migration of oil following an oil spill. In the past, this research has been incident-led, drawing conclusions of general application from the solution of a local problem at an existing pollution site (Hunter-Blair 1978, 1980). With the increasing emphasis on environmental impact assessment and environmental auditing, however, it is necessary to assess the risks that could arise at a 'clean' site following a hypothetical spillage.

It is this more recent emphasis that exposes the limitations of existing research into oil behaviour in the subsurface. In fact, the problem is not a deficiency of understanding of the theory of oil behaviour, but is rather concerned with the application of the theory to a specific site. Groundwater behaviour can often be satisfactorily predicted from the permeability, the storage coefficient, and a few geometrical parameters; prediction of solute behaviour requires in addition a knowledge (for example) of the aquifer's sorptive properties and coefficients of dispersion; the prediction of oil movement, however, requires local knowledge of the complex capillary interaction between oil, air, water and the aquifer's matrix, relationships that are difficult to determine either in the laboratory or in the field.

In the related problems of petroleum reservoir engineering, as in classical hydrogeology, small-scale, physically complex processes are described in terms of empirical properties that are applied on a statistically large enough scale to mask their imperfections. In pollution problems, however, the volume of aquifer in which the oil is present is small and changing, for example the unsaturated zone beneath a pipeline that has started to leak, or a thin pancake of oil on a seasonally fluctuating water-table. Under such circumstances the detailed physical characteristics of heterogeneous alluvium or fractured chalk become of greater significance than in the equivalent groundwater flow or solute transport problem.

Much of the foregoing applies equally to groundwater contamination by chlorinated solvents, with the difference that the latter are denser than water and sink through the water-table (Fig. 9.5). The other major difference is that whereas chlorinated solvents tend to be used and spilled as pure liquids, mineral oils are complex mixtures of hydrocarbons. Crude oils are naturally variable in composition from reservoir to reservoir; derived refined oils inherit this variability, constrained by the boiling point range and other characteristics of the specific refined product. Inevitably, the chemical variability of mineral oils affects their physical properties and their susceptibility to bacterial degradation.

In the UK, oil refineries are situated on the coast and cross-country crude-oil pipelines are few. Thus the following examination of the transport and attenuation of oil in aquifers concentrates on refined products. The products that are responsible for most urban pollution incidents are the light fuels: petrol, aviation fuels, diesel fuel (DERV), and other gas oils such as central heating oils. The heavier fuel oils and lubricating oils are much more viscous, hence less mobile, and less likely to reach the water-table from a surface spill.

The UK aquifers at greatest risk from oil spill incidents are those with only limited absorbent or low permeability superficial cover. This applies particularly to certain areas of the Chalk and Jurassic limestones. Other groundwater systems at serious risk, although to a lesser extent, include alluvial sands and gravels, and locally the Triassic sandstones.

Transport

The physical principles underlying oil migration in an aquifer are the same as those underlying groundwater flow. The flow of an oil phase may be described in similar terms to that of the water phase, with due allowance on the one hand for the differing and changeable physical properties of oil-derived liquids, and on the other hand for the interaction of oil, water, and possibly vapour/air phases.

Effect of varying oil phase properties

The simplified form of Darcy's law applied to saturated groundwater flow in a granular aquifer combines fluid and aquifer properties in a single parameter, hydraulic conductivity. For general application this term must be separated into the intrinsic permeability of the aquifer and the density and viscosity of the liquid:

$$v = -\sigma g k i / \mu$$

where v is velocity of liquid flow
σ is liquid density
g is gravitational constant
k is intrinsic permeability of the aquifer
i is pressure gradient
μ is liquid viscosity

As noted earlier, whereas density varies only slightly between different oils (roughly 0.68 g cm^{-3}

for petrol, and 0.84 g cm^{-3} for diesel fuel, for example), viscosity may vary greatly (from say 0.55 cp for petrol to 3 cp for diesel fuel, compared to about 1.3 cp for water). This will result in a correspondingly wide range in potential migration rates. Furthermore, changes in oil phase composition in the subsurface due to evaporation, solution, and degradation of light hydrocarbons would result in an increase of viscosity.

Effect of interaction of oil phase with water and vapour

For oil to migrate in an aquifer it must displace the water or vapour already present. Until the oil phase reaches maximum saturation, the effective intrinsic permeability of the aquifer is reduced, a reduction expressed mathematically by the 'relative permeability' concept familiar from unsaturated zone groundwater flow theory, but applied, if necessary, to three fluid phases. The relationship between relative permeability and water saturation in the water–air system has been examined for a range of soil textures in the field of soil physics (Rijtema 1969); three phase data are obtained, but rarely published, for petroleum reservoirs. The absence of two or three phase date (or a simple means of obtaining them) for the oil–water system or oil–water–vapour system for typical alluvial or sandstone strata is a constraint on the assessment of potential oil pollution incidents in the UK and elsewhere.

The relative permeability–saturation relationship is adequate for steady state situations, but for transient problems it is necessary to understand how oil–water–vapour saturations vary with the pressure differences between the phases. Again, such relationships for typical soil or aquifer materials are rarely, if ever, known, and there are no simple methods for their determination.

Fissured aquifers

In fissured and fractured aquifers, such as the Chalk, principles of multiphase fluid flow derived for granular aquifers cannot easily be applied at a local scale. The model experiments of Schwille (1984) have demonstrated that following an oil spill at the surface, the oil–vapour interface in the unsaturated zone is likely to move downwards at widely and erratically varying rates along 'fingers', even in a uniform fissure. At, or rather on, the water-table the lateral migration of oil is affected less by interaction with water and vapour phases, and more by the availability of fissure flow paths at this level.

In granular aquifers oil accumulates in the pore spaces; in aquifers dominated by fissure flow, the fissures can hold oil only in the layer on the water-table. In the unsaturated zone the fissures are usually too wide to retain oil by capillary action, but oil may accumulate in porous material between the fissures. Recent research has shown that in the Chalk, where pores are typically small and largely saturated with water even above the water-table, oil is absorbed only within about a centimetre of the fissure wall (Ashley 1990).

Approximation methods

In view of the difficulty of applying theory to practical problems, a number of simplifying assumptions have to be made, and some rules of thumb have been suggested that may permit an assessment of actual or potential oil spills to a first order of accuracy. In clean, coarse gravels the capillary forces that control the saturation-pressure relationships and the relative permeability–saturation relationships may be neglected, simple step functions may be inserted in their place.

In the specific case of a limited spillage of a known volume of oil, the maximum extent of migration of the oil body either vertically downwards from the surface or laterally on the water-table, may be determined from the oil retention capacity of the aquifer material and certain empirical factors that allow for capillary effects. A similar approach may be adopted for fissured aquifers, based on estimates of the absorptive capacity of fissure surfaces and the extent of fissuring (Ashley 1990).

Attenuation

There has been extensive research into the physical weathering and degradation of crude oils at sea; however, similar studies of the degradation of refined and waste oil in soils have been limited and most studies of deeper subsurface processes have provided only qualitative information on rates of degradation under natural conditions (Fried *et al* 1979).

Certain principles are well understood. For

example, small hydrocarbon molecules evaporate and dissolve in water more readily than larger molecules. Hence the volatility of petrol and its solubility of about 200 mg l^{-1} in water, compared to about 15 mg l^{-1} for diesel fuel. Similarly, smaller molecules are more readily degraded, although the molecular structure strongly influences the overall pattern of degradation. Hydrocarbons degrade much more rapidly under aerobic conditions than anaerobic conditions, although where an oil phase is present the oxygen available in the subsurface is utilized more rapidly than it can be replenished, leading to an early onset of anaerobic conditions.

Hunter-Blair (1978) conducted some limited experiments to compare laboratory degradation with observations of degradation at a site on the Chalk where diesel fuel had spilled. Re-examination of the same site after a ten year interval (Ashley 1990) revealed extensive degradation both in the unsaturated zone and in the oil still on the water-table. Of the oil initially absorbed into the chalk fissure walls in the unsaturated zone, typically 75 per cent has been lost, probably approximately divided evenly between physical weathering and degradation. The oil on the water-table appears to have undergone at least as extensive weathering. Oil that has penetrated into the body of the chalk away from fissures has been altered to a lesser degree, apparently due to more limited degradation.

Conclusions

Research into the occurrence of urban pollution only gained momentum in the UK during the late 1980s, although some aspects of the problem, such as heavy metal contamination of groundwaters, have been the subject of extensive investigation over many years. Specific examples of urban pollution related to inorganic compounds, including heavy metals, has attracted little research attention and the main emphasis has been directed towards organic pollutants with initial studies carried out to identify the existing concentrations in groundwaters. Although high concentrations have been found locally for some organic compounds, for example TCE, the evidence of widespread organic pollution beneath the urban areas that have been investigated is surprisingly small. Nevertheless,

local examples of pollution are such as to give cause for concern and clearly organic pollution of groundwater needs to be thoroughly investigated.

Two principal research needs are measurement and sampling techniques, and studies of attenuation processes. Because the permissible concentrations of many organic pollutants in groundwater are extremely low, precision measurement is fundamental for an understanding of their occurrence and the associated physical and chemical processes. Laboratory measurement techniques need refining and techniques for field measurement require development. Commensurate with the development of measurement techniques, are studies of the sampling procedures of which the isolation of the sample location and the use of non-contaminant sampling materials are of paramount importance particularly in studies of organic compounds. The manner in which organic compounds move through both the unsaturated and saturated zones is very poorly understood and is a major area for further research. The problems are manifold but efforts should be concentrated on understanding the attenuation processes for the principal identified pollutants in the major aquifers.

References

Ashley, R. P. (1990). Mechanisms of oil movement in the unsaturated chalk. PhD thesis. University of Birmingham.

Barker, J. A. (1982). Laplace transform solutions for solute transport in fissured aquifers. *Advan. Water Res.*, **5**, 98–104.

Barker, J. A. and Foster, S. S. D. (1981). A diffusion exchange model for solute movement in fissured porous rock. *Q. J. Eng. Geol.*, **14**, 17–24.

Chilton, P. J., Lawrence, A. R., and Barker, J. A., (1990). Chlorinated solvents in Chalk aquifers–some preliminary observations on behaviour and transport. In *Chalk*, Proc. Intern. Chalk Symp. (Brighton, 1989), pp. 605–10. Thomas Telford, London.

Dowse, L. H. and Selby, K. H. (1975). Groundwater pollution in an industrialised part of the Trent Basin. *Water Pollution Control*, **74**, 526.

Edmunds, W. M., Cook, J. M., Kinniburgh, D. J., Miles, D. L., and Trafford, J. M. (1989). *Trace element occurrence in British groundwaters*, Research Report No. SD/89/03. British Geological Survey, Keyworth.

Ford, M. (1989). Extent, type and sources of inorganic groundwater pollution below the Birmingham conurbation. PhD thesis. University of Birmingham.

Forsterner, U. and Wittmann, G. T. W. (1983). *Metal pollution in the aquatic environment*. Springer Verlag, Berlin.

Fried, J. J., Muntzer, P., and Zilliox, L. (1979). Groundwater pollution by transfer of oil hydrocarbons. *Ground Water*, **17**, 586–95.

Harrison, I. B., Parker, A., and Williams, G. M. (1981). *Investigation of the landfill at Eastfield Quarry, Fauldhouse, West Lothian, Scotland.* Report No. 81/13. Institute of Geological Sciences, London.

Hunter-Blair, A. (1978). *Oil pollution of a chalk aquifer–a case history*. Int. Symp. on groundwater pollution by oil hydrocarbons, Stavebni Geologie, Prague.

Hunter-Blair, A. (1980). Groundwater pollution by oil products. *J. Instn. Water Engs and Sci.*, **34**, 557–69.

Lawrence, A. R. and Foster, S. S. D. (1987). *The pollution threat from agricultural pesticides and industrial solvents*, Hydrogeology Report 87/2. British Geological Survey, Keyworth.

Lawrence, A. R., Chilton, P. J., Barron, R. J., and Thomas, W. M. (1990). A method of determining volatile organic solvents in Chalk pore-waters and its relevance to the evaluation of groundwater contamination. *J. Contam. Hydrol.*, **6**, 377–86.

Lloyd, J. W., Lerner, D. N., Rivett, M. O., and Ford, M. (1988). *Quantity and quality of groundwater beneath an industrial urban conurbation.* Proc. symposium urban water (Duisberg, FRG/April 1988), pp. 445–52. UNESCO, Paris.

Newman, R. and Ross, C. A. M. (1985). *Mineralogical and geochemical controls on heavy metal pollution in monolith lysimeters*, Report No. FLPU 85–5. British Geological Survey, Keyworth.

Price, M., Atkinson, T. C., Wheeler, D., Barker, J. A., and Monkhouse, R. A. (1989). *Highway drainage to the Chalk aquifer*, Hydrogeology Report Series 89/3, British Geological Survey, Keyworth.

Rijtema, P. E. (1969). *Soil moisture forecasting.* Instituut voor Cultuurtechniek en Waterhuishouding, Wageningen, Netherlands.

Rivett, M. O., Lerner, D. N., and Lloyd, J. W. (1990). Organic contamination of the Birmingham aquifer, UK. *J. Hydrol.*, **113**, 307–23.

Ross, C. A. M. (1985). The unsaturated zone as a barrier to groundwater pollution by hazardous wastes. In *Hydrogeology in the service of man*, pp. 127–41. Memoirs of the 18th Conference of the Int. Assoc. Hydrogeol., Cambridge, England.

Schwille, F (1981). Groundwater pollution in porous media by fluids immiscible with water. In *Proc. Int. Sym.* Noordwijkerhout, The Netherlands, 23–27 March 1981, (ed. W. van Duijrenbooden, P. Glasbergen, and H. van Lelvveld). *Studies in Environmental Science*, **17**, 451–63.

Schwille, F. (1984). Migration of organic fluids immiscible with water in the unsaturated zone. In *Pollutants in porous media*. Springer-Verlag.

Schwille, F. (1988). *Dense chlorinated solvents in porous and fractured media: model experiments.* Lewis Publishers Inc., Chelsea, Michigan.

Warwick, P., Anderton, W., Smith, B., and Williams, G. M. (1991). *Preliminary investigation of nickel speciation in polluted groundwater at Villa Farm.* Technical Report WE/91/12, British Geological Survey, Keyworth.

Williams, G. M., (1981). Underground disposal of wastes in Britain. In *Proc. Int. Sym.* Noordwijkerhout, The Netherlands, 23–27 March 1981, (ed. W. van Duijvenbooden, P. Glasbergen, and H. van Lelvveld.) *Studies in Environmental Science*, **17**, 421–6. Elsevier, Amsterdam.

10. Rural and agricultural pollution of groundwater

J. M. Parker, C. P. Young, and P. J. Chilton

Introduction

By the end of the 1970s, British hydrogeological research had established that changes in agricultural practice, aimed at increasing productivity, were having a profound impact on groundwater quality. The most notable effect was the rising level of nitrate in groundwater supplies. This was particularly evident in the aquifers of eastern and southern England where intensive arable cultivation is the dominant land-use.

The early investigations of the pollution of groundwater by nitrate, carried out in parallel by the Institute of Geological Sciences (now the British Geological Survey) and the Water Research Centre, concentrated on the unsaturated zone because of its critical role in the estimation of future trends in groundwater quality and because of the almost complete lack of published work on relevant processes in the unsaturated zone. The work consisted of extensive drilling programmes to sample the pore-water quality in the unsaturated zone of all the major aquifers, with emphasis on areas of intensive arable farming. Pore-water profiles of thermonuclear tritium were determined to help to understand the mode and rate of movement of solutes through the unsaturated zone.

A large body of data was collected which showed that nitrate losses from arable land were increasing, probably as a result of increasing applications of inorganic fertilizer, and that the concentration of nitrate in many groundwater sources was likely to rise for years to come (Foster and Young 1981). Grassland was not revealed to be a significant source of nitrate leaching to groundwater, except after ploughing when the accumulation of organic nitrogen in the soil is mineralized and susceptible to leaching. Although these general conclusions could be drawn, problems remained in the more detailed interpretation of the solute profiles because of uncertainties about the mechanism of solute transport in the unsaturated zone, particularly in the important Chalk aquifer. The contention surrounding this issue also hampered efforts to model the movement of nitrate through the unsaturated zone mathematically, although early attempts based on solute movement by a piston flow mechanism met with an encouraging degree of success (Young *et al*. 1976).

The research of the 1970s had also incorporated studies of the occurrence and behaviour of nitrate in the saturated aquifer of several research catchments. Although few data were available, these studies demonstrated the considerable variations in nitrate concentration that could occur both spatially and with depth in aquifers subject to diffuse pollution (Foster and Young 1981).

During the 1980s nitrate pollution has remained the most important aspect of groundwater pollution resulting from agricultural practices. Studies of the unsaturated zone have progressed to provide a better understanding of the solute flow mechanism and to give more precise information about nitrate leaching from different types of land-use. The issue of nitrate leaching from grassland has received more attention. Investigations of the saturated zones of unconfined aquifers have provided a fuller understanding of the behaviour of nitrate and other solutes in the groundwater system whilst studies of confined aquifers have attempted to determine the origin and security of their characteristically low-nitrate groundwater (Foster *et al*. 1986). The results of all these investigations have been incorporated into mathematical models to predict future groundwater nitrate concentrations. This chapter is an account of the key aspects of this research programme together with an account of more recent work on another aspect of rural pollution, that resulting from the use of pesticides.

Nitrate pollution

During the last ten years, nitrate concentrations have continued to rise in many groundwater sources in unconfined aquifers in Britain. However, there is evidence in water from some sources that nitrate levels are stabilizing and, in a very few sources, are even showing signs of declining (House of Lords 1989). In 1985, Britain adopted the EC Drinking Water Directive (80/778/EEC) which introduced a more stringent limit for the maximum acceptable concentration of nitrate in potable water. Compliance with this standard created problems for the water industry in many parts of the country and sparked a widespread debate, both political and scientific, on all aspects of nitrates in water (DoE 1986; House of Lords 1989).

Hydrogeological research and groundwater pollution modelling have been an important input to this debate, the most recent outcome of which has been the decision by the Government to imple-

ment a pilot scheme for groundwater quality protection through land-use controls in 'nitrate sensitive areas' (DoE 1988). The hydrogeological research is summarized in relation to the unsaturated and saturated zones of the aquifer.

Unsaturated zone research

A substantial proportion of the total research effort continues to be dedicated to the unsaturated zone. The research has been based primarily on the extraction of interstitial (pore) water from rock cores by centrifugation and other methods.

Arable land

Many of the investigation boreholes drilled under arable land during the 1970s revealed pore-water nitrate profiles with marked peaks in concentration in the upper part of the unsaturated zone. Maximum concentrations ranged from 40–70 mgN l⁻¹ compared with typical values of 10–20 mgN l⁻¹ at depths of 10 m or more. Implied leaching losses were in the order of 50–70 kgN ha⁻¹ a⁻¹ (Foster *et al.* 1982) representing an equivalent of up to 50 per cent of the applied nitrogen fertilizer.

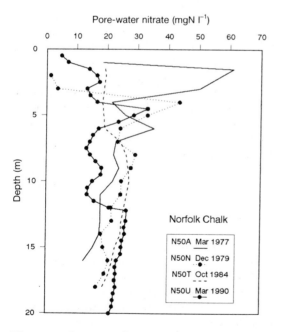
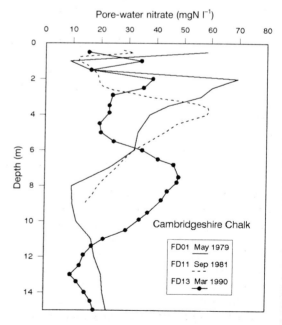

Fig. 10.1. Sequential pore-water nitrate profiles in the unsaturated zone of the Chalk beneath arable land in eastern England.

Periodic re-drilling has been carried out at a range of sites to monitor the changing nitrate concentrations in the unsaturated zone and determine trends in leaching losses. The major work in this area has been carried out by BGS at sites on the unconfined Chalk in Norfolk and Cambridgeshire (Parker and Chilton 1991). At least three sequential pore-water profiles have been obtained from each site in the period 1976–1990. Two examples are shown in Fig. 10.1.

There is evidence of downward movement of nitrate and other solutes through the unsaturated zone but not generally in a non-dispersive or systematic fashion. Nitrate concentrations in the upper parts of most of the recent profiles are lower than those observed in earlier profiles, despite similar or increased fertilizer applications. The marked peaks observed in the original nitrate profiles are seen to have moved down through the unsaturated zone with some attenuation and have generally been replaced by infiltration with a lower nitrate content.

It is possible that the high nitrate concentrations found at shallow depth in the mid-1970s may, in part at least, be a product of an atypical groundwater recharge and nitrate leaching regime arising from the 1976 drought and might not have been strictly representative of nitrate losses. It is probable, however, that the observed decrease in shallow unsaturated zone nitrate concentrations reflects a real reduction in leaching losses in recent years as a result of more frequent cultivation and earlier sowing of winter cereals using better crop strains. If this is indeed the case, then estimates of average leaching losses from arable land should be reduced. Current estimates derived from the most recent drilling in the unsaturated zone, suggest that average leaching from arable land at the research sites in eastern England is equivalent to about 25 per cent of the applied nitrogen (Parker and Chilton 1991). Previous calculations, based on hydrogeological studies, generally produced values of around 40 per cent (DoE 1986).

The re-drilling at existing research sites has indirectly demonstrated the benefits, in terms of the apparent reduction in nitrate leaching, of the move towards early sown winter cereals. Other work has been aimed at comparing the effect on leaching of different land-use management. An agricultural experimental site on the Chalk outcrop in Berkshire provided an opportunity to investigate the comparative effects of conventional ploughing and minimal cultivation (direct drilling) on leaching losses of nutrients from arable soils. Pore-water profiles under the two experimental plots showed that the latter led to generally lower leaching losses, presumably as a consequence of the reduction of autumn mineralization of organic nitrogen resulting from less aeration (Foster *et al.* 1986).

Grassland

During the 1970s investigations of the cause of groundwater pollution by nitrate concluded that the mineralization of soil organic matter following the ploughing up of established grassland was a major source of nitrate leached into the ground (Young *et al.* 1976). The greater part of the evidence for this was derived from agricultural field drainage, soil lysimeter, and crop uptake studies (Young 1981), with little corroborating data from studies of the underlying aquifers. To provide data linking the agricultural work with the hydrogeological studies, a five-year field experiment was undertaken to monitor the effects on nitrate leaching of changing the cultivation of established long-term grassland.

A site was identified on the unconfined Chalk of the South Downs in Sussex where no cultivation or other intensive agriculture had been practised for at least fifty years and probably much longer. The site is in a dry valley where 5 m of Coombe deposits overlie the Middle Chalk. The unsaturated zone is more than 40 m thick.

Pore-water nitrate profiling on the site during the mid-1970s had revealed uniformly low nitrate contents in the unsaturated aquifer, typical of other sites on unimproved and unfertilized grassland (Young and Gray 1978). Five, 12 m square experimental plots were established in 1978. One plot was maintained as a control of uncut, unfertilized grassland whilst the other four were converted to fallow, fertilized grassland, cereals, and a cereal/ley rotation and treated in accordance with local husbandry practices (Young 1986). The programme of agricultural practice is outlined in Table 10.1 and was accomplished by regular soil sampling to determine available nitrate over the period 1978–80. Periodic sampling of the underlying unsaturated zone of the Chalk was carried out from 1978–82.

The changes in available soil nitrate (Fig. 10.2) reveal the combined effects of rapid mineralization

Table 10.1   Agricultural regimes on experimental plots on the Chalk in Sussex

Year	Control	Fertilized grass	Fallow	Cereals	Cereal/ley
1978	Unfertilized uncut	Fertilized 63 kg N ha^{-1}, May. Cut 4 times, grass removed	First cultivated March 1978, rotovated and harrowed 4 times. No fertilizer	First cultivated March 1978, maize, 126 kg N ha^{-1} May	First cultivated March 1978, barley, 70 kg N ha^{-1} May
1979	As 1978	As 1978	Rotovated and harrowed 4 times. No fertilizer	Wheat, 70 kg N ha^{-1} April	Barley, 70 kg N ha^{-1}
1980	As 1978	As 1978	As 1979	Maize, 126 kg N ha^{-1} May	Grass, 63 kg N ha^{-1}, cut 4 times and removed
1981	As 1978	As 1978	As 1979	Wheat, 70 kg N ha^{-1} April	Grass, 63 kg N ha^{-1}, cut 4 times and removed
1982	As 1978	As 1978	Revert to grass	Revert to grass	Revert to unfertilized grass

and fertilizer application in the form of a sharp increase in the available soil nitrate on the fallow and cereal plots immediately after ploughing of the original grassland. The effects of crop uptake are also apparent from the subsequent marked fluctuations in available soil nitrate on the cereal plot compared with the fairly uniform level on the control plot whilst nitrate continued to be produced by mineralization on the fallow plot.

Unsaturated zone profiling showed the downward movement of the released nitrate in infiltration to the aquifer. On the fallow plot (Fig. 10.3), the first influx of nitrate following ploughing produced concentrations in excess of 100 mgN l^{-1}. These concentrations were reduced by the effects of substantial dispersion during the rapid downward movement of nitrate through the homogeneous Coombe deposits. Sequential profiles beneath

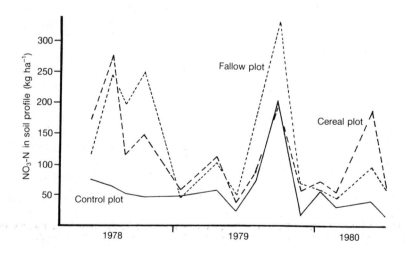

Fig. 10.2. Changes in available nitrate in soils following conversion of long-standing grassland on Chalk in Sussex.

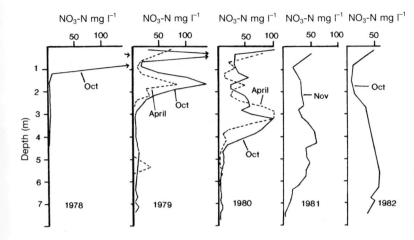

Fig. 10.3. Changes in porewater nitrate in the unsaturated zone beneath a fallow plot on Chalk in Sussex.

the cereal plot also exhibit progressive downwards migration of nitrate pulses released from the soil (Young 1986).

Integration of the masses of nitrate released into the Chalk suggested that no significant leaching of nitrate occurred from the control plot of unimproved grassland. During the period of monitoring (1978–1981), a total of about 200 kg N ha^{-1} was released from the cereal plot and 700 kg N ha^{-1} from the fallow plot. The latter demonstrates conclusively the substantial nitrate release that occurs as a result of ploughing established grassland. The results are consistent with those of other studies of nitrate release following ploughing of grassland on the Chalk in Berkshire (Ryden *et al.* 1984) and in France (Landreau and Morisot 1983). The nitrate loss to groundwater resulting from the mineralization of soil organic nitrogen was much less on the cereal plot due to the uptake of a large amount of the available nitrate by the crop. This illustrates the importance of planting immediately after ploughing established grassland, rather than allowing a fallow period.

Prior to 1980, it had been established that nitrate losses from permanent unfertilized grassland were small and gave rise to nitrate concentrations generally less than 6 mg N l^{-1} in the unsaturated zone (Young and Gray 1978). Pore-waters beneath long-term fertilized grassland had been found to contain greater concentrations, although not generally exceeding 20 mg N l^{-1}. High concentrations of nitrate were only thought to leach from grassland receiving excessive amounts of inorganic

fertilizer (> 500 kg N ha^{-1} a^{-1}) or as a result of ploughing-in long standing grass as has been described.

Research published by the Agriculture and Food Research Council (AFRC) (Ryden *et al.* 1984) focused attention on the possibility of significant nitrate leaching from intensively managed grazed grassland. Subsequent work by BGS, on a series of chalk sites in southern England, has confirmed that very high nitrate concentrations do occur in the unsaturated zone beneath grazed grassland (Parker and Perkins 1988; Parker and Chilton 1989). A total of 24 pore-water profiles have been drilled on grazed grass receiving 150–400 kg N ha^{-1} a^{-1}. Most of the sites are long term permanent pasture used for grazing dairy cattle and have annual mean effective rainfalls ranging from 170–360 mm a^{-1}.

In grazed grassland systems, the nitrogen in the grass is taken in by the animal and mostly re-cycled as excreta. The excreta is deposited in small areas where the concentration of nitrogen exceeds the capacity of the vegetation to absorb it so that much of the nitrate is leached out. It is evident from the results of investigations of the unsaturated zone that there is a close relationship between the amount of nitrate leached and the nitrogen fertilizer applied to grazed grassland. In the upper part of the unsaturated zone (10–15 m below the surface) nearly all the sites investigated have nitrate concentrations greater than 10 mg N l^{-1}. Beneath the more intensively managed grazed grass sites, the nitrate in the pore-water generally exceeds

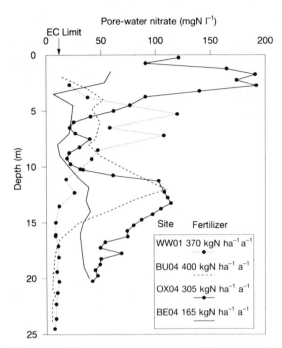

Fig. 10.4. Pore-water nitrate profiles in the unsaturated zone beneath grazed grassland on the Chalk aquifer.

50 mg N l⁻¹ and peak nitrate concentrations of more than 100 mg N l⁻¹ are present in many of the investigation boreholes (Fig. 10.4). At several sites, more than one borehole was drilled in each field to assess spatial variability. Despite obvious problems of non-uniform inputs of nitrogen from grazing animals, there was generally good agreement between the nitrate profiles (Parker and Chilton 1989).

The nitrate concentrations that have been found under grazed grassland are greater than is usually observed under intensively cultivated arable land. In broad terms, the data imply mean leaching losses of 75–150 kg N ha⁻¹ a⁻¹ from intensively managed, grazed grassland. Expressed as a percentage of the nitrogen inputs to the land, these values represent an average loss equivalent to at least 20 per cent of the applied nitrogen. The results suggest that grazed grassland is a significant source of nitrate leaching to groundwater and that leaching from grazed grassland receiving more than 100 kg N ha⁻¹ a⁻¹ is likely to give rise to nitrate concentrations above the EC recommended maximum. In contrast, direct leaching losses from

cut grass systems are low, typically around 10 kg N ha⁻¹ a⁻¹ except where significant quantities of nitrogen fertilizer are applied (> 250 kg N ha⁻¹ a⁻¹), because the crop has the ability to take up large amounts of nitrogen which are removed from the system when the grass is cut.

Mechanisms of solute movement in the unsaturated zone

The Chalk is the most important aquifer in the UK and has naturally been the main subject of research into solute movement, particularly since the process of solute movement through the Chalk is a complicated one. In order to resolve some of the uncertainties about the mechanism of solute transport through the unsaturated zone of the Chalk (Foster and Bath 1982), re-drilling was carried out at several of the original research sites to determine actual changes in the pore-water profiles of nitrate and thermonuclear tritium. The results have provided important new information but their interpretation presents some difficulties.

The Chalk is an unusual aquifer with high porosity but pore sizes so small that intergranular hydraulic conductivity is very low. Even in the unsaturated zone most of the pores remain saturated. The rock mass is traversed by numerous fissures and this secondary porosity provides the potential for relatively rapid flow through the unsaturated zone if the matrix tension falls sufficiently. Whether fissure or intergranular flow predominates at any time depends on the precise rock physical properties (see Chapter 5).

If it could be assumed that all nitrate and tritium in the pore-water profiles is moving down without dispersion, then it would be possible to interpret the history of recent nitrate leaching losses in relation to the cultivation history. Sequential tritium profiles through the Chalk's unsaturated zone have revealed the preservation of tritium peaks at many sites, which implies slow downward migration with only limited dispersion (Smith *et al.* 1970). Detailed analysis of the profiles in terms of their mass balance and peak movement reveals anomalies which suggest that the mode of transport may be more complex than the uniform 'piston-like' displacement that might be envisaged (Foster and Smith-Carington 1980; Geake and Foster 1989). The possibility of significant by-pass flow is an important factor.

At other locations, there is forward tailing and broadening of the tritium peaks in sequential tritium profiles, providing clear evidence of much greater dispersion (Fig. 10.5). The same is true of sequential nitrate profiles derived from re-drilling

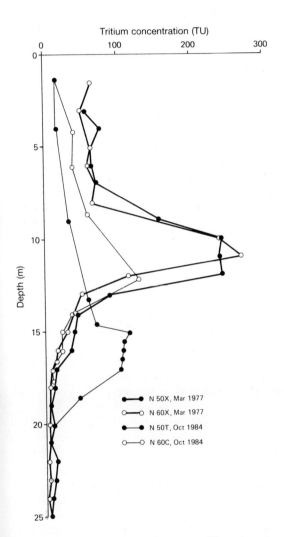

Fig. 10.5. Sequential tritium profiles in the unsaturated zone of the Chalk in west Norfolk.

at certain sites. For instance, re-profiling of four sites on the Chalk in Norfolk showed downward movement of nitrate through the unsaturated zone but not in a systematic, non-dispersive fashion

(Fig. 10.1). Dispersion is implied where there has been less downward movement than would be expected if piston flow was occurring and where there has been an apparent flattening and broadening of the original nitrate peaks. In consequence, there are higher nitrate concentrations than had been expected in the deeper part of the profile (Parker and Chilton 1991).

It has become apparent that different mechanisms of solute transport occur at different locations in the same aquifer. At some sites that have been re-drilled, there is evidence of downward movement of nitrate by apparently straightforward displacement, with little dispersion (Southern Water 1985). At other sites there is evidence of significant dispersion in the solute movement and in some locations the observed changes in form of the pore-water nitrate profiles cannot be explained (Parker and Chilton 1991). The variation is attributed to differences in the movement of water through the unsaturated zone. This is controlled by the physical properties of the Chalk and temporal variations in rainfall and antecedent moisture content and it has been shown that fissure or intergranular flow can predominate. The transport of any individual solute will also be influenced by its seasonal input distribution and its aqueous self-diffusion coefficient in porous media. Laboratory measurements recorded similar diffusion coefficients for tritiated water and nitrate (Geake and Foster 1989) which should, therefore, display similar movement, although differences may occur due to differences in their seasonal input.

There is not a simple universal model for solute transport through the unsaturated zone of the Chalk, nor many other aquifers. The preservation of tritium peaks at many Chalk sites clearly demonstrates that the bulk of solute transport through the unsaturated zone, by whatever mechanism, is slow. The evidence re-affirms the conclusion that, because of the slow downward transit, nitrate concentrations in groundwater supplies will continue to rise slowly in many unconfined aquifers for many years. The rate of movement in the Chalk, implied from the pore-water nitrate profiles is between 0.4 m a^{-1}–1.1 m a^{-1} (DoE 1986). In some locations, however, it may be difficult to simulate the changing form of the unsaturated zone profiles and to model the progress of nitrate pollution through the unsaturated zone accurately.

Saturated zone research

With the overall objective of establishing the future water quality trends in groundwater sources, research has been carried out to determine the processes of migration and attenuation of nitrate and other diffuse pollutants, in the saturated zone of the major UK aquifers. The work has included detailed catchment investigations in unconfined aquifers to establish the origin and distribution of diffuse pollutants within the aquifer and understand the factors controlling the concentration of these pollutants in pumped groundwater supplies. Investigations in the confined parts of aquifers have been aimed at understanding the origin and long-term security of the generally low nitrate groundwater that they contain. Such investigations have necessarily considered processes in the transition zone between the unconfined and truly confined parts of the aquifers.

The methods employed have included pore-water profiling, borehole flow logging, and laboratory core analysis to establish aquifer hydraulic properties and groundwater flow regimes, and installation and operation of monitoring/sampling networks to determine temporal variations in groundwater quality. Similar methods have been used in both the unconfined and confined aquifers but because the different zones pose different groundwater quality problems in relation to agricultural pollution, they are discussed separately.

Unconfined aquifers

Investigations of diffuse agricultural pollution in the saturated zone of unconfined aquifers have generally been carried out as a part of wider studies based on individual groundwater catchments, usually associated with a major public supply source.

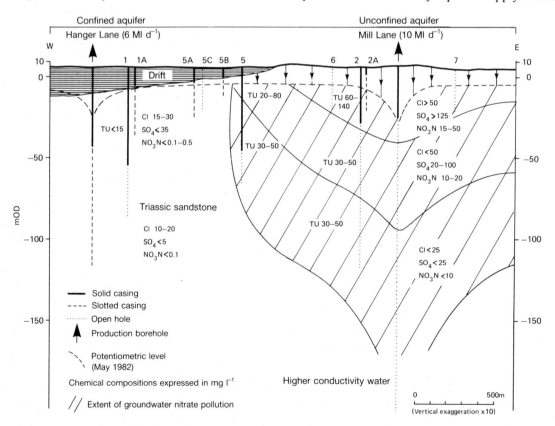

Fig. 10.6. Cross-section through the Triassic sandstones at Carlton in Yorkshire illustrating variations in groundwater quality.

Many of the findings are site specific, but some common features have emerged which can be illustrated by particular examples.

In most of the studies of nitrate pollution in the unconfined zones of British aquifers, it has been found that nitrate has penetrated to considerable depths in the aquifer. This can be seen in the Sherwood Sandstone at Carlton in Yorkshire (Fig. 10.6) where nitrate pollution is present to depths greater than 120 m in an aquifer from which abstraction commenced in 1968, initially producing low-nitrate water. The unsaturated zone here is thin (about 10 m) and the deep penetration of diffuse pollutants into the sandstone is thought to occur as a result of groundwater abstraction from boreholes that fully penetrate the aquifer. The pumping induces increased vertical groundwater flow which, if combined with high dispersivity, would enable the rapid penetration of pollutants. In the Chalk also, there are places where nitrate pollution has invaded much of the saturated aquifer (Parker *et al.* 1987), although in many Chalk catchments the impact of diffuse pollution on groundwater supplies is delayed by the slow movement of solutes (typically 1 m per year) through a thick unsaturated zone (often greater than 30 m).

Where diffuse pollution resulting from agricultural activity has already had a marked impact on an aquifer there is often a stratification of groundwater quality in which the most polluted water is present in the top of the saturated zone and concentrations of pollutants decrease with depth in the aquifer. Where the aquifer is relatively isotropic, this vertical stratification takes the form of a gradual change in groundwater quality but often there are abrupt vertical changes in groundwater quality resulting from lithological controls.

In the Sherwood Sandstone, the occurrence of major fissures and/or the presence of harder bands within the sandstone creates zones of preferential horizontal flow which may block or attentuate the downward movement of polluted water. At Carlton the lithological controls do not appear to have had much impact on the pore-water quality stratification but lateral flows above layers in the sandstone of lower permeability are important in terms of the movement of groundwater (Parker *et al.* 1985). A different lithology in the Triassic sandstones in Staffordshire gives rise to a zone of higher nitrate groundwater in the Bunter Pebble Bed

facies which carries the major groundwater flows within the aquifer (Young and Gray 1978). On a smaller scale, complicated areal and vertical variations in the nitrate concentration of groundwater in the Lower Greensand in the Woburn area occur because of marked anisotropy created by interstratification of clay and sand layers (Booker 1984).

Groundwater quality stratification controlled by lithology is also found in the unconfined Chalk which is often characterized by greater transmissivity, associated with better development of fissures, in the upper part of the saturated aquifer. In the Chalk at Colney in Norfolk, significantly higher nitrate concentrations in the upper high transmissivity zone provide a striking example of such stratification (Parker *et al.* 1983). In the Lincolnshire Limestone the Kirton Cementstones form an aquitard between the upper and lower limestones, creating stratification in the pore-water quality and complicated flow and quality variation in the mobile groundwater (Smith-Carington *et al.* 1983).

The widespread occurrence of groundwater quality stratification in unconfined aquifers affected by diffuse pollution from agricultural practices has raised many questions about the effectiveness of groundwater quality sampling for pollution monitoring (Parker *et al.* 1983; Parker and Foster 1986). Vertical variations in groundwater quality often become more significant than spatial variability. The comparison of samples from different boreholes is complicated because the chemistry of pumped samples can vary according to borehole depth, length of solid casing, intersection of fissure flow levels, etc. In the unconfined Sherwood Sandstone at Carlton, it has been observed that boreholes of very similar construction, in close proximity, produce water of quite dissimilar quality due to differences in the proportion of inflow from various depths in the aquifer (Parker *et al.* 1985). Similarly, where diffuse pollution has caused groundwater quality stratification, the quality of the discharge from a single borehole may vary significantly in response to changes in the pumping regime. Traditional monitoring methods employing pump or depth sampling of deeply penetrating boreholes are often inadequate for groundwater quality monitoring under these conditions. To be effective, monitoring networks and sampling methods must be designed to detect vertical as well as areal variations. For the early detection of diffuse agricultural pollution in unconfined aquifers, sampling from shallow depths in the saturated zone will be most sensitive.

In the majority of groundwater catchments that have been investigated, nitrate pollution has resulted from the normal agricultural practices of ploughing grassland and the application of inorganic fertilizers. The effect of irrigation of agricultural land with sewage effluent has, however, been the subject of one study on the Sherwood Sandstone (Harris 1985). The findings indicated that in certain parts of the catchment in question, high nitrate concentrations in the groundwater were a direct result of nitrification of ammonia present in the sewage irrigation.

Confined aquifers

To date, the impact of agricultural pollution on confined aquifers in the UK has been small and groundwater in these aquifers is generally characterized by low or zero nitrate concentrations. Although volumetrically the groundwater supplies derived from confined aquifers are quite small, they provide a potentially important source of low-nitrate groundwater for blending with polluted supplies from unconfined aquifers. Most confined aquifers have very limited natural groundwater circulation as a result of regional geological structure, but in some instances major flow has been induced by groundwater abstraction.

The principal focus of research on the likely impact of agricultural pollution on confined aquifers has been to establish the origin and long-term security of the low nitrate groundwater resource they contain. Major investigations have been carried out in four different confined aquifer environments; the Lincolnshire Limestone in south Lincolnshire (Lawrence and Foster 1985), the Chalk in Norfolk (Parker and James 1985), the Triassic sandstones in south Yorkshire (Parker *et al.* 1985, 1988) and the Lower Greensand in Bedfordshire (Booker 1984). Each of these investigations has considered the sequence of hydrochemical changes across the transition from groundwater polluted by nitrate to nitrate-free conditions within the aquifers.

At Carlton in Yorkshire (Fig. 10.6), there is an apparently abrupt change in groundwater quality marking the boundary between unconfined and confined aquifer conditions. Intensive arable agriculture on the thin soils of the sandstone outcrop has led to significant pollution of the aquifer by nitrate (Parker *et al.* 1985). Despite 10 years of abstraction from the adjacent area where the sandstone is confined by thick clay, there has been virtually no movement under the clay of water contaminated by nitrate in the recharge area of the outcrop. Because of the very large storage in the sandstone, the gentle regional gradient from west to east has not yet been reversed. The prospect for a continued supply of low-nitrate water from the confined aquifer will depend to some extent on the source of the water and, in the likely event that it is drawn from an unconfined part of the aquifer, on the effectiveness of the processes of oxygen removal and nitrate reduction. The prospects for bacteriological denitrification in the aquifer have also been considered (Parker *et al.* 1988).

The study in the Lincolnshire Limestone provided a better opportunity to investigate nitrate penetration into a confined aquifer. Since 1950, a number of major groundwater sources have been developed in the confined Lincolnshire Limestone in south Lincolnshire. The limestone is a relatively thin microporous aquifer in which flow is predominantly through fissures. The rapid movement of water through the limestone has meant that agricultural pollution has already had a significant polluting effect in the unconfined part of the aquifer and yet groundwater sources down-dip in the confined aquifer still abstract water with a very low nitrate content. The absence of nitrate from these sources, which are 10–15 km from the outcrop, either indicates retardation by diffusion in the limestone matrix and/or bacteriological denitrification.

The sequence of changes in groundwater quality along the groundwater flowpath from unconfined to confined conditions has been traced by sampling in existing open boreholes (Edmunds and Walton 1982) and detailed sampling in several cored investigation boreholes. The results, illustrated in Fig. 10.7, reveal that nitrate and dissolved oxygen diminish to negligible concentrations some 10 km from the edge of the aquifer outcrop, despite the penetration of thermonuclear tritium even further down-dip. The limestone varies in colour from grey to buff. This is attributed to oxygenated water flowing along fissures and penetrating the pores of the matrix, oxidizing finely disseminated pyrite in the limestone, and so causing the colour change. The proportion of buff limestone decreases down-dip.

Pore-water samples from a borehole some 4 km from the outcrop contain significant nitrate which demonstrates that diffusion from the fissure water to the limestone matrix does occur. Much lower nitrate concentrations were found in pore-water at greater distances from the outcrop area. However, significant tritium concentrations were found in the pore-water from boreholes with low or zero nitrate

concentrations. This is an important clue to the fate of nitrate once it has penetrated the confined limestone aquifer. The tritium concentrations indicate that the water which has diffused into the rock is of modern (post-1963) origin and should theoretically have contained nitrate also. The absence of this associated nitrate implies that denitrification is occurring. The hydrochemical evidence suggests that the process only occurs beyond the point where the dissolved oxygen content has been reduced to zero. Microbiological investigations confirmed the presence of potentially denitrifying bacteria concentrated on the fissure walls of the limestone.

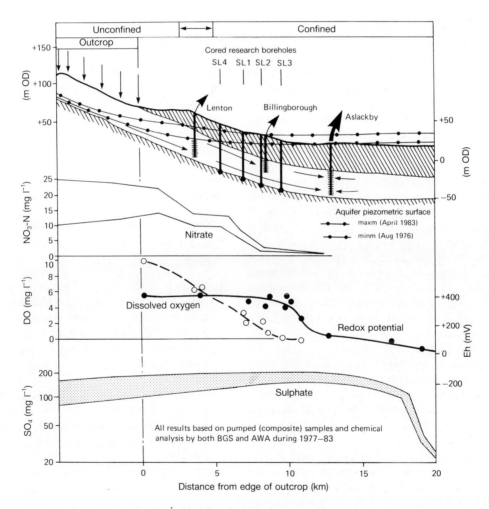

Fig. 10.7. Down-dip hydrogeological and hydrochemical section through the Lincolnshire Limestone of south Lincolnshire.

Mathematical models

The results of research on both the unsaturated and saturated zones of aquifers have formed the basis for a mathematical model which can be used to determine nitrate concentrations in groundwater from a knowledge of farming activities (Oakes 1982). The model consists essentially of two parts. In the first, the volumetric flow of water from the soil into the underlying rocks and thence into the water supply boreholes is calculated. In the second part, the quantity of nitrate that is leached from the soil and carried with the water into the aquifer and to the public supply boreholes is estimated.

The model employs a discrete, fully mixed cell formulation. The groundwater catchment is divided by a grid into cells, typically 250 m square and nitrate inputs to each cell from the overlying soil are estimated from land-use data with an appropriate lag to allow for transit through the unsaturated zone. Losses of nitrate through leaching are calculated from a simple set of rules which identify different types of crops with the average annual loss expressed as a percentage equivalent of the nitrogen applied (DoE 1986). The losses for any crop are spread over a three year period to simulate the mineralization of soil organic matter. The leaching rules are based on leaching losses determined from field measurements in the numerous boreholes drilled into the unsaturated zone of British aquifers. The validity of this method of determining nitrate losses has recently been the subject of criticism, principally by soil scientists (Addiscot and Powlson 1989), but the method has the advantage of simplicity which allows the model to be applied widely without requiring detailed field measurements and investigations. Confidence in the model predictions can only be improved if the database on leaching losses is itself improved by further research into the relationship between crop and fertilizer applications and nitrate leaching. One potential problem is the lack of adequate data about the effect on leaching losses of reductions in the use of fertilizers.

The model has been used extensively to predict future nitrate trends in groundwater sources and to assess the impact on those trends of changes in land-use. It is possible to examine various agricultural scenarios with the model, and hence to compare the economic implications of resolving the nitrate problem in a specific catchment by water treatment, by changes in land-use or some combination of the two. The Hatton Study (Severn Trent Water 1988) is a recent example of the use of the model in this way. The model is also to be employed in the assessment of aquifer protection measures in the recently designated nitrate sensitive areas (DoE 1988). It is evident from the wide range of catchments to which the model has been applied, that non-agricultural factors such as geology, depth to water-table, and effective rainfall have an important influence on the current severity of groundwater pollution by nitrate, the long-term outlook and the best means of constraining nitrate concentrations to acceptable levels.

Pesticides in groundwater

During the 1970s, the advances in the understanding of the processes responsible for the widespread increase in nitrate concentrations in groundwater led naturally to a consideration of the risk to groundwater quality of other diffuse-source agrochemicals (Royal Society 1981). If nitrate is leaching to groundwater, then it is likely that, with the intensification of agricultural use of pesticides, some of the more mobile compounds could also be leached.

Study of pesticides in groundwater does, however, present substantial technical problems (Lawrence and Foster 1987):

1. There is a wide range of pesticides in common agricultural use, many of which have equally toxic breakdown products. Analytical screening for all of them would be technically difficult and prohibitively expensive, and choices have to be made as to which compounds should be studied.

2. Many compounds are toxic at very low concentrations. Maximum allowable concentrations (the EC directive sets 0.1 μg l^{-1} for any compound and 0.5 μg l^{-1} for total pesticides) in drinking water are close to detection limits, and very sophisticated analytical procedures are required.

3. Relatively large volumes of water are required for analysis at such low concentrations, and

great care is required in sampling to avoid cross contamination and/or sample loss.

These considerable problems of devising reliable sampling techniques and analytical methods delayed confirmation of the presence of pesticides in British groundwater until the early 1980s. Between 1982 and 1984, surveys of pesticides in surface water and groundwater were carried out in the predominantly arable farming areas of East Anglia (Croll 1986). As had been anticipated, residues of the widely used phenoxyalkanoic group of agricultural pesticides were detected widely in surface waters in the region, with mecoprop, MCPA, and 2, 4-D being the most common. In addition, the triazine group of pesticides, especially atrazine, was detected in many surface waters, although this group was known to have limited agricultural usage. The survey reported for the first time that these same groups were present in groundwater, although with a lower frequency of occurrence. Concentrations were in the range 0.2–0.5 μg l^{-1}, and Croll (1986) reported similar findings from the North West, Severn Trent, and Thames regions.

To determine the transport and fate of selected pesticides in surface water and groundwater, a study was initiated by the Water Research Centre in 1986 in the upper catchment of the River Granta to the south-east of Cambridge. The catchment is underlain by chalk, with limited areas of boulder clay capping the high land and an alluvial tract following the course of the river.

The study has included sampling and analysis of surface water, groundwater, and rainfall, both on a regular basis and as special surveys, for example, during flood events. A survey of pesticide usage in the catchment during the early part of the study showed that the principal compounds employed in local agriculture were mecoprop, isoproturon, triallate, and chlortoluron. Atrazine and simazine represented only a small proportion of total agricultural pesticide applications.

The results of the survey so far provide general confirmation of Croll's (1986) findings, but with a wider range of compounds being found in surface water than in groundwater. A clear relationship has been found between both concentrations and numbers of compounds detected and river discharge, with maximum concentrations coming at the times of highest flow. Isoproturon, simazine, propyzamide, and chlortoluron correlated directly with stream flow, whereas atrazine and mecoprop consistently peaked at the times of heavy rainfall that generated the high river flows. The difference in behaviour is attributed to local variations in usage. The group which peak before the maximum flow are thought to be derived from rapid runoff from areas close to the river, whereas the second group are transported less rapidly, perhaps via field drains from more distant agricultural land. Comparison of concentrations measured in unfiltered and filtered samples suggests that a high proportion of the pesticide residues were in a dissolved state and not sorbed to solid particles. Maximum concentrations (0.44 μg l^{-1} isoproturon) were within the range reported by Croll (1986), and as the river became dominated by baseflow during the recession period, the number of compounds detected fell and concentrations were consistently below 0.1 μg l^{-1}.

In contrast, the most commonly recorded compounds in groundwater in the study were atrazine and simazine, although both isoproturon and chlortoluron were detected in samples from shallow observation wells. The latter compounds have not been detected at the larger public supply boreholes which penetrate deep into the Chalk. In general, triazine concentrations in the shallow boreholes were higher than those in the mixed samples from the deeper boreholes, suggesting a degree of vertical stratification. Overall, the concentrations conformed well to the results previously reported by Croll (1986), and were consistent with the concentrations found in surface water in the summer months.

The limited survey of rainwater showed that a number of species may be detected, at times consistent with their agricultural application, but identification of specific sources has not yet been possible and their transport potential and persistence in rain cannot be assessed. Rainfall concentrations were generally less than 0.1 μg l^{-1} and often below 0.05 μg l^{-1}.

Because of the many difficulties and high costs involved in field investigations of pesticides in groundwater, an essential prerequesite is to identify the most likely source of pesticide contamination and the most probable mechanisms of transport from the land surface to groundwater.

This information is needed to design investigation programmes, especially the sampling procedures and methods of monitoring.

A preliminary assessment of the transport and behaviour of pesticides (Lawrence and Foster 1987; Foster *et al.*, 1991), highlights some of the difficulties. Based on a knowledge of groundwater flow and pollutant transport in British aquifers and on the physicochemical properties of the pesticide compounds themselves, factors affecting leaching from the soil and transport through the unsaturated zone are discussed. Whilst a considerable body of data exists for adsorption and degradation of pesticides in 'standard', fertile organic clayey soils, these are not representative of the more permeable soils which are widely developed on aquifer outcrops. Even less is known about the mechanisms of pesticide retardation in aquifer materials. The hydraulic characteristics of the major British aquifers make the development of preferential flow in the unsaturated zone highly likely, and difficult to quantify. Preferential flow would be greatly favoured where surface drainage to the ground is by soakaways, which is likely to be the case in many situations where pesticides are used for non-agricultural purposes. Thus investigation of pesticides in groundwater is likely to be considerably more difficult than the studies of nitrate. The complex issues of transport and behaviour will be central to research on pesticides in groundwater which will be carried out over the next few years.

Acknowledgements

This chapter is based principally on studies by the British Geological Survey (BGS) and Water Research Centre (WRC), funded largely by the Department of the Environment. The assistance of many other organizations is gratefully acknowledged. The chapter is published by permission of the Directors of the BGS and the WRC. The authors wish to thank their many colleagues who have participated in this research during the past ten years.

References

Addiscott, T. and Powlson, D. (1989). Laying the ground rules for nitrate. *New Scientist*, 29 April, 28–9.

Booker, I. R. (1984). *Determination of the origin of nitrogen compounds in confined aquifers*. Final Report to the CEC: Contract Reference ENV-586-UK (H). Commission of the European Communities, Brussels.

Croll, B. T. (1986). The effects of the agricultural land-use of herbicides on fresh waters. In *Effects of land-use on fresh waters* (ed. J. F. de L. G. Solbe), pp. 201–9. Ellis Horwood, Chichester.

Department of the Environment. (1986). *Nitrate in Water*. Pollution Paper No. 26. HMSO, London.

Department of the Environment. (1988). *The Nitrate Issue*. HMSO, London.

Edmunds, W. M. and Walton, N. R. G. (1982). The Lincolnshire Limestone—hydrochemical evolution over a ten-year period. *J. Hydrol.*, **61**, 201–11.

Foster, S. S. D. and Smith-Carington, A. K. (1980). The interpretation of tritium in the Chalk unsaturated zone. *J. Hydrol.*, **46**, 343–64.

Foster, S. S. D. and Young, C. P. (1981). Groundwater contamination due to agricultural land-use practices in the United Kingdom. In *A survey of British hydrogeology 1980*, pp. 47–60. Royal Society, London.

Foster, S. S. D. and Bath, A. H. (1982). The distribution of agricultural soil leachates in the unsaturated zone of the British Chalk. In *Impact of agricultural activity on groundwater*. Memoirs XVI, **2**, pp. 109–22, Prague. Int. Assoc. Hydrogeologists.

Foster, S. S. D., Cripps, A. C., and Smith-Carington, A. K. (1982). Nitrate leaching to groundwater. *Phil. Trans. R. Soc. London*. **296**, 477–89.

Foster, S. S. D., Bridge, L. R., Geake, A. K., Lawrence, A. R., and Parker, J. M. (1986). *The groundwater nitrate problem*. Hydrogeology Report No 86/2. British Geological Survey, Keyworth.

Foster, S. S. D., Chilton, P. J., and Stuart, M. E. (1991). Mechanisms of groundwater pollution by pesticides. *J. Instn Water Env. Mgmt.*, **5**, 186–93.

Geake, A. K. and Foster, S. S. D. (1989). Sequential isotope and solute profiling in the unsaturated zone of the British Chalk. *Hydrol. Sci. J.*, **34**, 79–95.

Harris, R. C. (1985). *An investigation into differences in nitrate concentrations of groundwater abstractions in the Smestow Valley, Staffordshire*. Severn Trent Water Authority Report No RP 85-069. Severn Trent Water, Birmingham.

House of Lords. (1989). Select Committee on the

European Communities, Nitrate in Water, Session 1988–89, 16th Report, HL Paper 73. HMSO, London.

Landreau, A. and Morisot, A. (1983). *Évaluation de la vulnérabilité des aquifères libres aux nitrates d'origine agricole*. Report No. BRGM 83 SGN 026 EAU. BRGM, Orléans, France.

Lawrence, A. R. and Foster, S. S. D. (1985). The security of low-nitrate groundwater resources in the confined Lincolnshire Limestone aquifer. *J. Instn Water Engrs. Sci.*, **40**, 159–73.

Lawrence, A. R. and Foster, S. S. D. (1987). *The pollution threat from agricultural pesticides and industrial solvents: a comparative review in relation to British aquifers*. Hydrogeology Report No 87/2. British Geological Survey, Keyworth.

Oakes, D. B. (1982). Nitrate pollution of groundwater resources—mechanisms and modelling. In *Non-point nitrate pollution of municipal water supply sources: issues of analysis and control* (K. H. Zwirnmann). International Institute for Applied Systems Analysis. Collaborative Proceedings Series CP-82-S4, pp. 207–30. Laxenburg, Austria.

Parker, J. M. and Chilton, P. J. (1989). *Nitrate leaching to groundwater from grassland on permeable soils*. Unpublished Report WD/89/40C. British Geological Survey, Wallingford.

Parker, J. M. and Chilton, P. J. (1991). *Pore-water nitrate profiles in the Chalk unsaturated zone—results of 1990 re-drilling*. Unpublished Report WD/91/13C. British Geological Survey, Wallingford.

Parker, J. M. and Foster, S. S. D. (1986). Groundwater monitoring for early warning of diffuse pollution. In *Monitoring to detect changes in water quality*, pp. 37–46. IASH Publication No. 157.

Parker, J. M. and James, R. C. (1985). Autochthonous bacteria in the Chalk and their influence on groundwater quality in East Anglia. In *Microbial aspects of water management*. J. App. Bact. Symp. Series, pp. 15S–25S.

Parker, J. M. and Perkins, M. A. (1988). *Investigations of nitrate leaching in a Chalk catchment in Wiltshire*. Unpublished Report WD/88/12C. British Geological Survey, Wallingford.

Parker, J. M., Perkins, M. A., and Foster, S. S. D. (1983). Groundwater quality stratification—its relevance to sampling strategy. In *UNESCO TNO Symposium MIIGS*, pp. 43–54. Noordwijkerhout: UNESCO TNO.

Parker, J. M., Foster, S. S. D., Sherratt, R., and Aldrick, J. (1985). *Diffuse pollution and ground-water quality in the Triassic Sandstone aquifer in southern Yorkshire*. Report. **17**, (5). British Geological Survey, Keyworth.

Parker, J. M., Booth, S. J., and Foster, S. S. D. (1987). Penetration of nitrate from agricultural soils into the groundwater of the Norfolk Chalk. *Proc. Instn Civ. Engrs*, Part 2, **83**, 15–31.

Parker, J. M., Mason, J., and Kelly, D. P. (1988). *A study of denitrification in the Triassic Sandstone aquifer in Yorkshire, England*. Unpublished Report WD/88/23. British Geological Survey, Wallingford.

Ryden, J. C., Ball, P. R., and Garwood, E. A. (1984). Nitrate leaching from grassland. *Nature*, **311**, 50–3.

Royal Society (1981). *A survey of British hydrogeology 1980*. Royal Society, London.

Severn Trent Water (1988). *The Hatton catchment nitrate study*. Severn Trent Water, Birmingham.

Smith, D. B., Wearn, P. L., Richards, H. J., and Rowe, P. C. (1970). Water movement in the unsaturated zone of high and low permeability strata by measuring natural tritium. *Proc. Symp. Isotope Hydrology*, pp. 73–87. IAEA, Vienna.

Smith-Carington, A. K., Bridge, L. R., and Robertson, A. S. (1983). *The nitrate pollution problem in groundwater supplies from the Juriassic Limestones in central Lincolnshire*. Report No. 83/3. Institute of Geological Sciences, London.

Southern Water (1985). *Report on Thanet nitrate investigations. A study of the occurrence and cause of high concentrations of nitrate in groundwater on the Isle of Thanet and their future trends*. Southern Water Authority, Worthing.

Young, C. P. (1981). The distribution and movement of solutes derived from agricultural land in the principal aquifers of the United Kingdom with particular reference to nitrate. *Water Sci. Tech.*, **13**, 1137–52.

Young, C. P. (1986). Nitrate in groundwater and the effects of ploughing on release of nitrate. In *Effects of land-use on fresh waters*, (ed. J. F. de L. G. Solbe). Ellis Horwood, Chichester.

Young, C. P. and Gray, E. M. (1978). *Nitrate in groundwater—the distribution of nitrate in the Chalk and Triassic sandstone aquifers*. Technical Report No. TR 69. Water Research Centre, Medmenham, Bucks.

Young, C. P., Oakes, D. B., and Wilkinson, W. B. (1976). Prediction of future nitrate concentrations in groundwater. *Ground Water*, **14**, 426–38.

11. Microbiology of aquifers[1]

L. Clark, N. C. Blakey, S. S. D. Foster, and J. M. West

Introduction

At the beginning of the 1980s knowledge of natural microbial activity at depth in aquifers was 'roughly equivalent to our state of knowledge of the microbiology of the planet Mars' (Wilson *et al.* 1983). Only very limited investigations had been undertaken and their results were equivocal.

In the absence of pollution, the concentration of dissolved organic carbon in groundwater is typically only 1–2 mg l^{-1} and microbiologists reasoned that this was probably too low to support significant life. Moreover, early studies of the vertical distribution of micro-organisms indicated that their numbers dropped off very sharply at the base of the soil. The efficiency of sand filters, used for water treatment to remove micro-organisms from waters, also suggested that they could not enter or move through porous aquifers. There was, however, definite evidence of sulphate reduction in deep confined groundwater systems (from the presence of hydrogen sulphide in wells) and this process can take place through bacterial metabolism requiring the presence of some form of readily-assimilable carbon substrate (Erlich 1981; Matthess 1982).

Much original research was initiated in Britain in the 1980s on indigenous subsurface bacteriological activity. This research was stimulated by

1 the possibility of biodegradation of organic compounds that entered aquifers as a result of accidental spillage or casual discharge;

2 the need to understand more about the process of denitrification in groundwater as a result of the pressing requirement to evaluate the security of water supplies that have low concentrations of nitrate in confined aquifers;

[1] With glossary at chapter end.

3 the significance of bacteriological activity in relation to the possible migration of radionuclides from nuclear waste respositories.

The two main fields of work were, therefore, in situations where no artificial organic loading from surface pollution was occurring (2 and 3 above), and those where organic loading may have been significant (1 above).

At present there is a great deal of research into the microbiology of processes at shallow depths such as those associated with agricultural activities, waste disposal practices, and the reclamation of derelict land. Work is also being undertaken to study the influence of microbiological processes on borehole degradation through clogging by iron-based precipitates. These subjects are not considered in this chapter because, although the processes involved may have similarities with processes in deep aquifers, the two situations are distinct.

Fields of study

Disposal of organic wastes

As part of a nationwide investigation into the behaviour of hazardous wastes in landfills (Chapter 8), a comparative study of the influence of industrial wastes and domestic wastes on groundwater quality was carried out in the Chalk aquifer. One of the principal aims of the study was to determine the role of bacteria in the attenuation of the organic fraction of the leachate from both types of waste, and to ascertain the microbial population throughout the depth profile of uncontaminated chalk as well as in adjacent, polluted strata. Improvements in techniques adopted for both sample handling and enumeration of micro-organisms were established at an early stage of these investigations and then adapted and modified for use in studies of the Triassic Sherwood Sandstone aquifer in later stages of the programme. The field studies of aquifer microbiology have been supported by biodegradation tests in the laboratory, where the ability of indigenous microbes to degrade organic compounds, generally found in landfill leachates, has been investigated by activity tests using aquifer material taken from both the Chalk and the Sherwood Sandstone.

Diffuse nitrate pollution

The mechanisms of transport and attenuation of nitrate, derived from agricultural activities, as groundwater moves from the surface through the soil, the unsaturated zone, and finally, the saturated aquifer, have been the subject of intensive research for the past two decades (Chapter 10). The role of microbiological processes in the attenuation of nitrate in groundwater has been the focus of much work in the 1980s. This concentrated on the processes in the unsaturated zone and in the vicinity of the 'nitrate front' in the saturated aquifer. The 'nitrate front' is the zone in a confined aquifer where the nitrate concentration drops markedly to almost zero (Foster *et al.* 1985).

Radionuclide transport

Sulphur and iron bacteria are both believed to be important in the uranium biogeochemical cycle and have been studied because of their potential role in the uptake and transport of radionuclides.

Methodology

The methods of sampling in the field and enumeration of bacteria in the laboratory have much in common over the three distinct fields of study. The methods developed in the studies on organic wastes are discussed in detail while those developed in the nitrate and radionuclide investigations are discussed only where they differ significantly.

Organic waste disposal studies

Sampling

Borehole construction at the principal sites chosen for these studies has been described in detail by Towler *et al.* (1985), and Blakey and Towler (1988), for the Chalk and Sherwood Sandstone sites respectively. In summary, undisturbed samples from landfills and rock formations were collected in U100 barrels or core barrel liners. Surface contamination and cross-contamination between samples from different zones were prevented as far as possible, by steam cleaning and flame-sterilizing/ autoclaving the U100 core barrels, liners, and end caps before sealing in sterile polythene bags. Once recovered, the core material (still inside the core barrel or liner) was sealed in sterile polythene and stored at 4°C prior to dispatch to the laboratory for analysis.

Following removal of the end caps, each U100 core barrel was placed in a manual extruding device and about 10 cm of material removed with a sterile knife. A sterile stainless steel tubular corer was then pushed into the centre of the exposed face, and the main core further extruded so that waste material could be removed from around this small core. This small sample was then extruded from the corer into a pre-weighed sterile polythene bag for viable bacteria counts, and a sample for comparison was placed in a sterile pot for total bacteria counting with a microscope. Material from each end of these small cores was discarded during extrusion to avoid contamination.

Enumeration

The small core sample required for counting viable bacteria was weighed in the polythene bag and a volume of sterile diluent (0.1 per cent peptone) was added to give a 1:10 suspension by volume. The sample was then blended in a Colworth Stomacher for 30 seconds, by which time it could be transferred by pipette.

Aseptic decimal dilutions of this preparation were made and aliquots of 0.1 ml spread onto yeast extract agar (Department of the Environment *et al.* 1983). This growth medium consistently gave counts similar to, or higher than, other media tested including diluted soil extract agar (Wilson *et al.* 1983), and an artificial soil medium (Skinner *et al.* 1952), and was, therefore, adopted. Plates were incubated at 22°C and counted after 7 days.

Viable counting usually gives an under-estimate of the true population of the organisms present. In particular, sensitive bacteria (strict anaerobes for example) may be lost owing to inappropriate sample handling. Additional information can often be gained by using a *direct* microscopic counting method, although the viability of the microbes observed remains in doubt.

A simple staining technique, originally developed for the bacteriological examination of soils by Jones and Mollison (1948), was adapted for both the Chalk and Sherwood Sandstone studies. Aniline blue in acetic acid with aqueous phenol solution was used as the stain since it produced a clear microscopic contrast between the stained cells and the background. Acetic acid dissolved the bulk of the chalk matrix, clarifying the microscopic field of view. For sandstone samples, the bacteria had to be separated from the bulk of the solid matrix by agitation. Samples from both formations were prepared for examination following the methodology summarized in Table 11.1. Accuracy of the methodology was assessed by inoculating chalk with bacterial cell suspensions. A linear relationship between the average bacterial count and the volume of added cell suspension was identified, showing that a quantitative determination of bac-

Table 11.1   Outline methodology for the total microscopic counting of microbes in the Chalk and Sherwood Sandstone aquifers (after Jones and Mollison 1948)

Step	Chalk	Sherwood Sandstone
1	Random, aseptic sampling (1–10 g from recovered core material).	
2	—	• Grind sample in a pestle and mortar with filter-sterilized phenol/glacial acetic acid solution (PGAA)*
3	Place sample in a pre-weighed sterile plastic pot and reweigh. Mix with filter-sterilized distilled water and PGAA solutions for 30 minutes—	
	1 g chalk: 10 ml distilled water and 9.5 ml PGAA	• Make up to 22.5 ml
4	Allow large particles to settle (approximately 2 minutes)	
5	Take 1 ml of acid-insoluble residue in PGAA solution and place in small vial with 25 μl of 1% aqueous aniline blue solution and stain for 1 hour.	• Take 5 ml of insoluble residue in PGAA solution and place in small vial with 25 μl of 1% aqueous aniline blue solution and stain for 1 hour
6	After thorough mixing (Whirlymix) take a 200 μl subsample, add to 10 ml of filter-sterilized distilled water and pass through a Swinnex filter containing a 0.22 μg Millipore membrane supported on a glass-fibre filter pad.	
7	Repeat steps 1–6 in triplicate	
8	Air dry each membrane on a glass slide, and impregnate the membrane with a few drops of immersion oil rendering the membrane transparent. Cover with a cover slip.	
9	Count at $\times$ 1000 magnification	

* PGAA solution – 15 ml 5% wt/vol aqueous phenol solution to 4 ml glacial acetic acid.

teria in both Chalk and Sherwood Sandstone samples was possible.

Intensive sampling

From the outset of the studies described above, it was suspected that the distribution of bacteria in core material would not be uniform, with higher numbers on fissure surfaces than in the matrix of the aquifer. An experiment was performed to assess the variation between small chalk samples taken close together from the same core and to determine how representative were single samples taken from similar material. Sampling was carried out as described above, with the exception that five small mini-cores were taken in a radial pattern around the centre of the exposed core instead of one, and at four further locations down a 25–30 cm length of core.

Table 11.2 Viable bacterial counts obtained from intensive sampling of core from the Chalk in Suffolk (Depth 10.0–10.5m)

Sample	Geometric mean (CFU g^{-1})	Minimum count (CFU g^{-1})	Maximum count (CFU g^{-1})
1	1.5×10^3	3.4×10^2	5.0×10^3
2	9.6×10^2	6.0×10^1	5.8×10^3
3	1.3×10^4	1.5×10^3	7.2×10^4
4	4.5×10^5	1.8×10^5	1.4×10^6

The main counts at each of the four positions down the core are shown in Table 11.2, together with the respective maximum and minimum counts. The samples were examined using the viable count procedure and showed considerable variation. This variability in bacterial counts over relatively small distances indicates that reliance on single samples as being representative of a core as a whole should be treated with caution.

Effects of prolonged storage

Three separate cores were examined to assess the survival of micro-organisms in chalk during storage between core recovery and analysis. A half-metre length of core was extruded into polythene tubing, sealed, and stored at 4°C. Each core was examined in three sections denoted as top, middle, and bottom. Periodically, small cores were taken from each section using the techniques previously described, and examined using the viable counting technique.

The three areas of the three cores gave nine sampling positions at which the bacterial population was evaluated over several weeks.

Samples from five positions initially gave very low counts (about 20 colony forming units (CFU) per gram) but after a week these were seen to rise dramatically, with typically around 10^5–10^7 CFU per gram present after 14–18 days. After this time a slow decline in counts was observed. The other four sample positions had high counts at the start of storage, which rose slightly over a period of 1–2 weeks, then declined slowly.

The tests provided results which matched to some degree the way the cores were handled, between recovery from the borehole and the start of the storage trials. One core was examined within a day of recovery and the initial counts were very low in all three areas sampled. Another core had been stored in its aluminium U100 liner for 28 days at 4°C before examination, and low initial counts from two areas were observed along with high counts from a third. The final core had been stored in its U100 liner for 40 days prior to sampling and showed high counts from all three areas.

Although not conclusive, these tests provide circumstantial evidence that slow changes in bacterial population occur during storage and that these changes can be accelerated on exposure of the cores to air. Cores stored in aluminium liners tend to have low counts initially but as, or if, oxygen diffuses into the core, progressive bacterial activity can be stimulated giving higher populations with time. This problem can be exacerbated further by extruding the sample on recovery into polythene bags prior to storage. The relatively high viable counts in samples from boreholes A01C and B02C (Fig. 11.1) compared with the other boreholes, is attributed to enhanced growth after storage in polythene tubes compared to aluminium U100 liner tubes.

Therefore, microbial populations under storage at 4°C are dynamic and not static. The degree and type of change appears to be dependent on storage conditions, particularly on the accessibility of atmospheric oxygen to the sample. Whether the observed increases were due to multiplication of

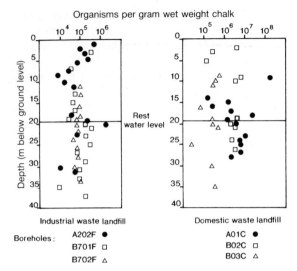

Fig. 11.1. Microbial population of the Chalk's unsaturated/saturated zones, in boreholes at landfill sites in Suffolk, determined as the total microscopic count.

bacteria or gradual resuscitation of damaged or dormant organisms remains to be investigated.

Laboratory activity tests

Microbial activity in the Chalk and Sherwood Sandstone cores was studied under aerobic conditions in the laboratory. Volatile organic acid solutions were added to samples from both formations held in autoclaved 2 litre culture flasks, and the subsequent removal of the acids from the solution was measured by a gas chromatographic technique.

The organic acid solutions were prepared using sterilized native groundwater from supply boreholes close to each site. Ethanoic, propanoic, butanoic, and pentanoic acids were added so that the total volatile acid (TVA) concentration of these stock solutions ranged from 250–10 000 mg C l^{-1}. Different proportions of individual acids were chosen to simulate the composition of landfill leachate expected at the TVA concentrations adopted (Blakey and Towler 1988).

The Chalk samples were chosen from a relatively uncontaminated borehole at an industrial waste disposal site in Suffolk. The Sherwood Sandstone samples were selected from the borehole profile

beneath a domestic waste landfill in Nottinghamshire.

TVA solution (500 ml) was added to approximately 250 g of chalk or crushed sandstone contained in 2 litre culture flasks. These were then placed on an orbital shaker. Each test was duplicated as a minimum requirement.

The flasks were incubated under constant temperature at 10°C. Prior to sampling with sterile disposable pipettes, the flasks were weighed in order to compensate for liquid volume loss by evaporation. Samples were prepared for analysis using methods described by Towler *et al.* (1985).

The results are summarized in Fig. 11.2 which shows that the volatile acids in landfill leachate are easily degradable by indigenous bacteria in both aquifers under aerobic conditions. The degradation in the chalk samples (Fig. 11.2(a)) begins immediately with the rate dependent on the concentration of TVA or the consequential pH. The rate of degradation is almost zero at a pH below 7.3 (TVA concentration 8210 mg C l^{-1}), the relatively high pH being maintained by buffering of the chalk ($CaCO_3$) aquifer material. In the sandstone samples (Fig. 11.2(b)), the degradation onset is delayed with the delay greater at high concentrations of TVA. The delay is believed to be due to the reduced buffering capacity in the sandstone compared to the chalk with a resultant delay in raising the pH of the solution above 7.3. This delay retarded the microbial population growth but, once microbial growth began at higher pH values, the activity, measured as the rate of TVA removal, was much higher than in chalk because of their previous acclimation to landfill leachate. This acclimation was because the sandstone samples were taken from directly beneath a landfill.

Diffuse nitrate pollution

Investigations have been undertaken near 'nitrate fronts' in parts of the Jurassic limestones in South Lincolnshire (Lawrence and Foster 1986), the Chalk of central Norfolk (Parker and James 1985) and the Triassic sandstones of South Yorkshire (Parker *et al.* 1988). In all these investigations samples were collected for bacteriological assay by boring into the centres of soil and rock cores recovered from investigation boreholes. The soil

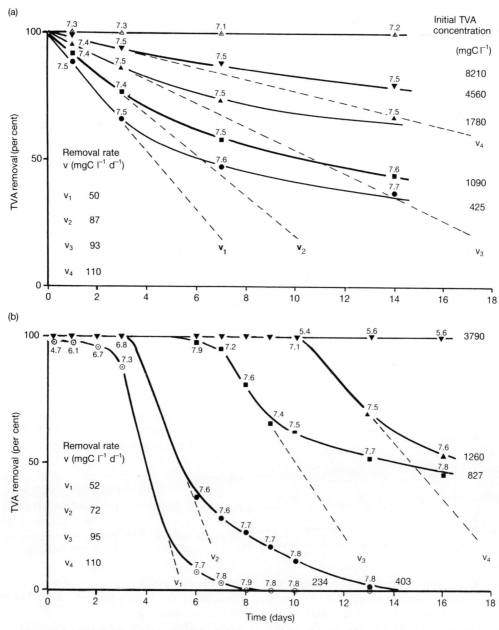

Fig. 11.2. Microbial activity measured as total volatile acids removal in (a) the Chalk and (b) Sherwood Sandstone underlying a landfill; incubated under aerobic conditions at 10°C.

Notes:
    i. TVA. Total Volatile Acids (mgC l⁻¹) in solution. Initial concentrations of test solutions shown on right of figure.
    ii. Dashed lines are tangents to data curves; they show initial rates of TVA removal from each test solution.
    iii. ⊙ 7.7 pH of solution at each data point.

Table 11.3 Characteristics of selected representative bacterial colonies showing denitrification potential. The colonies were recovered from a site on the Chalk in central Norfolk where the aquifer is semi-confined (after Foster *et al.* 1985).

	Colony number (Glacial sand, 16m)		Colony number (Chalk, 34m)
	7	9	2
Characteristics			
Nitrate broth gas	+	+	+
Pigment			
fluorescence	+	+	+
Gram stain	−	−	−
Oxidase	+	+	+
Growth at 4°C	+	+	+
Growth at 41°C	−	−	−
Gelatinase	+	+	+
Catalase	+	+	+
Arginine			
dehydrolase	+	+	+
Motility at 22°C	+	+	+
Citrate	+	+	+
Urease	−	−	−
PHB granules	−	−	−
Substrate utilization			
Glucose	+	+	+
Sucrose	+	+	+
Maltose	−	+	+
Ethanol	−	+	+
Sorbitol	−	+	+
Acetic acid	+	+	+

to standard tests for the determination of total aerobes and anaerobes. The presence of denitrifying bacteria was established by the enrichment culture technique. A range of different culture media was used, from weak organic acids and sugars to thiosulphates and relatively rich organic agars and broths. Incubations were carried out at temperatures ranging from 10–30°C for periods of up to twenty days under microaerophilic to totally anaerobic conditions, and the extent of growth on the culture and nitrogen gas production were observed.

In one instance, confirmed denitrifying bacteria were isolated and characterized by morphological examination, by biotyping through a battery of biochemical tests and by their utilization of a range of organic substrates during anaerobic incubation at 22°C in a minimal salts medium (Table 11.3).

Radionuclide transport

With respect to the studies of the role of bacteria in the transport of radionuclides, sulphate-reducing, and iron-oxidizing bacteria have been isolated from deep geological formations in mines, and studied by epifluorescence microscopy. Their potential for growth under optimum and extreme conditions has been studied in the laboratory using adenosine triphosphate assay (West *et al.* 1985*a*, 1989).

Assays of bacterial populations in aquifers

Organic waste disposal

Micro-organism populations were enumerated over the full depth of the investigated profiles in three boreholes from a domestic waste landfill site and three boreholes from an adjacent industrial waste landfill site on the Upper Chalk in Suffolk. The results of a direct counting technique, based on the phenol aniline blue (PAB) method (Table 11.1) for both sites are shown in Fig. 11.1. The prefix A in the borehole identification code shows that the borehole was drilled directly through the landfill while the prefix B shows that it was drilled adjacent to the landfill.

Bacteria were found distributed throughout the

or rock sample, with its retained pore-water, was subjected to some on-site testing, with follow-up work at the base laboratory. In the case of the Jurassic limestones, which are considerably harder than the other two formations, samples were obtained both by scraping the surface of identifiable fissures in a sterile environment and by crushing the limestone in a sterile bag, adding a dispersant, and extracting pore-water by spinning in sterilized centrifuge liners.

Sub-samples were brought into the form of homogeneous suspensions and normally subjected

Chalk in each borehole and to a depth of about 40 m in borehole B701F. The total counts for A01C and B02C were higher and probably reflect the different handling and sampling procedures, as well as the prolonged storage used for material from these boreholes when compared with subsequent determinations on the other four boreholes (Towler *et al.* 1985). The presence of leachate from the domestic waste landfill appears to result in an increased bacterial population (compare A01C with B02C), with the sample from the base of the landfill (9 m) having the highest total count of 1.9×10^8 organisms per gram (wet weight).

Likewise, similar work on the Sherwood Sandstone was carried out using an adaptation of the PAB direct counting method (Table 11.1) and the results are given in Fig. 11.3. Bacteria were identified to a total depth of 40 m but the population was not uniformly distributed. Over half the 30 samples examined had a population below the limit of detection, with the rest averaging 10^5 organisms per gram wet weight of sample. The presence of quantifiable populations coincided with profile zones where chemical evidence of leachate biodegradation had been identified; highest counts were found at the landfill/unsaturated zone interface, whereas little or no evidence of bacterial presence was identified over a zone of low pH coinciding with a landfill leachate plume. Bacteria only reappeared from 17 m downward where the pH returned to background (neutral to alkaline)

levels. Precise reasons for this are unknown but it may be due to the presence of different trophic types in this zone, which may have been difficult to identify using the direct counting technique.

Diffuse nitrate pollution

Assays

At the sites investigated, which were close to the 'nitrate fronts', bacteria capable of denitrification were found to be widely distributed in the subsurface to depths of from 40–80 m, with highly significant populations in certain depth intervals. However, the total populations (generally 50–500 bacteria per gram wet weight) are many orders of magnitude less than those present in agricultural soils. Typical profiles of the nitrate and tritium in pore-water, and of bacterial populations or activity in confined limestone and sandstone aquifers are shown in Fig. 11.4 and 11.5. The failure to detect bacteria from some depths in these profiles is believed to reflect the absence of fissuring in the rock core samples concerned. All samples taken from the heart of non-fissured Jurassic limestones (Fig. 11.4) were found to be sterile, while those scraped from fissure walls had significant populations. It is, therefore, suspected that indigenous bacteria are attached to the walls of micro and macrofissures in the form of biofilms (Lawrence and Foster 1986).

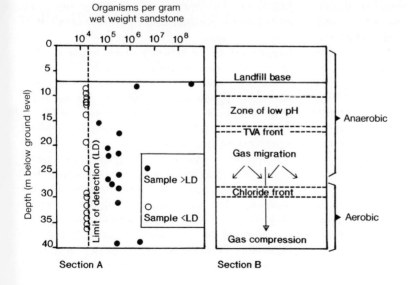

Fig. 11.3. Microbial population of the unsaturated zone of the Sherwood Sandstone determined as the total microscopic count for a landfill site in Nottinghamshire, Section B shows the geochemical profile of the unsaturated zone.

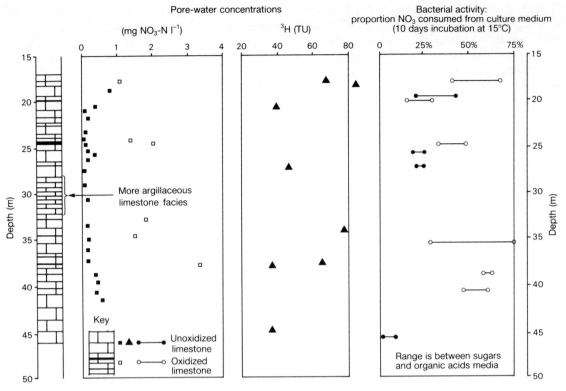

Fig. 11.4. Profiles of pore-water chemistry, tritium, and bacteriological activity in fissure-wall samples in the confined Lincolnshire Limestone (after Lawrence and Foster 1986).

At the Chalk site in Norfolk, the denitrifying bacteria that were confirmed formed a narrow and specialized group (that was motile and gram-negative) suggesting selective development during migration from the soil in groundwater. Among representative colonies, most showed characteristics typical of *Pseudomonas fluorescens* (Table 11.3), demonstrating the ability to grow on various organic substrates (weak sugars, alcohols, and acids) as a sole carbon source (Parker and James 1985). This group, like most other denitrifying bacteria, consists of facultative aerobes, which would only reduce nitrate when growing in the near-absence of oxygen.

Microbial activity

The fact that bacteria are shown to be present in geological samples recovered from depth in aquifers does not in itself constitute evidence of their activity. They could indeed be present in a dormant form.

The possibility of bacteriological denitrification in the unsaturated zone of the Chalk beneath arable land has been investigated at sites in Norfolk and Hampshire (Foster and Young 1980), but no consistent evidence of the presence of potential denitrifiers below depths of one metre or so was found, despite the presence of a possibly suitable organic substrate. In contrast, nitrifying bacteria, which are believed to be chemolithotrophic and not requiring an organic substrate, were much more widely distributed, although their activity is likely to be limited by the very restricted supply of oxidizable organic nitrogen.

Independent evidence that significant bacteriological denitrification is occurring comes from isotopic studies of the groundwater systems concerned. The pore-waters of both the upper part of the semiconfined Chalk in Norfolk and the more porous facies of the confined Lincolnshire Limestone (Fig. 11.4) contain thermonuclear tritium at concentrations (in the range 10–80 tritium units)

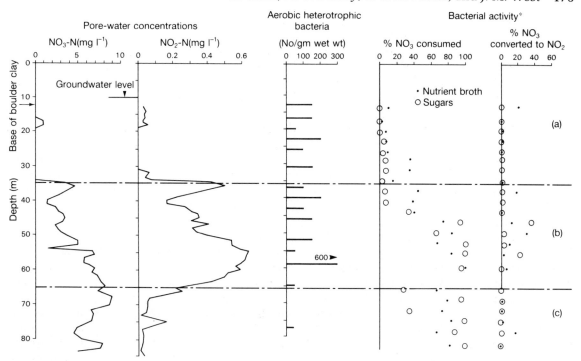

Fig. 11.5. Pore-water composition and bacterial activity in confined Triassic sandstones in South Yorkshire (after Parker *et al.* 1988). (a) Low biodenitrification potential in zone of recent groundwater with originally low-nitrate content; (b) evidence of biodenitrification (possibly incomplete resulting in nitrite accumulation), probably limited by insufficient organic carbon, in groundwater of recent origin and high nitrate content; (c) biodenitrification potential but little evidence of activity in somewhat older groundwater with high nitrate content.

which imply that a significant proportion of modern (probably post-1964) recharge has penetrated into these systems. Since the soils of the recharge areas of these aquifers have been used for arable farming since at least 1930, all infiltration during the subsequent period will have contained significant nitrate. The fact that thermonuclear tritium has been detected where nitrate levels are negligible, is strong evidence of active bacteriological denitrification. Moreover, in the case of the Lincolnshire Limestone, it is possible to undertake reasonably accurate hydrological and nitrate balances in some subcatchments. Such balances suggest that nitrate contamination of the confined aquifer would have been far more extensive than is observed in the field if no denitrification was occurring.

The amount of nitrate that can be reduced (to

nitrogen) will depend on the type and quantity of oxidizable organic substrate available, on the establishment of microaerophilic conditions and on the availability of other minor nutrients. Different biological oxidations would provide different amounts of reducing equivalents; for example, 1 g of oxidized carbohydate would be sufficient to reduce almost 0.4 g of nitrate nitrogen. To calculate *in situ* biodenitrification rates, it is necessary to know the total number of denitrifying bacteria per unit area of active surface, the type and concentration of organic substrate being utilized, and the metabolic rate of bacteria at the prevailing nutrient concentrations. These are virtually impossible to establish in practice, but the potential range of activity can be estimated from the results of enrichment cultures.

Almost all the bacteriological samples taken

from fissures in limestones exhibited a minimum denitrification capacity in excess of 150 mg NO_3 $-N$ m^{-2} d^{-1} based on a reduction of 10 mg NO_3 $-N$ l^{-1} over a 2 km flow distance traversed in 100 days, the aquifer having a volume–fissure wall area ratio of 0.015 m. This is sufficient to account for the observed field situation, which requires a reduction of 1.5 mg NO_3 $-N$ m^{-2} d^{-1} (Foster *et al.* 1985). Similar conditions apply at the site on the semi-confined Chalk in Norfolk.

The question arises as to the processes controlling oxygen consumption in the mobile groundwater. These processes appear to be related to the up-gradient oxidation of finely-disseminated pyrite (iron sulphide) in the rock formation (especially in the case of the Lincolnshire Limestone) and of organic matter of geological origin. Another key question is the nature and origin of the organic substrate being used for biodenitrification. Monitoring of water supplies pumped from the Lincolnshire Limestone and the Chalk shows that the mobile fissure water generally contains less than 2 mg DOC l^{-1}, and this, whatever its form, is unlikely to be sufficient to support the observed denitrification. However, preliminary work suggests that the pore-waters of both formations in the relatively unweathered state often contain 5–25 mg DOC l^{-1} and the rock matrix in excess of 2000 mg DOC kg^{-1} dry weight (Lawrence and Foster 1986), with 30–150 mg kg^{-1} dry weight of total carbohydrates in chalk (Whitelaw and Rees 1980).

Radionuclide transport

The movement of radionuclides in the subsurface can be considerably influenced by microbial activity. Many reviews are now available on the microbiology of radioactive waste disposal in geological formations and their role in radionuclide movement (eg Mayfield and Barker 1982; West *et al.* 1982, 1985*b*; Chapman and McKinley 1987). Radionuclide sorption experiments (West *et al.* 1987) with sulphate-reducing bacteria (isolated from a relevant geological formation), Fullers Earth (a potential backfill material) and ^{137}Cs (a common radionuclide in radioactive waste) showed that microbial presence decreased the amount of retardation of the radionuclide on the solid phase. However, the role of microbes is very complex and their action may enhance or retard migration

depending on the biogeochemical conditions (Fletcher 1985).

The role of microbes in radionuclide movement can also be studied in the natural environment by the use of natural analogues. The Poços de Caldas project is such a study and has sought to investigate a number of areas of concern with regard to the performance assessment of the disposal of radioactive waste. The Osamu Utsumi uranium mine, Poços de Caldas, Brazil is a roll front uranium deposit and, as part of a multi-disciplinary study (Smellie *et al.* 1989), an examination of the movement of natural series radionuclides across redox fronts was made. These processes at such fronts are analogous to those which could occur at radiolytic redox fronts found in the near-or far-field of deep repositories for high level waste or spent fuel.

Microbiological analyses of the redox front showed that microbial populations were found in the rock and groundwater at whatever depth sampled, and that these populations could be supported by chemolithotrophic organisms which derive their energy from redox front reactions. A simple model was used to assess the significance of these microbial populations in redox front processes. The model was based on the assumptions that any microbial population in a deep geological formation requires a supply of energy and nutrients (notably C, N, P, and S) for their metabolism; it can be assumed that hydrogen and oxygen are not metabolism-limiting in an aqueous environment. The average composition of a microbe is assumed to be in the ratio C:H:O:N:P:S = 160:280:80:30:2:1 (dry weight) (West *et al.* 1989). It is evident that in this environment, with a restricted nitrogen supply, nitrogen is the limiting nutrient preventing biomass production from exceeding about 0.1 g m^{-3} a^{-1} dry weight. The principal reaction providing energy from microbial conversion of nutrients into biomass in this environment is the oxidation of pyrite, which occurs at a redox front where inflowing groundwater saturated with oxygen reacts with the non-oxidized rock mass. Calculations based on the known kinetics of this reaction and various other assumptions show that enough energy is available for microbial growth solely from the oxidation of iron from the ferrous to the ferric state, without consideration of the oxidation of sulphide to sulphate. Although simplistic, and not accounting for nutrient recycling, the model calculations are compatible with field observations of the redox front, which is moving faster than one would expect on the basis of a simple dissolved oxygen inventory. Simultaneously, microbial activity appears to be involved in the mobilization of uranium and its reprecipitation at the redox front.

Other analogue studies are in progress around the world. Most of these include a microbiological component as the importance of geomicrobiological processes is now recognized.

Conclusions

Bacteria have been found in the saturated and unsaturated zones of the Chalk, Triassic sandstones, and Jurassic limestones to depths of over 40 m in the Chalk and limestones and over 80 m in the sandstones. Population distribution is uneven in all formations, influenced to some degree by handling, sampling, and prolonged storage of samples as well as lithological variations.

Organic-rich leachate from domestic waste appears to influence the bacterial population in the Chalk positively with increased numbers being counted directly below a landfill site. In the Triassic sandstones increases in population only correlate with horizons where optimum pH (neutral or basic) conditions are maintained.

Preliminary laboratory studies have indicated that the kinetics of microbial degradation of the organic constituents in landfill leachate are greater in an acclimated system—a system in which the microbial population has become accustomed to a stable physicochemical environment. The presence of bacteria capable of denitrification of groundwater has been proved. However, there is a lack of direct evidence of denitrification in aquifers but there is substantial circumstantial evidence that such processes are taking place. Bacterial activity is also believed to be important in the mobilization and precipitation processes involved in the transport of radionuclides in groundwater.

The research programmes discussed here were mainly undertaken in the first half of the 1980s. Research activity into the microbiology of deep aquifers is now dormant. This is unfortunate because it is extremely important to pursue actively the entire field of research into the microbiology of aquifers in order to understand fully the processes involved and their consequences. There is an emphasis in hydrogeological studies on the pollution and quality of groundwater and a fundamental understanding of aquifer microbiology is necessary to be able to forecast and model the effects of pollution, and to design and implement bio-remedial works.

GLOSSARY

Activity tests	Laboratory test to measure metabolic activity of micro-organisms.
Colworth Stomacher	Equipment to homogenize samples.
CFU	Colony forming units.
Enrichment culture technique	Method for increasing microbial population or certain elements in that population.
Trophic types	Forms of bacteria distinguished by their nutritional needs.
Chemolithotrophic bacteria	Bacteria which use inorganic compounds as energy sources and employ carbon dioxide as their principal carbon source.
Heterotrophic bacteria	Bacteria which obtain their energy from organic compounds (Autotrophic bacteria *can* obtain their energy from sources other than organic compounds).
Oligotrophs	Micro-organisms that can survive on very low levels of nutrients.
Gram negative or positive	Classification of bacteria based on their cell wall structure, tested using Gram's staining method.
Aerobic bacteria	Bacteria requiring oxygen for survival.
Anaerobic bacteria	Bacteria which can live in the absence of oxygen.
Facultative anaerobes	Aerobic organisms which can also grow under anaerobic conditions. (Obligate anaerobes can *only* grow under anaerobic conditions.)

References

Blakey, N. C. and Towler, P. A. (1988). The effect of unsaturated/saturated zone property upon the hydrogeological and microbiological processes involved in the migration and attenuation of landfill leachate components. *Water Sci. Tech.*, **20**, 119–28.

Chapman, N. A. and McKinley, I. G. (1987). *The geological disposal of nuclear waste*. Wiley, Chichester.

Department of the Environment, Department of Health and Social Security (1983). *The bacteriological examination of water supplies 1982*. Reports on Public Health and Medical Subjects No 71, Methods for the Examination of Waters and Associated Materials. HMSO, London.

Ehrlich, H. L. (1981). *Geomicrobiology*. Marcel Dekker, New York.

Fletcher, M. (1985). Effect of solid surfaces on the activity of attached bacteria. In *Bacterial adhesion* (ed. Savage, D. C. and Fletcher, M.), pp. 339–62. Plenum Press, New York.

Foster, S. S. D. and Young, C. P. (1980). Groundwater contamination due to agricultural land-use practices in the United Kingdom. *UNESCO-IHP Studies and Report in Hydrology*, **30**, 268–82.

Foster, S. S. D., Kelly, D. P., and James, R. C. (1985). The evidence for zones of biodenitrification in British aquifers. In *Planetary ecology: selected papers from 6th International Symposium on Environmental Biogeochemistry*, pp. 356–69. Santa Fe, USA, October 1983. Van Nostrand Reinhold, New York.

Jones, P. C. T. and Mollison, J. E. (1948). A technique for the quantitative estimation of soil microorganisms. *J. Gen. Microbiol*, **2**, 54–69.

Lawrence, A. R. and Foster, S. S. D. (1986). Denitrification in a limestone aquifer in relation to the security of low-nitrate groundwater supplies. *J. Instn. Water Engrs Sci.*, **40**, 159–72.

Matthess, G. (1982). *The properties of groundwater*. Wiley, New York.

Mayfield, C. I. and Barker, J. F. (1982). *An evaluation of the microbiological activities and possible consequences in a fuel waste disposal vault. A literature review*. Atomic Energy of Canada Ltd, TR-139.

Parker, J. M. and James, R. C. (1985). Autochthonous bacteria in the Chalk and their influence on groundwater quality in East Anglia. In *Microbial aspects of water management* J. App. Bact. Symp. Series. pp. 15S–25S.

Parker, J. M., Mason, J., and Kelly, D. P. (1988). *A study of denitrification in the Triassic Sandstone aquifer in Yorkshire, England*. Hydrogeology Report WD/88/23. British Geological Survey, Keyworth.

Skinner, F. A., Jones, P. C. T., and Mollison, J. E. (1952). A comparison of a direct and a plate counting technique for the quantitative estimation of soil micro-organisms. *J. Gen. Microbiol.*, **6**, 261–71.

Smellie, J., Chapman, N. A., McKinley, I. G., Penna Franca, E., and Shea, M. (1989). Testing safety assessment models using natural analogues in high natural-series groundwaters. The second year of the Poços de Caldas project. *Proc. Mat. Res. Soc. Symp.*, **127**, pp. 863–970.

Towler, P. A., Blakey, N. C., Irving, J. E., Clark, L., Maris, P. J., Baxter, K. M., and Macdonald, R. M. (1985). A study of the bacteria of the Chalk Aquifer and the effect of landfill contamination at a site in eastern England. In *Hydrogeology in the service of man*, **3**, 84–97. Memoirs of 18th Congress of Int. Assoc. Hydrogeol.

West, J. M., McKinley, I. G., and Chapman, N. A. (1982). Microbes in deep geological systems and their possible influence on radioactive waste disposal. *Radioactive waste management and the nuclear fuel cycle*, **3**, 1–15.

West, J. M., Christofi, N., and McKinley, I. G. (1985*a*). An overview of recent microbiological research relevant to the geological disposal of nuclear waste. *Radioactive waste management and the nuclear fuel cycle*, **6**, 79–95.

West, J. M., McKinley, I. G., Groga, H. A., and Arme, S. C., (1985*b*). Laboratory and modelling studies of microbial activity in the near field of HLW repository. *Proc. Mat. Res. Soc. Symp*, **50**, 533–8.

West, J. M., Haigh, D., Hooker, P. J., and Rowe, E. J. (1987). *Radionuclide sorption onto Fullers Earth (calcium montmorillonite)—the influence of sulphate reducing bacteria*. Report FLPU 87–7 British Geological Survey, Keyworth.

West, J. M., McKinley, I. G., and Vialta, A. (1989). The influence of microbial activity on the movement of uranium at Osamu—Utsumi mine, Poços de Caldas, Brazil. *Proc. Mat. Res. Soc. Symp.*, **127**, 771–7.

Whitelaw, K. and Rees, J. F. (1980). Nitrate-reducing and ammonium-oxidizing bacteria in the unsaturated zone of the Chalk aquifer of England. *Geomicrobiol. J.*, **2**, 179–87.

Wilson, J. T., McNabb, J. F., Balkwill, D. L., and Ghiorse, W. C. (1983). Enumeration and characterization of bacteria indigenous to a shallow water-table aquifer. *Ground Water*, **21**, 134–42.

12. Geological and hydrogeological aspects of the deep disposal of nuclear wastes in Britain

N. A. Chapman and T. J. McEwen

The concept of geological disposal

Radioactive wastes, although often less harmful than many other toxic industrial and domestic by-products, get the Rolls-Royce treatment when it comes to disposal. Research and engineering studies over the past twenty years have led to the development of sophisticated techniques for immobilizing, encapsulating, and handling radioactive wastes of all categories. In parallel, a whole new field of detailed safety assessment has grown up, mainly derived from approaches used in predictive performance assessment of nuclear reactors, and the associated field of radiological protection. No other type of waste has been subject to such intense characterization, such detailed analysis of its long-term behaviour, and such careful comparison of the health effects of various solutions for its disposal.

Despite many options being advanced for disposing of the relatively small volumes of solid radioactive waste arising from the world's nuclear power and weapons programmes, the most feasible solution still remains burial in the ground. Whether this is in near-surface trenches and bunkers, or in deep, mined repositories, the basic principle underlying long-term safety is the same. This has been termed the 'multi-barrier' concept, whereby reliance is placed on the containment capacity of a number of nested barriers which act, individually or in concert, to limit the rate at which radionuclides from the waste can be mobilized into groundwaters in the rocks surrounding the repository. These barriers comprise the solid matrix in which the waste materials are set, the containers

that are used for moving the waste, any overpacks that surround these, mineral or cementitious buffer materials used to fill up void space around buried containers, and the rock itself. The man-made components are termed 'engineered barriers', and lie in the 'near-field' of the disposal system, while the rock lies in the 'far-field' that connects the repository to the 'biosphere'. A more detailed description of the background to geological disposal is found in Chapman and McKinley (1987).

Over very long periods, internal processes in the waste (some contain unstable organic materials, others produce substantial radioactive decay heat), and attack by pore-waters (e.g. in cements) and, most significantly, groundwaters moving through the near-field, cause slow degradation of the wastes and the engineered barriers. Not only does the rate of groundwater flow, and the chemistry of the waters, affect mobilization rates and chemical speciation of radionuclides, it also controls the rate at which they are transported through the far-field to the biosphere. As a consequence, hydrogeology and hydrochemistry have found themselves at the centre of research and safety assessment modelling for radioactive waste disposal. Since one of the most obvious ways of ensuring adequate containment of wastes is to bury them in geological formations with very low throughflow of water, this research has led hydrogeology down the unusual path of developing techniques to examine non-aquifer formations with very low permeabilities. As a consequence, our understanding of very low-flow environments has increased considerably during the 1980s.

In this chapter we do not address the considerable effort that has gone into the development of hydrogeological understanding and techniques (for example Black *et al.* 1985; Chapman *et al.* 1987), but concentrate instead on the broader issues underlying deep disposal of radioactive wastes in Britain, for which hydrogeology provides the common backdrop. As will be seen, our main emphasis is on the issues that arise when we attempt to incorporate our concepts of groundwater flow and solute transport into very long-term predictive models.

The wastes and the respository

In March 1989 UK Nirex Ltd, the company responsible for disposing of Britain's radioactive wastes, announced that it intended to carry out geological investigations of the Sellafield and Dounreay nuclear sites to determine their suitability to host a deep repository for low (LLW) and intermediate level (ILW) wastes. At both sites the investigation would be concerned with a repository located in basement rocks underlying sediments, at depths of about 500–1000 m. Sellafield, renowned for its nuclear fuel reprocessing facilities, and Dounreay for the experimental fast-breeder reactor, are both coastal sites (Fig. 12.1) whose predicted deep geological structures are now being confirmed in detail by drilling. The nomination of these two sites took place after a two year site selection programme which included geological comparisons and predictive safety assessments based, in part, on geological information. This resulted in a short-list of preferred sites, from which Sellafield and Dounreay were selected for initial scrutiny. If both these sites prove unsuitable, other sites will be considered for investigation.

A programme of exploratory site investigations began in mid-1989, and will provide information for the selection of a single site for detailed, long-term characterization during the 1990s. This process will include the construction of an underground research laboratory, and the progressive exploration of the rock volume in conjunction with the excavation of the repository, which is planned to be operational by 2005. During this period, Nirex will undertake continuous iterations of its safety assessment programme, which is designed to analyse the long-term performance of a reposit-

Fig. 12.1. Location of the two sites scheduled for preliminary investigation for a deep radioactive waste repository.

ory, and consequently to focus both research and site investigation on issues that might be critical to the radiological safety of disposal. The principal regulatory authority in Britain is Her Majesty's Inspectorate of Pollution within the Department of the Environment, and its equivalent body in Scotland. When Nirex submits its proposals to the regulators they will be considered by HMIP and the other authorizing departments, using their own, independent, assessment programme. Although the methodologies differ somewhat, both Nirex and HMIP will make use of similar conceptual approaches for predicting future geological evolution of a repository, and both will have recourse to the same site-specific geological data for their models. These approaches have been developed via research programmes sponsored by both organizations, and by participation of the many UK scientists involved in international projects. Much of the conceptual basis for predicting the geological behaviour and long-term evolution of deep repositories is now held in common by all national authorities managing nuclear wastes, and relies on levels of understanding that are often at the forefront of our knowledge in the earth sciences. This chapter examines the geological background which has led up to the present situation, and looks at the main geological issues which arise now that firm proposals exist for disposal of Britain's nuclear wastes at specific locations.

In 1981 the UK Government decided that high-level waste (the vitrified by-product of reprocessing nuclear fuel) should be stored for about fifty years to allow much of the activity to decay prior to disposal. However, there was no call for storing the very much larger volumes of lower activity wastes. Until 1987, the disposal option was to have been shallow land burial for the lower level component, which represents the majority of these wastes, and deep disposal for the more active of the intermediate level wastes, but in May 1987 Nirex cancelled site investigations at four potential repository sites in England, and UK policy became that of eventual deep disposal of all nuclear industry commercial and research wastes.

The wastes concerned can be categorized approximately as follows:

1. Operational low level waste (gloves, filters, laboratory equipment, and the like) and reactor decommissioning low-level waste (concrete and steelwork). This will be packaged in 200 litre steel drums, or in larger steel boxes.

2. Operational intermediate level waste (fuel cladding, fuel element debris, solidified sludges, and ion-exchange materials from clean-up of liquid effluents, items of redundant plant, and equipment, etc.) and reactor decommissioning intermediate-level waste (items from within reactor cores, including graphite and pressure vessel steel). This

will mostly be packed into 500 litre steel drums, or steel boxes. Very large items may be contained in self-shielded concrete boxes.

The repository will be designed by Nirex to hold about 2×10^6 m^3 of waste by the time it is sealed around 2055, assuming that much of the low-level waste can be compacted prior to disposal (UK Nirex Ltd 1987,1989). Approximately one third of the waste will arise from the decommissioning of nuclear reactors. The combined effects of radioactive decay heating from some of the wastes, chemical reactions in the wastes and barriers, and the geothermal gradient may lead to parts of the repository being subjected to average temperatures of up to 80°C. In the Nirex proposals, wastes will be disposed of in mild steel containers which will be vented as necessary to permit the release of gases generated by the degradation of organic materials in the wastes, and by anaerobic corrosion of metals in the waste and containers. The proposed design of the repository will make use of considerable volumes of cementitious material both as grout in and around waste containers, and as bulk backfill. This will provide a controlled high pH chemical environment around the waste which ensures that the solubility of many of the radionuclides is restricted to low levels, and provides a large surface area in the 'near-field' zone of the repository for sorption of radionuclides (Hodgkinson and Robinson 1987; Saunders 1988; UK Nirex Ltd 1988). The cement backfill will be designed to be sufficiently porous and permeable to allow relatively rapid chemical mixing within the near-field, to take full credit for its buffering potential. Although the full reference inventory of radioactivity in the wastes is substantial (Table 12.1) only a limited number of radionuclides contribute to the overall activity of the repository (Fig. 12.2). These are ^{90}Sr, ^{137}Cs, and ^{63}Ni in the period up to about 1000 years, ^{59}Ni and ^{93}Zr in the intermediate timescale, and ^{238}U and ^{226}Ra at times greater than a few million years (Billington *et al.* 1989). However, performance assessments which account for the mobilization and migration rates of the full spectrum of radionuclides in the repository show that the significant nuclides of concern are (depending on groundwater transit times from the repository to the surface) ^{36}Cl and ^{129}I in the short term, ^{126}Sn in the period between 10^4 and 10^6 years,

Table 12.1   Reference assessment inventory. Radionuclide content of the 2 000 000 m^3 of wastes assumed for disposal over 50 years (from UK Nirex Ltd, 1989)

Nuclide	Half-life	Total inventory		Operational waste		Decommissioning	
		LLW (TBq)	ILW (TBq)	LLW (TBq)	ILW (TBq)	LLW (TBq)	ILW (TBq)
H-3	12.35y	1.3×10^2	6.8×10^3	3.4	6.3×10^3	1.3×10^2	5.4×10^2
C-14	5730y	4.8×10^1	6.6×10^3	4.5×10^1	6.5×10^3	3.8	7.9×10^1
Cl-36	3×10^5y	5.7×10^{-2}	2.6	3.0×10^{-3}	1.3	5.4×10^{-2}	1.3
Ca-4l	1.4×10^5y	5.8×10^{-1}	1.3	0	3.0×10^{-2}	5.8×10^{-1}	1.3
Co-60	5.27y	3.7×10^1	2.9×10^6	3.3	2.4×10^6	3.3×10^1	4.3×10^5
Ni-59	7.5×10^4y	6.0×10^{-2}	6.1×10^4	3.1×10^{-3}	6.1×10^4	5.7×10^{-2}	4.4×10^2
Ni-63	100y	8.3	4.2×10^6	2.7	4.2×10^6	5.6	4.4×10^4
Se-79	6.5×10^4y	3.2×10^{-2}	1.1×10^1	1.1×10^{-5}	1.1×10^1	3.2×10^{-2}	9.4×10^{-2}
Sr-90	29.1y	1.0×10^1	1.4×10^6	8.4	1.3×10^6	1.9	8.7×10^4
Zr-93	1.5×10^6y	8.6×10^{-2}	2.0×10^3	0	2.0×10^3	8.6×10^{-2}	7.0
Nb-93m	13.6y	0	1.6×10^3	0	1.6×10^3		
Nb-94	2×10^4y	5.4×10^{-2}	7.7×10^3	2.5×10^{-9}	7.7×10^3	5.4×10^{-2}	1.0
Tc-99	2.1×10^5y	4.3×10^{-1}	5.7×10^2	1.1×10^{-1}	5.3×10^2	3.2×10^{-1}	4.7×10^1
Pd-107	6.5×10^6y	4.3×10^{-3}	5.7	1.1×10^{-3}	5.3	3.2×10^{-3}	4.7×10^{-1}
Ag-108m	127y	3.0×10^{-2}	4.2×10^{-1}	3.0×10^{-2}	4.2×10^{-1}		
Sn-126	10^5y	3.2×10^{-2}	2.0×10^1	1.8×10^{-5}	2.0×10^1	3.2×10^{-2}	9.4×10^{-2}
I-129	1.57×10^7y	3.6×10^{-5}	1.5	3.4×10^{-5}	1.5	1.6×10^{-6}	4.7×10^{-2}
Cs-135	2.3×10^6y	3.2×10^{-1}	3.1×10^1	1.3×10^{-5}	2.9×10^1	3.2×10^{-1}	1.7
Cs-137	30y	2.9×10^1	2.3×10^6	2.4×10^1	2.2×10^6	4.5	8.7×10^4
Sm-151	90y	4.3×10^{-3}	1.4×10^4	4.3×10^{-3}	1.4×10^4		
Pb-210	22.3y	1.6×10^{-2}	1.9	1.6×10^{-2}	1.9		
Ra-226	1600y	1.6×10^{-1}	2.0	5.1×10^{-2}	1.2	1.0×10^{-1}	8.7×10^{-1}
Ra-228	5.75y	3.7×10^{-3}	1.9	3.7×10^{-3}	1.9		
Ac-227	21.77y	5.4×10^{-6}	6.2×10^{-2}	5.4×10^{-6}	6.2×10^{-2}		
Th-228	1.9y	3.7×10^{-3}	1.9	3.7×10^{-3}	1.9		
Th-229	7.34×10^3y	0	3.6×10^{-1}	0	3.6×10^{-1}		
Th-230	7.7×10^4y	4.8×10^{-5}	4.0×10^{-2}	4.8×10^{-5}	4.0×10^{-2}		
Th-232	1.41×10^{10}y	3.7×10^{-3}	3.0	0	2.1	3.7×10^{-3}	8.7×10^{-1}
Pa-231	3.28×10^4y	1.0×10^{-5}	9.9×10^{-2}	1.0×10^{-5}	9.9×10^{-2}		
U-233	1.58×10^5y	0	9.5×10^{-2}	0	9.5×10^{-2}		
U-234	2.45×10^5y	1.1×10^1	5.8×10^1	1.6	3.5×10^1	9.6	2.4×10^1
U-235	7.04×10^8y	5.9×10^{-1}	8.1×10^1	2.7×10^{-1}	7.8×10^1	3.2×10^{-1}	2.5
U-236	2.34×10^7y	8.8×10^{-5}	3.7	8.8×10^{-5}	3.7		
U-238	4.47×10^9y	2.5×10^1	2.0×10^2	1.6×10^1	1.4×10^2	9.6	5.9×10^1
Np-237	2.14×10^6y	3.2×10^{-2}	3.4×10^1	8.8×10^{-5}	3.4×10^1	3.2×10^{-2}	1.7×10^{-1}
Pu-238	87.7y	1.9	1.7×10^4	8.5×10^{-1}	1.5×10^4	1.1	1.3×10^3
Pu-239	2.41×10^4y	2.9	1.7×10^4	1.7	1.1×10^4	1.2	6.0×10^3
Pu-240	6537y	3.0	1.5×10^4	1.7	1.1×10^4	1.3	3.9×10^3
Pu-241	14.4y	2.1×10^1	5.7×10^5	3.1	2.5×10^5	1.8×10^1	3.2×10^5
Pu-242	3.76×10^5y	1.0×10^{-1}	1.1×10^2	3.5×10^{-4}	1.1×10^2	1.0×10^{-1}	3.8
Am-241	432y	2.8	5.0×10^4	1.8	4.8×10^4	9.9×10^{-1}	1.5×10^3
Am-242m	152y	2.3×10^{-3}	7.6×10^1	2.3×10^{-3}	7.6×10^1		
Am-243	7370y	8.8×10^{-4}	9.7×10^1	8.8×10^{-4}	9.7×10^1		
Cm-244	18.11y	2.5×10^{-2}	1.9×10^3	2.5×10^{-2}	1.9×10^3		

Notes: 1. Certain nuclides (eg Mn-54 half-life 312 days, Ru-106 half-life 367 days) are not included in this list because they decay to insignificant levels by the time of repository closure.
2. Short Life Daughters are not listed.

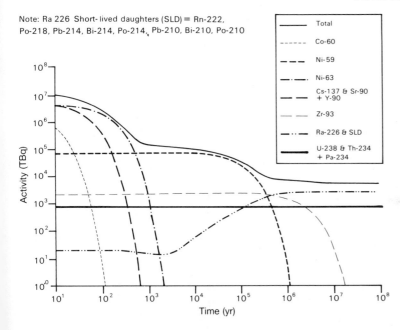

Note: Ra 226 Short-lived daughters (SLD) = Rn-222, Po-218, Pb-214, Bi-214, Po-214, Pb-210, Bi-210, Po-210

Fig. 12.2. Radioactivity in the repository as a function of time after disposal (from UK Nirex Ltd, 1989).

and ^{226}Ra (from the decay of ^{238}U) at very long times into the future.

Both sites now being investigated would have a repository situated in hard, fractured basement rocks. Designs for such a facility (Fig. 12.3) envisage that most of the wastes would be emplaced in large caverns, 25 m wide, 35 m high, and about 250 m long. A total of 26 such vaults would be required which, together with service galleries and four access shafts, would require the excavation of more than 6×10^6 m^3 of rock (UK Nirex Ltd 1989). Prior to the selection of sites, design work was also assessing options for repositories in anhydrite, stiff clays, and chalk. Options for siting the repository offshore in deep sediments and hard rocks below the seabed were also assessed.

The options for deep geological disposal in the UK

Understanding of the types of low groundwater flow regime found in deep clay formations and massive fractured rocks has increased immensely in the last ten years, almost entirely as a result of research into radioactive waste disposal. Coupled with a greater confidence in modelling such low-

energy, low-flow environments, this has caused a reappraisal of the salt-clay-granite approach to selecting deep repository sites, built on requirements for disposal of high-level, heat-emitting wastes. This had placed great emphasis on the need for a low permeability, thermally stable host-rock. It became apparent that the selection of sites for deep disposal of less thermally active wastes needed to be based more on defining suitable *large-scale hydrogeological environments*, with the focus on long groundwater return times, a very slowly evolving regional groundwater regime, and a reduction in emphasis on the host rock itself (Chapman *et al.* 1986, 1987). The features of the geological environments thought to be most suitable are characterized by:

- a high level of confidence in the predictability of the local and regional hydrogeology, with the minimum of geological complexity;

- long groundwater 'return' paths to the surface, preferably resulting in progressive mixing with older, deeper waters, or leading to discharge to the sea. Such slow groundwater movements are usually associated with areas of low regional hydraulic gradient and/or low hydraulic conductivity;

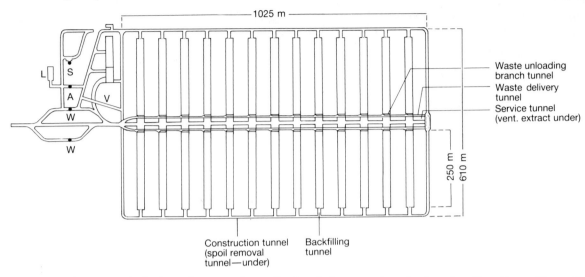

Fig. 12.3. Conceptual design of a repository for hard, basement rock showing the disposition of disposal caverns and ancillary works. W: Waste delivery shaft (8m i.d.); A: Access shaft (6m i.d.); S: Spoil removal shaft (6m i.d.); V: Ventilation tunnel; L: Underground laboratory.

- ease of construction to allow for economic repository design;

- meeting many other widely accepted guidelines regarding the regional and local significance of mineral deposits, geothermal gradients, seismicity, formation depth, etc.

In the UK, the types of geological environment containing rock formations which meet the above requirements can be broken down as follows (Fig. 12.4; Chapman *et al.* 1986):

1. **Inland basinal environments:** Deep sedimentary basins containing mixed sediments with a high proportion of low permeability formations (mudstones, evaporites, etc.). Regional groundwater flow would be mainly confined to any aquifer units, and would tend to be down dip with sub-vertical fluxes across the low permeability units at very low advection rates or, where there is little or no advection, at rates dominated by diffusion.

2. **Seaward dipping and offshore sediments:** Similar in concept to (a), with ground-water movements expected to be very slow towards and under the coast. The lack of any significant head variations in sub-seabed formations would result in almost zero flow.

3. **Low permeability basement under sedimentary cover:** Basement rocks of low intrinsic permeability (principally hard shales, mudstones, slates, quartzites, or volcaniclastics with some crystalline rocks) beneath more recent sedimentary cover. Groundwater movement would occur dominantly in the cover, with little anticipated hydraulic connection with the basement.

4. **Hard rocks in low relief terrain:** Low relief environments, such as those currently being developed for waste disposal facilities in Sweden, have little driving potential for groundwater movement, although the scale of the groundwater flow systems is small compared with the previous environments, owing to the control by frequent major fracture zones.

5. **Small islands:** Almost regardless of rock-type.

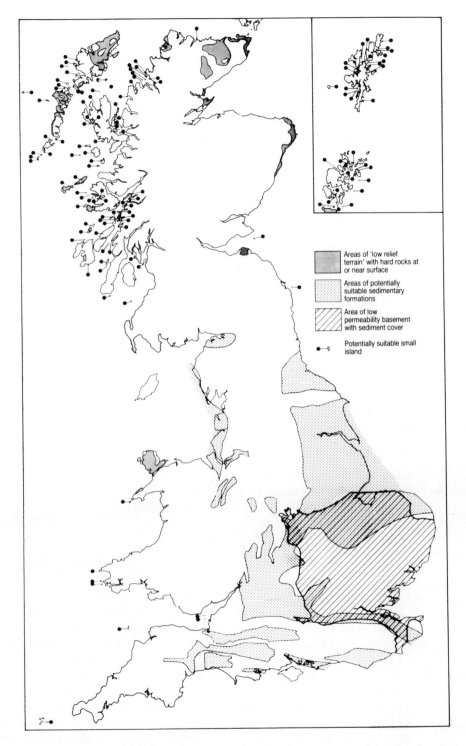

Fig. 12.4. Geological environments in Britain considered to have potential for the development of a deep repository.

A repository might be sited below the sea water to freshwater interface, where groundwater fluxes are thought to be extremely low. Island environments have the additional advantage of the massive dilution capacity of the sea in respect of eventual releases of radionuclides.

Five stratigraphic intervals were considered for environments (1.) and (2.): Kimmeridge Clay (including the Upper Corallian and Ampthill Clay), Oxford Clay, Lias, Mercia Mudstone Group (including the Penarth Group), and the complete Permian sequence including the basal clastic sediments. These formations were selected because their lithologies are mainly argillaceous or evaporitic. Areas of interest have been defined for each interval using boundaries which are the vertical projections to surface of the following structural contours:

(1) where any part of the formation is more than 200 m below the surface;

(2) where the base of the formation is 1000 m below the surface;

(3) where the formation thins to less than 50 m or 100 m in thickness, depending on the formation:

Much of southern and eastern England contains potentially suitable disposal environments. The area known as the Eastern England Shelf, contains all the stratigraphic intervals considered. Here the geological complexity is minimized by the absence of significant faulting and folding.

The 'basement under sedimentary cover' environment (3.) is represented by a large area of Precambrian and Palaeozoic basement present at relatively shallow depths as part of the London Platform. This is covered by up to several hundred metres of Mesozoic sediments. The boundaries of the area of interest were initially defined such that the basement was not overlain by any significant aquifers, but were subsequently revised when it was appreciated that this may not be a critical safety issue.

Low-relief crystalline, igneous and metamorphic rocks, well indurated argillaceous rocks, and some clastic sediments (4) are found mainly in the north and west of Britain. They do not occur in England, and in Wales are only found in Anglesey and the Lleyn Peninsula. In Scotland such areas are mainly restricted to the east, north-east, parts of the west coast and the Outer Hebrides. The geology of these areas varies considerably from peneplained Lewisian gneiss, through granites intruded into Moine metasediments, to extremely thick, highly indurated sandstones.

More than one hundred small islands (5) with areas greater than 0.5 km² were identified. Those with extreme topography and those not sufficiently far from the shore to have independent hydrogeological regimes were not included. The majority of the islands lie off the west coast of Scotland, or in the Orkneys and Shetlands, although a few are found around the coasts of England and Wales.

Since these five geological environments cover such a large part of the UK it was considered reasonable to try to reduce the initial 'area of search' by concentrating at the outset on areas where the geological structure and predictability was simplest. The environments were thus grouped with respect to their relative geological complexities and the consequent degree of difficulty thought likely to be encountered in both investigating and assessing them:

(1) types 4, 5, and 2 (in areas of low geological complexity);

(2) type 1, and 2 (in areas of greater geological complexity);

(3) type 3.

It was decided to investigate the potential of Group 1 first: hard fractured rocks in low relief terrains, small islands, and offshore dipping sediments in areas of simple structure only. The safety assessment modelling teams felt that the basement under sedimentary cover environment (Group 3) offered such potential that it too was investigated from the start, regardless of potential complexities. It was also considered to be conceptually quite different, and consequently important to assess in more detail. Two additional 'offshore' concepts were included at this stage; disposal beneath the continental shelf in either hard, fractured basement rocks (mainly beneath the Atlantic Ocean around

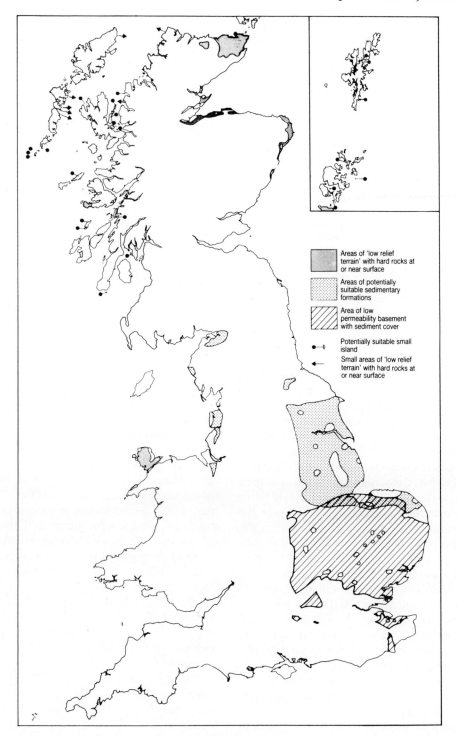

Fig. 12.5. Reduced area of search following geological ranking and superposition of planning and environmental factors.

the north-west of Scotland), or in the sediments which dip seawards beneath the North Sea. Both concepts involved gaining access to a repository via an oil-platform type strucure. In the hard basement rocks the repository would be little different in form to one situated below the land, but for the sediments a matrix of wide shafts was proposed to contain the waste.

Reducing the still considerable area of search on the UK mainland down to a shortlist of sites involved applying a variety of non-geological factors to the assessment both of areas, and the initial list of several hundred potentially available sites in those areas. Planning and non-radiological environmental impact considerations had an immediate effect in considerably reducing the geological area of interest (Fig. 12.5), and matters such as the size and shape of sites further reduced the list of sites. Following this initial sifting the iterative process of arriving at a shortlist of about ten sites eventually made use of a multi-attribute decision analysis system very similar to that already demonstrated in the choice of sites for disposal of high-level waste in the USA (Keeney 1987; Merkhofer and Keeney 1987). Further attributes considered covered issues such as post-closure radiological safety, safety and costs of transport of wastes to the repository, overall costs, operational safety of various repository designs, and the many facets of local and national impacts. The geological input to this model involved assessing sites in terms of their predictability (basically, how simple it would prove to characterize them adequately, and answer the most probable critical questions arising from the safety assessment) and the availability of proven techniques to obtain the relevant data in the rock types and environments concerned. The latter factor obviously militated against the offshore concepts, where detailed characterization would be both difficult and expensive. Additional geological input was incorporated in attributes dealing with construction costs and the general 'robustness' of a concept; essentially reflecting the level of confidence in being able to develop and operate a repository in a particular environment using current well-tested technology. Again, these attributes militated strongly against the offshore concepts in comparison with the land-based alternatives.

When a final shortlist emerged from this analysis, Nirex decided to concentrate its initial field investigations on the two sites with the most nuclear experience, reserving its investigation of any other shortlisted sites for the event that neither Dounreay nor Sellafield prove suitable.

The geology of Dounreay and Sellafield

Both sites now selected for preliminary investigation are on the coast, and both lie on the very edges of sedimentary basins. Consequently both sites are underlain by sediments which display quite rapid thickening towards the basin and the characteristic development of growth faulting parallel to the basin margins, and both have basement rocks within reasonable depths as a target host formation for a waste repository.

Dounreay

Dounreay, on the north coast of Scotland, lies close to the western margin of the Orcadian Basin in which the Devonian age Caithness Flagstones were deposited. To the immediate south and east lie the Reay diorite and the Strath Halladale granite, which formed part of the eroded mountain complex from which the continental Devonian sediments were formed in the intermontane Orcadian Basin. A cross-section through the site in a north-west–south-east direction (Fig. 12.6), based partly on a recent seismic reflection survey, illustrates the general structure in the immediate area of the site. The Strath Halladale granite crops out 3 km to the south of the site, and is present as an easterly dipping sheet within Moine metasediments, and the Reay diorite is likely to have a similar form. The Moine is most likely to form the basement directly beneath the site. It consists dominantly of psammite and semi-pelite with local calc-silicate bands and rare quartzites, and may be migmatized.

The site itself is underlain by rocks near to the base of the Middle Devonian, Upper Caithness Flagstone Group. These rocks comprise an alternating sequence of calcareous pale grey to greenish-grey siltstones and fine-grained sandstones with rarer mudstones, of believed lacustrine origin (Donovan 1980). They have a characteristic flaggy

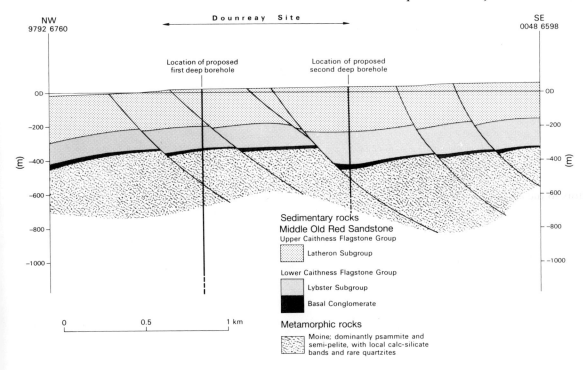

Fig. 12.6. Cross-section showing the predicted geology of the Dounreay site.

appearance, are cut by well developed, broadly spaced orthogonal and rhombohedral joint sets and have a regional dip of about 10⁰. Two divisions of the Caithness Flagstones are likely to be present beneath the site, the Latheron Subgroup of the Upper Caithness Flagstone Group overlying the Lybster Subgroup of the Lower Caithness Flagstone Group. There is likely to be a relatively thin basal conglomerate.

Figure 12.6 shows that the sediments are cut by normal faults, throwing down to the east and dipping at 40–60⁰, and exhibiting syndepositional movement. Throws generally decrease upwards, from as much as 130 m at the base of the Devonian to perhaps not more than 10 m at the surface. Small scale reverse faulting can also be seen in coastal exposures. These faults can be traced into the basement where they may coincide with pre-existing shear zones, reactivated during Devonian basin development. The sediments beneath the site in the proposed location for the first deep borehole are approximately 350 m thick, but they thin rapidly towards the west, such that the basement is exposed 1.5 km south-east of the site.

The groundwater flow pattern in the area of the site will be dominated by the location, orientation, and properties of the faults, especially where they are present within the basement. The flagstones have relatively high hydraulic conductivities, but the area is of low relief and groundwater heads are expected to be small throughout the formations of interest. The most probable host rock for the repository is the Moine at depths of 700–1000 m.

Sellafield

Sellafield lies on the Lake District coastal plain, which is almost completely covered by superficial deposits consisting of a complex admixture of many recent sediment types, up to 55 m thick. A cross section through the site in an east–west direction, based on recent seismic reflection data, illustrates the structural complexity of the region, which lies on the eastern margin of the Irish Sea Basin (Fig. 12.7). The superficial deposits overlie the Ormskirk Sandstone and the St Bees Sandstone, of Permo-Triassic age, which dip at 10⁰ to

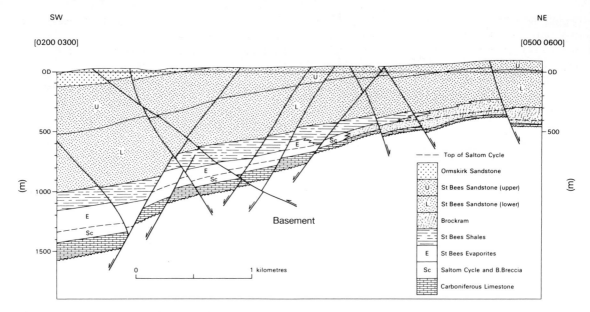

Fig. 12.7. Cross-section showing the predicted geology in a north-easterly direction from Sellafield.

15° to the west and south-west and thicken rapidly in the same direction. The St Bees Sandstone contains more shaley and mudstone partings towards its base, and overlies the St Bees Shales which comprise two principal facies; in the lower part, blocky-weathering siltstone and silty mudstone, commonly gypsiferous, with coarse sandy bands and calcareous concretions and, above, laminated micaceous siltstone, mudstone, and subordinate sandstone with load casts, desiccation cracks, and mud-flake breccias. Below the St Bees Shales is a thick sequence of St Bees Evaporites containing three cycles. Thick anhydrites may be developed, as may halite, and the lower parts of the St Bees Evaporites may contain thick-bedded dolomites. The evaporites rest sharply on basal breccia, which lies on an irregular surface of Lower Palaeozoic and Carboniferous age. The base of the Permo-Triassic sequence is represented in places in the east of the area by Brockram, a marginal conglomerate of the St Bees Sandstone which, in its lower parts, may be the lateral equivalent of the St Bees Shales and St Bees Evaporites.

Whilst inland from Sellafield the Lower Palaeozoic basement rocks are predicted to belong exclusively to the Borrowdale Volcanic Group lying unconformably on the Skiddaw Group, beneath and to the west of the site it is thought that Silurian greywackes are more likely to be present. The Borrowdale Volcanic Group is made up of intermediate to acid lavas and pyroclastic rocks, associated sedimentary, and volcaniclastic strata, and the Skiddaw Group of slates and greywackes. Both groups are intruded by igneous rocks including, locally, two major granitic bodies, the Eskdale Granite and the Ennerdale Granophyre. Carboniferous Limestone lies with a strong unconformity on the basement over parts of the area, and can be as much as 200 m thick. Due east of Sellafield it is not present. Shales and mudstones with evaporites of the Mercia Mudstone Group are present a short distance offshore, and high angle normal and, occasionally, reverse faults are ubiquitous throughout the whole of the area. Three distinct phases of faulting have been recognized, with vertical displacements of up to 250 m.

The basement rocks in the region would allow for repository construction at depths of around 1000 m to the east of Sellafield. They may have suffered two phases of deformation prior to the syndepositional Permo-Triassic faulting. Groundwater flow takes place dominantly within the St Bees Sandstone, and is probably directly towards the sea, being recharged in those regions further inland where the superficial deposits are either thin or absent. The pattern of groundwater flow is likely to be strongly controlled by the faulting, by the large vertical contrasts in hydraulic conductivity caused by the varied composition of the sediments and, in part, by the thickness, hydraulic properties, and extent of the superficial deposits which are also present offshore with similar thicknesses to those on land. Flow within the basement will be determined by the extensive fault system which is thought to be present, and by the continuity and properties of the overlying low permeability shales and evaporites.

The site investigation programme

The programme of site characterization that was just commencing at the time of writing requires planning consent from local and regional authorities at several stages of the work. Obtaining planning consent for the eventual construction of a repository is likely to result in a public inquiry before any significant work can start. The repository will have to be licensed by the national regulatory authorities before it can commence operation. Under present procedures, consent to drill the first exploratory boreholes also requires planning approval from the local authorities.

Current proposals for the investigations of Dounreay and Sellafield envisage a two stage programme of work at each site prior to the selection of one of them. If the anticipated public inquiry is in favour of development at one of the sites, then a third stage of site characterization will commence, in parallel with the progressive construction of the repository.

The first stage of work, currently underway, is a preliminary exercise to provide the designers and the safety assessment teams with a 3-D structural model of the sites. This will be achieved largely by a programme of regional geophysical surveying, supported by the drilling of two deep boreholes (800–1800 m) at each site. These boreholes will be used to calibrate the seismic and gravity surveys, and to provide first information on basic hydrogeological properties of the rocks. Head, flux, and transmissivity measurements will be supported by preliminary hydrogeochemical evidence for the residence times of deep groundwaters. The core

material will provide data on geotechnical properties for the design team, as well as material for experimental studies of rock/radionuclide interactions.

During this work, which will take 12–18 months, the safety assessment teams will be constructing refined groundwater flow models both to characterize the sites and eventually, to test their potential response to future perturbations caused by the repository and climatically induced hydraulic fluctuations. This approach will allow very specific questions to be raised for consideration in the second stage of the work. For example, it is envisaged that predictions made by the flow models can be tested by making hydraulic measurements in zones specified as particularly sensitive by the models. The potential significance of various features (such as fault zones, lithological variations, and other heterogeneities) can be tested by the model using estimates of parameter values, before specifying the most useful means of providing supporting field data, and the best locations to make the measurements. The second stage, which will overlap with the first, will thus comprise a longer period of drilling, hydraulic testing, and groundwater sampling, continuing until about 1993.

The first stage in developing the repository, which will take place after the public inquiry, will be the sinking of an exploratory shaft, and the commencement of underground exploration via trial adits. The objective of this third stage of work will be to characterize in detail the precise volume of the rock in which the repository will lie, and to allow for a period of experimentation focused on safety or construction issues, and the demonstration of the technological aspects of waste emplacement. In this stage it will probably be necessary to make detailed changes to repository design or to details of the engineered barrier design in order to accommodate specific features of the rock. It is intended that as much as possible of the exploratory work be conducted from the underground facilities using remote sensing geophysical techniques and, where possible, to ensure that exploratory boreholes are subsequently mined out during construction. This is to minimize damage to the rock surrounding the repository, and avoid significant disruption of the natural groundwater flow system.

The geological issues arising from safety assessments

Assessment methodology

Performance assessment of a disposal system is an iterative process that guides the design of the repository by defining the most sensitive components and issues, and the physicochemical processes that are of prime importance to safety. It also highlights areas of critical data uncertainty and so helps to focus site characterization work. In common with other national programmes, the assessment models used in the UK tend to be based on conservative assumptions; that is they take a pessimistic view of future performance and will consequently tend to overestimate the radiological impacts of disposal.

Both Nirex and DoE have been developing methodologies for radiological safety assessment which made use of both deterministic analyses, and of probabilistic risk assessment (PRA). With PRA, the overall performance of a repository is calculated using a series of connected models which describe, in a simplified manner, the behaviour of critical components or processes in the repository, the geosphere, and the biosphere. The data for these models are in the form of probability distribution functions (PDFs); the estimated or measured frequency of occurrence of all possible values of each parameter. The approach with deterministic assessments is to look at the detailed performance of the whole, or parts, of the system when subject to a variety of evolutionary 'scenarios'. These use best estimates of parameter values, and are particularly useful in understanding the impact of an event or process. The results of PRAs, although logically more comprehensive, can be less easy to interpret directly. Consequently, both approaches are required in order to understand future performance, and the impacts of uncertainties in the data used.

A major component of the Nirex site selection exercise was the information arising from a two-stage 'comparative assessment of concepts and areas for deep emplacement': CASCADE (Billington *et al.* 1989), which involved modelling the performance of basic repository designs situated in generic geological sites, loosely based on the most promising areas of interest. This work was aimed at assessing how well concepts performed against the yardstick of the assessment principles laid down in 1984 by the Department of the Environment (DoE 1984). These state that, among other guidelines, the appropriate radiological safety target for a repository, at any time, is a risk to an individual in a year equivalent to that associated with a dose of 0.1 mSv; about one chance in a million.

Prior to the CASCADE exercise, the DoE had been carrying out its own 'dry run' performance assessment exercise for a hypothetical deep repository situated in clays at approximately 400 m below the Harwell site in Oxfordshire. The basic geological information had been provided by some limited exploratory drilling carried out by the British Geological Survey in 1980–1 (Black *et al.* 1985). All lithological and hydrogeological data were taken from a group of very closely spaced boreholes, together with some regional spring and well sampling in the recharge area of some of the aquifers penetrated. Consequently, the data were considerably limited compared to what would be expected from a real site investigation. The results of this modelling exercise (Gralewski *et al.* 1987) highlighted the need for regional data, rather than information only from the site itself, and pointed to the importance of understanding groundwater flow at a site using 3-D as well as 2-D hydraulic models. Even with 2-D models it was possible to show the potential significance to radionuclide releases to the biosphere of multiple return pathways through the rocks. Although neither the Dounreay nor the Sellafield sites possess the same type of alternating aquifer/aquitard lithology as Harwell, these results are equally valid. Both sites will require 'off site' drilling, and both will require 3-D modelling to fit together a picture of possible radionuclide migration paths in a complex interconnected network of major fracture zones and lithological boundaries. 3-D modelling is thus a complex exercise which is most useful for developing a working understanding of a site whereas, for more straightforward illustration of the results of certain processes or options 2-D, and even 1-D, radionuclide transport models are often most appropriate.

From the above examples there are clearly many issues related to the geological input to safety assessments that could be selected for discussion.

However, a number of extremely important points have emerged from both the CASCADE study and the parallel research programmes of both Nirex and the DoE, and are currently the focus of interest. These are highlighted in the following sections.

Groundwater flow modelling and the problem of geological 'probability distribution functions'

The importance of being able to predict groundwater residence and transit times is made clear in all the British assessment work to date. Much interesting discussion has taken place on the topic of selecting values for critical parameters, such as hydraulic conductivity, for application to the porous-medium flow models which will almost inevitably be used for regional scale groundwater flow models in all types of rock. This has been focused by probabilistic models, which demand PDFs for such a parameter. While it is quite possible to give estimates of upper and lower bounds of, for example, hydraulic conductivities of fractured rocks, application of the extreme values to very large columns of the rock mass when making steady state predictions of flow can produce very short apparent return times that are inconsistent with the perception of real flow regimes, where any relatively rapid movements are restricted to very discreet zones. This issue, together with others related to flow in fractured rocks, is being investigated through UK involvement in projects such as the Stripa test mine in Sweden. The best resolution will always be found when flow models are applied to real sites, where there is some constraint on flow predictions provided by detailed hydraulic testing and hydrochemical evidence of groundwater residence times. This was certainly a valuable experience found during the Nirex shallow site investigations in 1987 (e.g. Bath *et al.* 1988). For the time being, however, despite the overall very conservative approach taken in safety assessment, we believe that it would be unwise to place too much emphasis on the extreme results of generic flow models.

The lack of well-documented databases for many geological processes with observations over adequate time periods makes it difficult to interpret the results of PRAs that use PDFs to attempt predictions over millions of years. Risk analysis works properly when the models used are well validated, and the PDFs have been obtained by extensive testing and observation of materials in use. One might consider, for example, the stress-strain behaviour of engineered components used in the construction industry. Prediction over geological timescales shares the same problems as very long-term global climate prediction; most processes occur very slowly, or events occur very irregularly with long periodicity and frequently go unobserved. We only infer their nature by their effects. While geologists can generate valid PDFs for some geological parameters, it is important to be quite clear how, exactly, the data are going to be used. For example the expense and effort of obtaining fracture densities and apertures across a whole site could be enormous and may be quite unnecessary for the type of modelling which might be most appropriate. For other data it may not be possible to gather them at all, and the uncertainties may be irreducible. For these, the alternative is to set the PDF bounds very wide indeed. This can produce very pessimistic results or, in some cases, optimistic results in which the 'expectation values' of risk are reduced owing to the wide spread of input parameter values. PRAs of very long term geological behaviour, although very useful in developing understanding, should be treated with great caution if decisions are to be based on their quantitative results.

Time dependency of natural processes

In the UK, major disruptive geological events are unlikely to disturb a repository to a significant extent, owing to the predicted large-scale tectonic stability of the British Isles for the coming 10^8 years. However, even on the one million year timescale climatic change is likely to have a considerable effect, not only on the biosphere, where the recipients of any future radiation doses could be living in anything from a greenhouse effect super-interglacial to a full glaciation, but also on the geosphere, where groundwater fluxes could be substantially modified.

When considering the effects of climatic change in Britain, the principal geological features of concern are the changes in sea-level and drainage that could occur, together with erosion rates and the stress and hydraulic response of the rock to ice-loading and unloading. Safety assessments show the benefits of having a repository under the sea bed or on a small island in order to take credit for the massive dilution potential of the oceans for any eventual releases. This was taken into account in defining hydrogeological environments in the UK in which to seek repository sites, and both Dounreay and Sellafield are, indeed, on the coast. However, studies of climatic change suggest that the current sea-level in Britain is relatively very high, that beyond about 10 000 years into the future periglacial conditions are likely to dominate for 70 per cent of the next one million years, and that sea-levels may consequently fall by about 30–140 m. The topographic consequences of this can be seen in Fig. 12.8 which demonstrates that

many current coastal and island locations could be far removed from a marine environment.

The central issue with respect to time dependent processes is not so much the ability to model what occurs at any particular 'snapshot' in time, as the problem of how to cope with their cumulative effects, particularly when a process is cyclical with a relatively short periodicity (say of the order of 10 000 years). In some areas this is probably not a significant concern; for example, hydrochemical evidence suggests that in both thick sequences of sediments and in some deep fractured rock environments, the response time of fluxes in the deep groundwater regime to transients in hydraulic

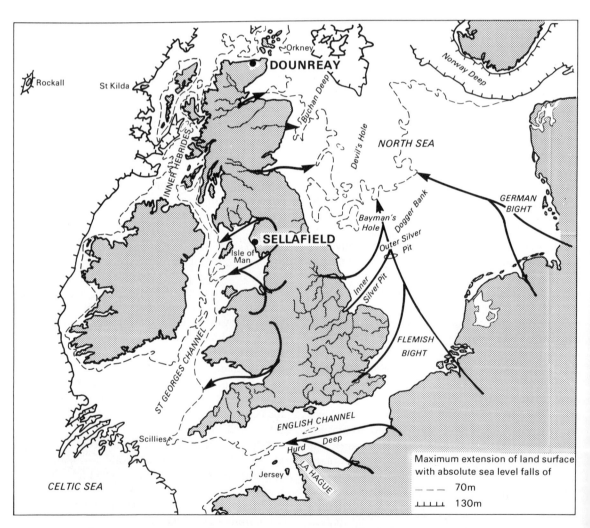

Fig. 12.8. Extension of land surface in Britain under periglacial conditions.

head and recharge water chemistry as a result of sea-level changes, is extremely long (Neuzil 1986). The recovery time of the saline/freshwater interface beneath a small island in hard fractured rock, when returned to a mainland environment by a drop in sea-level, has been estimated to be of the order of tens of thousands of years. These observations do not, however, justify the unqualified application of steady state flow conditions to times beyond about 10^5 years and, in fracture zones in the upper few hundred metres of the UK sites, steady state models are unlikely to be valid on still shorter timescales. Indeed, it may be useful to uncouple the sluggishly responsive zones of the deep groundwater regime from the rapidly cycling regions both at depth and near the surface when considering the estimation and significance of radionuclide concentrations and radiation doses. This means that considerable thought has to go into interpreting dose or risk versus time graphs at extreme times into the future. Much thought is going into the issue of time dependency in both the DoE and Nirex research programmes, and this also has an important input into the topic of timeframes in which to consider the results of performance assessment.

During the course of the future site investigations it is anticipated that much useful information on time dependent processes will arise from palaeo-hydrogeological studies of the sites concerned, combined with studies of the effect of cyclical climatic change on the properties of specific conductive fracture zones.

Is reversible sorption a geologically reasonable concept?

The simplest models of radionuclide transport in advecting groundwater make use of laboratory derived distribution coefficients which describe the uptake of radionuclides from the water onto rock surfaces by a variety of processes commonly termed 'sorption'. These coefficients vary with the physical and chemical state of the water, the radionuclide, and the rock surface and there is always some uncertainty in selecting ranges of values in geological environments which are inevitably spatially and temporally inhomogeneous. A complimentary issue is the very long time periods treated in groundwater transport calculations. Laboratory sorption kinetics appear to be very fast initially,

and settle down to an apparently steady state in a matter of hours. Sorption is taken to be reversible. It is difficult to believe that this reversibility persists over geological timescales, and that second or third order kinetic processes do not lead to irreversible fixation within the surfaces of minerals. Where this could be shown to be the case, by studies of naturally occurring radionuclides for example, then a sorption model is clearly giving faster radionuclide transport times (i.e. the safety case is making a generally pessimistic assumption).

The alternative approach is to use our knowledge of the solution chemistry of the radionuclides and components of the mineral phases in the rocks to produce purely thermodynamic models of radionuclide/rock interaction in which the effects of physicochemical changes can, in theory, be predicted. The problem here is that many of the chemical species which might exist are exotic, and thermodynamic data are only slowly becoming available. As with sorption, this approach is also unable to address comprehensively the kinetics of radionuclide/rock interactions over very long timescales, when reactions take place in aqueous pore and micro-fracture environments where diffusion, structured water layers, and complex mineral surface electrochemistry may dominate transfer processes, with consequent effects on the stability of radionuclide complexes. In some deep geological environments only metastable equilibria appear to exist between pore-waters and rock, despite the age of the rock formation, and these are difficult to model.

The thermodynamic approach does come closest to reality, and current research is enhancing the databases used considerably. In the meantime, for assessment purposes, reversible sorption remains a usefully conservative assumption, despite the fact that we are unconvinced that it can always maintain reversibility over very long timescales. The issue is that of defining under what circumstances conversion to an irreversible fixation mechanism occurs in the rock formations of concern in order to allow the safety assessments to be less pessimistic and more realistic. There is clearly much that could be learned from more detailed studies of natural systems.

Gas release and migration from the repository

Gas is produced in the repository during the degradation of the waste and its containers. A wide variety of gaseous species will result from the biodegradation of some of the wastes during the first hundred years or so, although the principal product will be carbon dioxide. However, the main source of gas is likely to be the production of hydrogen during the aerobic corrosion of aluminium and magnox in ILW and, more importantly, the anaerobic corrosion of steel in both LLW/ILW. This hydrogen production is likely to be significant during the first few hundreds or thousands of years after disposal, with a potential production rate of about 2×10^6 m^3 H yr^{-1} (at STP) during the first 500 years (Rees and Rodwell 1988). Were the repository completely gas-tight this could lead to disruption of the engineered barriers, and even the rock, by overpressure. However, it is expected that the gas will be able to escape relatively easily through the fracture network in the rock. The effect of this progressive gas release on radionuclide transport and groundwater flow is currently being assessed, and models are being developed for two-phase flow in fracture networks. These are being tested by experiments in boreholes to measure the rate of pressure dissipation after gas injection into deep, fractured rock.

Human intrusion into the repository

Perhaps the least predictable and most philosophical aspect of safety assessments is the treatment of future unintentional intrusion into a lost and forgotten repository during drilling or mining. In the UK, attempts are being made to address this issue in as scientific a manner as possible, largely because early assessments indicated that such activities could constitute one of the principal release mechanisms. The main risk is seen to arise to a geotechnical worker who handles contaminated core material from a borehole through or close to a repository. Consequently the question arises as to the probability of such drilling taking place.

It has been a relatively simple matter to relate past drilling practice and frequency to the type of rock formation in which a site lies, and so produce an annual frequency of borehole intrusion in terms of holes yr^{-1} km^{-2}. In the sites of interest, such drilling has been either for water or mineral exploration. Such an approach makes some very fundamental assumptions about unchanging exploration

drilling, and mineral or hydrocarbon extraction techniques. In particular, the basic assumption of total loss of all records of a repository, which makes such intrusion inevitable when very long periods are considered, raises many questions about the future nature of society and the way in which the results of safety assessments should be interpreted.

Work has thus been carried out to put exploratory drilling into the wider context of how exploration programmes work at present. In particular, interest is being shown in the type of remotely sensed geophysical anomaly that might be presented by a repository, and whether this would alert any inadvertent intruders to potential danger. Fig. 12.9 shows the calculated magnetic anomaly that would be produced by a waste repository at Dounreay containing about three million tons of iron in the waste and engineered barriers, and an even more distinctive anomaly for a repository in a deep clay formation. The apparent man-made regularity of these features and their persistence (even following corrosion of steel, the green magnetite resulting from the anaerobic process remains ferromagnetic) may, however, attract as much as deter future prospectors. An analysis of the likely risks from human 'intrusion' into a repository situated in the geological environment discussed earlier has been made by Nirex (Jowett and Chapman 1989). This indicated the risks to be comparatively lower at sites such as Dounreay and Sellafield than in the sedimentary rocks of eastern England. Although geologists can contribute much to this debate, its final resolution will depend largely on the perceptions and attitudes of current generations towards the significance of risks to unknown future societies.

Long-term excavation response of the rock

The construction of the repository may significantly influence the hydraulic conditions in the host rock and surrounding formations. For this reason, as well as to model the effect of stresses on regional groundwater flow, it will be important to characterize the *in situ* stress regimes during site investigations. Excavation response around openings in the Moine metasediments at Dounreay and Borrowdale Volcanic Group rocks at Sellafield will depend on the excavation method and on the stress distribution in the rock mass during and after excava-

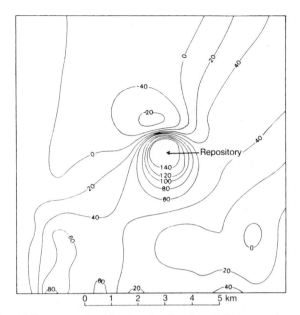

(a) Dounreay: Combined regional and modelled anomaly

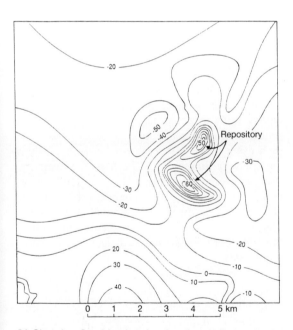

(b) Clay site: Combined regional and modelled anomaly

Fig. 12.9. Calculated magnetic anomalies produced by radioactive waste repositories (contours in nT).

tion. The disturbances of interest are the creation, extension, or reactivation of fractures that could result in either a modified structural condition, or modify the hydraulic conductivity of the rock close to cavern and shaft walls. This could result in short-circuiting of hydraulically active zones in the rock along 'skin' regions of enhanced conductivity, which, if significant, would require special grouting and sealing techniques during construction. Again, UK involvement in the research at the Stripa mine is providing valuable information on such techniques.

A further issue of concern is the very long-term response of the rock to the presence of voids in the repository. Although it is the intention to backfill the whole repository completely, cavern crowns are very difficult to fill totally, and are the zones in to which any growth of voids in the wastes, caused by corrosion and degradation, would eventually migrate. Although extension of such voids upwards into the overlying rock is considered unlikely from preliminary assessments by Nirex, the long-term mechanical and hydraulic behaviour of such features will need further study when access to the potential repository rock is available during detailed underground investigations.

An additional excavation response which must be accounted for is the de-watering of overlying rock during pumping to keep the repository dry during a 50 year operational life. From the viewpoint of establishing stable geochemical conditions in the engineered barriers, it would be beneficial if the resaturation time after completion of the repository were short, and that any air in the partially saturated rocks dissipated or was consumed by oxidation reactions quickly in order to minimize the aerobic period. Modelling of the effects of various modes of resaturation has been an important aspect of the UK programme of research into corrosion and engineered barrier design, and both resaturation time and re-establishment of reducing conditions are estimated to be of the order of a few tens to a few hundreds of years.

Time frames for performance assessment.

The DoE assessment principles indicate that the individual risk target is appropriate at any time. This could be interpreted to mean that individual risk ought to be calculable at all future times. At present, whilst DoE await a submission from Nirex which they can assess, the exact nature of what the safety case should contain in terms of projected long-term risks remains a rather grey area. Consequently, there is a continuing debate underway on the possibility of using various time frames for performance assessment, and on the type of calculations that might be most appropriate for them. Geological advice is very likely to be at the base of these discussions, particularly in respect of the longest time-frames being considered.

In our view, the confidence that safety assessment teams can place in the numerical results of calculations of performance at long times into the future does not permit the presentation of very specific figures such as risk to individuals. This view is based, first of all, on the reasons given in the prior discussion of time dependency and cyclicity of natural processes in the shallow geosphere and biosphere. Second, as with human intrusion, the exact meaning of risks and doses to populations and individuals that may or may not exist in the future seems very philosophical. Geologists are required to provide evidence of the longevity of the radionuclide transport and retardation processes being modelled by means of data from geochemical analogues in rock formations. These 'natural analogues' (see, for example, Côme and Chapman 1985, 1986, 1988; Chapman and McKinley 1990), tend to reinforce the view given earlier that we can uncouple deep processes which are only slowly affected by global environmental change, from processes in the upper tens or hundreds of metres, which respond as readily as the biosphere to such changes. The case is often made that we should not consider individual doses to humans calculated beyond the next ice age (say 10 000 years hence). We would certainly support the view that any quantitative radiological predictions which require more than the very simplest of assumptions about the biosphere/geosphere interface zone after unknown cycles of environmental change, have little real meaning. Consequently, a requirement to present, with a high level of confidence, radiation dose calculations, or risk figures, for times beyond a few thousands of years would seem to place an unreasonable burden of credibility on safety assessment groups. Conversely, our understanding of deep groundwater flow systems suggests that is is quite reasonable to make quantitative predictions about the likely dispersion and disposition of radionuclides throughout the rocks around a repository for very much longer periods.

The question thus arises as to whether alternative comparisons could be useful when considering the acceptability of a waste repository in terms of its very long term 'geological timescale' impacts. In seeking appropriate yardsticks we suggest that, although a dose/risk target is unreasonable for times beyond a few thousands of years, a comparator might be adopted for longer periods that relates the availability (location, chemical form, mobility, concentration, and fixation) of waste radionuclides to other radioactive or toxic species naturally dispersed through the rock volume. Assessment of such availability might, for example, consider the rates at which such species are naturally released to the environment by erosion or mobilization in groundwater. In other words, while targets based on radiological protection may be reasonable for as long as we can reasonably predict how radiation doses might be received, the assessment of longer term behaviour might be related to our more justifiable confidence in predicting the nature of the deep geological environment over such timescales.

Validation of predictive geological models

The strand that connects all the issues discussed above is that of using geological knowledge and observations to construct models of the physical processes that affect repository performance, and then using these models to predict future behaviour for very long periods. Perhaps the overriding issue at present is whether the predictive models used in performance assessment can be *validated*. Does the model provide a reasonable representation of reality, and is the representation valid for the time periods for which the model is making predictions?

Much effort is being put into model validation, not only by DoE and Nirex, but throughout the international community. The INTRAVAL project, organized by the Swedish Nuclear Power Inspectorate (Andersson 1988), and in which both DoE and Nirex take part, is a quantitative attempt to validate radionuclide transport models. The current consensus of this group is that it will take a combination of many types of evidence, laboratory and field experiments, site characterization, and natural analogue studies to provide convincing proof that our models and computer codes provide a fair representation of reality. A developing view within INTRAVAL is that it is difficult to validate any geochemical transport model for general use. A model can only really be considered to have been demonstrated as either a valid description of a process, or as a valid description of the effects of a group of processes at a specific site. The former 'process validation' can be achieved largely by field or laboratory experiment, whereas the latter 'site validity' is likely to rely on natural geochemical analogues of multiple processes, or on a thorough geological characterization of the site based on palaeo-hydrogeological interpretation of the evolution of the groundwater regime.

Conclusions and prospects

In Britain geologists have been preparing to carry out a thorough site characterization programme for a deep waste repository for the last ten years, during which time considerable development of intellectual concepts and models has taken place, but little field testing has occurred. This chapter has focused previous assessments of the 'state of the art' in the techniques of site characterization (e.g. Chapman *et al.* 1987) on to the likely requirements at two specific sites.

In our view, geologists now have the techniques and methodologies available to measure the parameters that are likely to be required in repository design and safety assessment. While there is still a lot to be learned about how to apply these techniques most effectively during site characterization, almost all the critical issues to have emerged from recent work on performance assessment are largely unconnected with either the data or the data gathering process. By looking at the results of recent research and safety assessments we have highlighted some more general geological issues which go beyond the confines simply of technique. The majority of the issues discussed above is, in fact, concerned with:

1. the considerable pessimism built into some assessment models as a response to data or conceptual uncertainties (such as how best to use information on hydraulic properties or sorption in models);

2. demonstrating confidence in predictions made for processes that occur over 'geological timescales' into the future;

3. how geological data can be used to compare the behaviour of natural radioactivity to that mobilized from a repository at very long times into the future as an alternative to making dose or risk calculations for these 'geological' timescales.

In all these areas geology still appears to hold the key to the future.

Acknowledgements

This chapter is a slightly revised version of a paper originally presented at the International Geological Congress, Washington DC, in July 1989, and describes work carried out whilst the authors were with the British Geological Survey. The views expressed have developed during the course of work carried out for both the Department of the Environment and UK Nirex Ltd, but are those of the authors, and do not necessarily reflect current or future policy in either organization. Both are thanked for their continuing support of the studies. We also wish to thank our many colleagues working in the British radioactive waste management programme, much of whose work has been used in preparing this review.

References

Andersson, K. (1988) International co-operation in the field of verification and validation of radionuclide transport models; experiences and need for further work. In *Geoval 1987*, pp. 179–87. Swedish Nuclear Power Inspectorate.

Bath, A. H., Entwistle, D., Ross, C. A. M., Cave, M. R., Falck, W. E., Fry, M., Reeder, S., Green, K., McEwen, T. J., and Darling, W. G. (1988). Geochemistry of pore-waters in mudrock sequences. Evidence for groundwater and solute movements. In *Hydrogeology and safety of radioactive and industrial hazardous waste disposal*. Report No. 160, pp. 87–98. BRGM

Billington, D. E., Lever, D. A., and Wisbey, S. J. (1989). *Comparative assessment of concepts and areas for deep emplacement* (Phase 2): Summary. Report No. NSS/A301. UK Nirex Ltd.

Black, J. H., Holmes, D. C., Alexander, J., and Brightman, M. A. (1985). The role of low permeability rocks in regional flow systems: the Harwell area study. In *Hydrogeology of rocks of low permeability*, pp. 107–117. Memoirs 17th Cong. Int. Ass. Hydrogeol.

Chapman, N. A., and McKinley, I. G. (1987). *The Geological disposal of nuclear waste*. Wiley, Chichester.

Chapman, N. A., and McKinley I. G., (1990) Radioative waste: back to the future? *New Scientist*, **1715**, 54–8.

Chapman, N. A., McEwen, T. J., and Beale,H., (1986). Geological environments for deep disposal of intermediate level wastes in the United Kingdom. In *Siting, design, and construction of underground repositories for radioactive wastes*, pp. 311–28. IAEA, Vienna.

Chapman, N. A., Black, J. H., Bath, A. H., Hooker, P. J., and McEwan, T. J. (1987). Site selection and characterization for deep radioactive waste repositories in Britain: Issues and research trends into the 1990s. *Rad. Waste Manage. Nucl. Fuel Cycle*, **9**, 183–213.

Côme, B. and Chapman, N. A. (1985, 1986, and 1988). CEC Natural Analogue Working Group. Reports of the first three meetings. Commission of the European Communities Reports Nos. EUR 10315, 10671 and 11725 (pp. 223, 279, and 231). CEC Luxembourg.

DoE. (UK Department of the Environment) (1984). *Disposal facilities on land for low and intermediate-level radioactive wastes: principles for the protection of the human environment* HMSO, London.

Donovan, R. N. (1980). Lacustrine cycles, fish ecology, and stratigraphic zonation in the Middle Devonian of Caithness. *Scott. J. Geol.*, **16**, 35–50.

Gralewski, Z. A., Kane, P., and Nicholls D. B. (1987). *Development of a methodology for post-closure radiological risk analysis of underground waste repositories*. Illustrative assessment of the Harwell site. Report No: DOE/RW/87.034. UK Department of the Environment.

Hodgkinson, D. P. and Robinson, P. C. (1987). *Nirex near-surface repository project; Preliminary radiological assessment: Summary*. Report No. NSS/A100. UK Nirex Ltd.

Jowett, J. and Chapman, N. A. (1989) UK Nirex studies of intrusion frequency. In *Risks associated with human intrusion at radioactive waste disposal sites*, pp. 142–158. OECD/NEA, Paris.

Keeney, R. L. (1987).An analysis of the portfolio of sites to characterize for selecting a nuclear repository. *Risk Analysis*, **7**, 195–218.

Merkhofer, M. W. and Keeney, R. L. (1987). A multiattribute utility analysis of alternative sites for the disposal of nuclear waste. *Risk Analysis*, **7**, 173–194.

Neuzil, C. E. (1986). Groundwater flow in low-permeability environments. *Water Resour. Res.*, **22**, 1163–95.

Rees, J. H. and Rodwell, W. R. (1988). *Gas evolution and migration in repositories—current status*. Report No. NSS/G104, UK Nirex Ltd.

Saunders, P. A. H. (1988). *Research and safety assessment*. Report No. NSS/G100. UK Nirex Ltd.

UK Nirex Ltd. (1987). *The way forward. A discussion document*.

UK Nirex Ltd. (1988). *Presentation of the Nirex disposal safety research programme*. Report No. NSS/G108.

UK Nirex Ltd. (1989). *Deep repository project*. Preliminary environmental and radiological assessment and preliminary safety report. Nirex Report No. 71.

13. Transport in fractured rock

J. A. Barker

Introduction

Over the last decade the challenge of understanding transport in fractured rocks has been taken up with a commitment of manpower and funds which is probably unprecedented in the field of hydrogeology. The main impetus for this undertaking has come from the pressing need to predict the safety of radioactive waste repositories in fractured rock. Much of the work has been carried out by scientists and engineers who would not regard themselves as hydrogeologists, yet their work must been seen as representing a significant contribution to this field of study. At the same time the petroleum industry has given greater attention to the development of fractured reservoirs, and exploration and evaluation studies have been important topics during the last 10–15 years (Nelson 1987).

While significant contributions have been made by Britons, both within and beyond their own particular interests (e.g. the Chalk, Triassic sandstones, and granite batholiths), those contributions need to be put in the context of international progress; indeed, much British work has been carried out within the framework of international programmes.

A significant bias in this chapter is towards modelling; there are a number of reasons for this. Modelling has tended to be a focus of research into fractured rocks: much of the field and laboratory work has been aimed at providing data upon which transport models can be formulated and calibrated. For radioactive waste disposal, predictions of behaviour must extend over thousands of years; such predictions cannot be based just on field or laboratory experiments lasting at most a few decades and, therefore, mathematical models are essential. Also, as with any field of study, the sophistication and degree of success of the models in use provide a good measure of scientific progress.

The chapter begins with a brief description of the major factors and processes affecting transport in fractured rock. That is followed by a fairly detailed description of the models in use. Finally, a selection of relevant field and laboratory studies is outlined. The term transport is taken in a broad sense to include the transport of groundwater itself, heat, and solutes (including pollutants).

Factors and processes affecting transport

Fracture morphology

The term fracture will be used here to represent any discrete discontinuity, natural or induced, which provides a conduit for the movement of fluid. In any study of fractured rock the genesis and morphology of the rock is of fundamental importance; one of the most recent treatments was given by Bles and Feuga (1986). No attempt will be made to provide a detailed geological description of fractured-rock systems; however, certain characteristics can have very significant effects on transport and are therefore given specific mention.

A system of fractures may not be continuous and thus may not provide a significant pathway for regional transport. Fracture zones are complex and include shear fractures, extension fractures, brecciated zones, and stress dissolution features, together with various filling and alteration products. Most systems exhibit a dominant orientation which results in very significant anisotropy. Shear zones appear to control the hydrology of many sites in argillaceous rocks.

Recently, fractal statistics have been employed in the description of fracture patterns seen in outcrops. Fractal dimensions between 1.12 and 1.16 were obtained by Barton and Larsen (1985) for three tuff outcrops; Jacquin and Adler (1987) obtained a value of 1.8 for an outcrop in a quarry. One approach to understanding these statistics being considered is to model the actual fracturing process.

Metamorphic and igneous rocks generally have low primary (matrix) porosity and permeability so fractures may provide most of the storage capacity and bulk permeability. Sedimentary rocks generally have a much higher primary porosity which dominates that of the fractures; however, that storage may not be accessible for the transport phenomenon within the time-scale of interest.

Where a layer of material has been deposited on the walls of the fracture this may impede the exchange of water, chemicals, and particles between the fracture and the rock matrix. Such a layer is sometimes referred to as a fracture *skin* (e.g. Moench 1984). When the material completely blocks part of a fracture, channelling of the flow is enhanced. These layers are possibly the site of much biological activity.

Fracture aperture size is very important in determining the rate of flow as, for a given hydraulic gradient, the flux is normally found to be proportional to the cube of the aperture. Apertures close up as the stress increases, so the gross permeability of a fractured rock normally decreases significantly with depth. Equivalently, fracture permeability varies with fluid pressure; this has important implications for the hydraulic testing of rocks if large pressure changes are involved.

In recent years, a great deal of evidence has been collected, mostly in crystalline rocks, which shows that fracture apertures are spatially very variable, so the water flow is concentrated in sets of channels. These variations probably result from the mechanical nature of the fracturing. Some of the evidence for *channelling* is given below (see also the section on hydraulic testing).

An experiment was carried out in a tunnel in the Stripa mine in Sweden, where about 400 individual measurements of flow rate, covering an area of 800 m² indicated that flow is very unevenly distributed (Neretnieks 1985). Tracer experiments in the mine (Abelin *et al*. 1985) showed that 90 per cent of the water is carried by about 5–20 per cent of the fracture planes. Similar observations of varia-

bility have been observed in tunnels and other excavations; the points where water emerges are often marked by the precipitation of iron hydroxides. Many experiments to measure fracture permeabilities in boreholes have found that only a small proportion of the visible fractures carry any significant amount of water. Bourke *et al*. (1985) described an experiment where five parallel holes were drilled in the plane of a fracture in granite in Cornwall. Flows between sections of the boreholes, isolated by packers, showed that flow took place in only a few channels occupying about 10 per cent of the fracture plane.

Perhaps the most direct evidence for channelling, on a small scale, comes from metal injection experiments described, for example, by Pyrak-Nolte *et al*. (1987): a low melting-point metal is injected into a fracture which is taken apart and photographed when the metal has cooled.

Channelling is important because:

(1) it determines effective porosity and hence velocities;

(2) the contact area controls the amount of sorption and diffusion into the matrix; and

(3) flow interaction between intersecting fractures is sensitive to the density of the channels.

Flow and advection

The term advection (or convection) is used to refer simply to something being carried along in a flow. This transport mechanism is therefore essentially determined by the distribution of the flow velocity.

Darcy's law is valid in fractures when the flow is linear-laminar: when the Reynolds number is inversely proportional to the friction factor. The transition from laminar to turbulent flow in fractures appears to occur when the Reynolds number is in the range 2300–2400.

When Darcy's law is valid, each fracture can be ascribed a transmissivity, T_f. For a perfect planar fracture of uniform aperture, a_f, and with smooth surfaces the transmissivity is given by the 'cubic law':

$$T_f = \frac{g a_f^3}{12 \nu}$$

where v is the kinematic viscosity of the fluid. Note the sensitivity of transmissivity to the size of the aperture implied by this equation. The validity of the equation has been discussed by Witherspoon *et al.* (1980) and others.

Theoretical considerations (e.g. Tsang 1984) of the effect of fracture-wall roughness on transport indicate that the aperture density distribution, rather than just the mean aperture, is required to give an adequate description of both flow and advection.

Adsorption

Adsorption of contaminants on fracture surfaces can be an important retarding mechanism. It is greatly enhanced when those contaminants diffuse into a rock matrix which can present a very large active area. The fracture-surface adsorption can then often be ignored, or can be regarded as part of the internal matrix surface.

To date almost all transport models consider the process to be fully reversible and described by a linear adsorption isotherm, ignoring kinetics; an adsorption coefficient, K_d, is used to summarize and characterize the process.

Diffusion

It is now well known that diffusion from fractures into immobile water is often the dominant factor retarding and dispersing solutes and heat in fractured rocks (e.g. Neretnieks 1980).

Table 13.1 summarizes the factors relating the diffusion process. At the bottom of the table typical times (in seconds) are given for diffusive transport over a distance of one metre (to extrapolate to any other length simply multiply the values given by the square of the length (in metres)). The same information is given visually in Fig. 13.1 which shows how the time for diffusion depends on distance. For any process (natural or induced) a point on the graph can be identified; the position of the point in relation to the band for the diffusion process of interest is important in determining the characteristic behaviour and the type of model applicable to the situation. (This will be discussed later in relation to quasi-steady-state models.)

The effects of matrix diffusion on contaminant transport in fractured rock was noted by Foster (1975) in relation to the observed migration of tritium in the Chalk. Adopting the same mechanistic view, Young *et al.* (1976) provided an interpretation of the distribution of nitrate in the Chalk.

There has been some interest in the process of diffusion by migration on the surfaces of pores (e.g. Rasmuson and Neretnieks 1983). However, if adsorption is either linear or negligible, such a phenomenon only serves to modify the apparent diffusion coefficient.

Dispersion

The term dispersion refers to the process of spreading during transportation: solutes, particles, and heat are all dispersed in groundwater. Hydrodynamic dispersion refers more specifically to spreading due to velocity variations in the fluid.

Table 13.1 Diffusive transport processes and parameters

Entity being transported	Water	Solute	Heat
Transport potential	Potentiometric head	Concentration	Temperature
Conductivity parameter	Hydraulic conductivity	Diffusion coefficient	Thermal conductivity
Storage parameter	Specific storage	Porosity	Specific heat × density
Matrix flux law	Darcy	Fick	Fourier
Characteristic time (s)	$10–10^4$	$10^9–10^{11}$	$10^6–10^7$

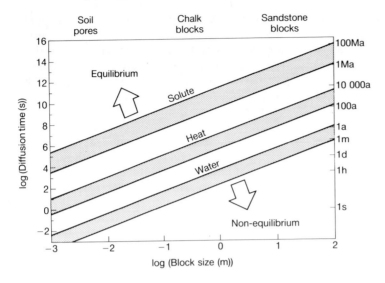

Fig. 13.1. Approximate dependence of times for diffusion through a granular matrix on the block size and the diffusion mechanism.

The widely observed and discussed scale-dependence of *dispersivity* (a parameter used to quantify hydrodynamic dispersion) applies to fractured rock as well as to porous media. Various explanations have been proposed (for example based on channelling and fractal geometry), but the issue remains a very open one.

The variation of velocity across a single fracture will lead to some dispersion; indeed, if there were no mixing mechanism, such as lateral diffusion, the velocity variations would yield an effective dispersion coefficient which would grow continuously with time. However, such dispersion is usually regarded as negligible in comparison with dispersion due to the heterogeneity of the fracture: when fractures branch and combine the amount of dispersion increases greatly. The effect of longitudinal molecular diffusion in a fracture is almost invariably negligible.

Colloids

Colloids occur in groundwater as a result of nucleation of dissolved species and release from the bulk material. Three factors make their presence important in understanding transport:

(1) they offer a large surface area (for adsorption) in relation to their mass;

(2) they diffuse much more slowly than molecules and ions;

(3) they tend to remain in midstream (for example in a fracture as described by Bonano and Bayeler (1985)) and therefore have larger average velocities than molecules and ions.

There is a particular concern about the possibility of radionuclide transport by colloids from repository sites (Advogadro and De Marsily 1984). The transport process is not simply described, however, because of the complex interaction of surface, hydrodynamic, and gravitational forces. Also it is not possible to make very general comments about the relative importance and magnitude of these phenomena as the sizes of colloids span a very large range (about 1–2000 nm). For example, diffusion coefficients are roughly inversely proportional to particle diameter, and therefore vary greatly from one colloid to another. A model of colloid transport in fractured rock has, however, been formulated by Grindrod (1989).

An investigation by Bradbury and Green (1986) included a study of the accessibility of the pore space of granite to colloidal particles in the size range 90–300 nm; these particles neither entered the pore structure nor blocked the pore apertures.

Multiphase flow

Many contaminants, such as organic solvents and oil, enter groundwater systems in the immiscible phase. Although multiphase systems have long been studied in the petroleum field, in the hydrogeological field progress has only just reached the point where models are being produced.

In fractured media capillary effects can be dominant. The wetting phase (often water) tends to occupy the smaller pores while the non-wetting phase tends to occupy the larger pores and fractures, and is thus susceptible to a larger permeability. The physical parameters affecting such flows are: density; viscosity; interfacial tension; wetting properties; and fracture widths, wall roughnesses, and dips. There are as yet inadequate data on the characteristic parameters for real fracture systems (e.g. Chilton *et al.* 1990).

The qualitative behaviour of dense immiscible liquids has been investigated by Schwille (1988) using laboratory models. It seems unlikely that the complex and fascinating behaviour he observed will be accurately modelled in the near future.

Modelling

To a great extent the differences between the various models in use reflect the different conceptualizations of the geometry of the fractures under consideration. Although the conceptual models are normally quite simple, the details of the mathematical and numerical formulation are formidable for fractured media. Only the general conceptual ideas are described and discussed here.

An attempt has been made to divide models into distinct categories. This is only partly successful since many combinations of the various conceptual features of the models are compatible, so a specific model may fall into two or more of the categories specified below.

Models of fractured rock, both analytical and numerical, generally fall into one of two classes: spatially deterministic or spatially stochastic (random). In the former class all geometrical parameters (fracture positions, sizes, etc.) are given specific values, but in the latter class only distributions of values with specified statistics are given.

Equivalent porous medium (EPM) models

When the fracture density is high on the scale of interest, fractured rocks are often assumed to behave as unfractured systems. Then models that have been used for homogeneous rocks can be employed provided appropriate parameter values are used. Such models are termed *equivalent porous* (or *homogeneous* or *continuous*) *medium* models. This approach seems reasonable when it is recalled that even sedimentary aquifers can be regarded as *fractured* at the scale of the grains. However, Sagar and Runchal (1982) provide a cautionary analysis: in particular, a system of fractures might be well represented in terms of flow by an EPM model, but such a model may give very poor estimates of solute travel times. So these models are more likely to be effective in predicting groundwater flows than other transport processes.

Khaleel (1989) considers fractured basaltic rocks where the fractures are defined by the columns of matrix material. For uniform fractures he concludes that EPM models are applicable on a scale about six times the column diameter, but for a lognormal aperture this can be 20–30 times the diameter. Long *et al.* (1982) proposed that the EPM is acceptable provided a plot of the measured directional permeability has the form of an ellipsoid. Troisi *et al.* (1989), who considered only hydrodynamics, proposed a criterion for accepting the EPM approximation based on a point of discontinuity in a plot of Reynolds number against friction factor. Pankow *et al.* (1986) compared two contaminated sites in fractured rock and concluded that the combination of characteristics made only one suitable for modelling using an EPM model.

Typical EPM models are the NAMMU package developed by the Theoretical Physics Division at Harwell (Herbert *et al.* 1986) and the British Geological Survey models (Noy 1985).

Network models

Various geometric forms are used when modelling fracture networks, including: infinite fractures, finite fractures (in two and three dimensions), fractures on a grid (e.g. Smith and Schwartz 1984), and even self-similar (fractal) networks. However, the vast majority of work has been performed using randomly generated networks, and only these are considered below.

When the number of fractures is relatively large it can be appropriate to regard the system as being characterized by various statistics (e.g. the average fracture length and aperture width). Stochastic network modelling involves the generation of fracture networks conforming to these statistics in a random manner. Specific realizations are created using computer programs with random number generators (the Monte Carlo method). The transport processes of interest are then modelled on each realization, and the results from many realizations are combined to form a stochastic description of the behaviour of the system.

One such model was developed at the Harwell Laboratory and is called NAMNET (Robinson 1982). The Stripa Project has funded the development of the NAPSAC code, also developed at Harwell. Perhaps the widest studies of these models has been carried out at the Lawrence Berkeley Laboratory in California (e.g. Long *et al.* 1982; Long and Witherspoon 1985). Chiles (1989) reviewed the models from the point of view of the statistical description of a fracture system.

One problem that can be investigated using these models is that of the relationship between fracture density and the large-scale connectivity of the fractures (such work can be regarded as an aspect of the branch of physics known as percolation theory (Robinson 1983)). One result is that connected flow paths appear in model simulations when, in two-dimensions, the number of intersections per fracture is between three and four. Also, Robinson (1984) typically found that variability in permeability between different realizations became small on a scale about ten times larger than the mean fracture length.

Once the flow pattern has been established, solute transport can be simulated (various numerical methods can be applied but particle tracking is particularly convenient). It has been found that dispersion does not conform to a Fickian-type description (characterized by a dispersion coefficient).

The great difficulty with stochastic network models is that of calibration against field results. This is a topic into which much research effort is being directed at present (for example Long and Billaux 1987).

Andersson *et al.* (1984) presented a model which is partly deterministic and partly stochastic: observed fracture interceptions by boreholes are employed but additional fractures are added with a density determined by the observed fractures. This approach serves to quantify uncertainty in flow problems and also provides guidance on how to acquire additional information.

The behaviour of flow at fracture junctions is important in determining the transverse or lateral spread (dispersion) of solutes or particles being transported through a network of fractures. Most simulation models assume complete mixing at the intersections, while others route the streamlines across junctions. Laboratory studies have led Hull and Koslow (1986) to propose rules for the routing of streamlines in models.

Channel models

Motivated by evidence of channelling, outlined earlier, specialized channel models have been developed. Neretnieks *et al.* (1982) used such a model to reproduce experimentally measured breakthrough curves. Abelin (1986) interpreted a tracer test over 5 m in an *in situ* fracture using both channel and advection-dispersion models; they gave equally good comparisons with the data. Tsang and Tsang (1987) introduced a variable-aperture channel model. Characterization of the roughness of fracture surfaces is important in applying such models and is a topic of recent interest; studies have included descriptions in terms of fractal dimensions (see Brown 1989 and references therein).

An important effect of channelling is that individual flows may not meet and their waters may not mix over significant distances, consequently a Fickian description of dispersion may not be appropriate even on a large scale.

Double porosity (DP) models

When a significant proportion of the total storage capacity of the system is provided by the rock matrix, while the fractures provide the dominant path for regional transport, a detailed study of the interaction of the fracture and matrix phases is

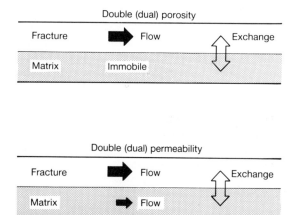

Fig. 13.2. Schematic representation of the difference between double porosity and double permeability models.

required. Perhaps the simplest conceptualization of such systems is provided by double porosity models (sometimes confused with dual permeability models, Fig. 13.2). One of the several origins of the concept appears to have been the paper by Barenblatt *et al.* (1960) who, in modelling pumping tests in fractured porous rock, envisaged two overlapping continuous media (the matrix and fracture phases corresponding to the primary and secondary porosity, respectively) with an exchange mechanism.

Fractured rocks have often been represented by sets of identical, equally-spaced parallel conduits (e.g. Grisak and Pickens 1980; Barker 1982). The actual positions of the fractures are normally of no significance, so these models are mathematically equivalent to and correctly described as double porosity models.

In the case of solute transport, for which the matrix phase contains stagnant, or relatively stagnant, fluid, it is not always clear what part of the system is regarded as being stagnant. In crystalline rocks microfissures clearly play a role (Fig. 13.3), but there may also be stagnant zones within the main fractures as a result of the variations in aperture width and fracture filling materials. Note the dead-end porosity depicted in Fig. 13.3.

In general, a multiple permeability model would be preferable but, given the quality of data and

various uncertainties, a double porosity approximation is practical and appropriate. It is probably the simplest conceivable model which has a distribution of velocities (two delta functions).

Diffusive type

Diffusive models are those for which the transport in the blocks of rock matrix can be described by one of the flux laws given in Table 13.1. The potential (head, concentration, or temperature) at the surface of the blocks is equal to that in the fracture system (provided there is no fracture skin).

The blocks are taken to exist in a (mathematical) space outside the normal three-dimensional transport space: the only case that can be visualized easily is of one-dimensional advection in the fractures with one-dimensional diffusion into slabs of matrix material. The shapes of the matrix blocks are important in determining the behaviour and Barker (1985b) was able to show that all of the information concerning this shape could be described by a single mathematical function (termed the block-geometry function). This function can be extended to describe fracture skin (Barker 1985c). The block shapes normally treated are: slabs (also known as slides), cylinders, and spheres; Barker (1985c) gave a mathematical formulation encompassing all three types.

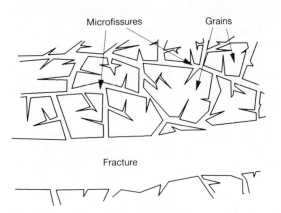

Fig. 13.3. Schematic diagram of a fractured crystalline rock (not to scale). Typical sizes: fracture aperture, 0.01–1 mm; microfissure aperture, 0.01–10 μm; grain sizes, 0.1–10 mm.

Quasi-steady-state (QSS) type

Probably the simplest physically reasonable description of interchange between two phases is when the flux, F, between them is proportional to the difference in their potentials (heads, temperatures, or concentrations):

$$F = \alpha(\phi_f - \phi_m)$$

where α is an exchange coefficient and ϕ_f and ϕ_m are the potentials of the fracture and matrix phases respectively. This is sometimes referred to as a quasi-steady-state approximation, as the equation describes diffusive steady-state transport between two systems at fixed potentials. (Some writers prefer to reserve the term dual-porosity for systems with this simple exchange process.)

This type of model was proposed by Barenblatt *et al.* (1960) for pumping tests, by Coats and Smith (1964) for solute diffusion into dead-end pores, and by O'Neill (1977) for thermal transport. The model can be recommended for its simplicity, particularly when details of the exchange process are not well understood. It has served well to describe transport in soils and even weathered chalk (Black and Kipp 1983). It also appears to be appropriate when fracture skin dominates the exchange process between phases (Moench 1984). However, when viewed as an approximation to a diffusive model, the QSS model is only appropriate when changes within the fractures are slow in relation to the time for diffusive equilibrium (Table 13.1) across a matrix block or other immobile fluid zone. The exchange coefficient can then be related to the block geometry (Barker 1985*b*, 1985*c*), but the diffusive and quasi-steady-state models are incompatible for relatively fast changes within the fractures.

Figure 13.1, introduced earlier, indicates the relationship between the transport process, the matrix block size, and the time-scale of changes. The QSS model can be used for systems (including tracer tests and pumping tests) represented by points which are above the appropriate band for the diffusion process. The figure shows that soil tracer experiments are well represented by the quasi-steady-state model because they are carried out on a time-scale which is long compared to the characteristic times for diffusive equilibrium across

a zone of stagnant pore-water (this is not the case for aggregated soils).

There is some evidence that the effective value of the coefficient, α, varies with the flow velocity, and this has been taken to indicate that the exchange mechanism is, at least partly, advective rather than diffusive (Raven *et al.* 1988).

The well-known Water Research Centre model of nitrate transport (Oakes 1982, 1989) includes the effect of matrix diffusion by an exchange term of the quasi-steady-state type.

Dual permeability models

The term dual (or double) permeability is sometimes used synonymously with double porosity. However, it is better to reserve the term for the case where there is significant regional transport in the rock matrix (Fig. 13.2). Unfortunately, the term has also been adopted to describe a specific type of numerical model (Miller and Clemo 1988) where large fractures are modelled discretely and smaller fractures are modelled as a continuum. Baca *et al.* (1984) presented a method for modelling double permeability rock with discrete fractures; they use finite elements to represent the blocks but line elements for the fractures. A similar boundary integral formulation was presented by Rasmussen *et al.* (1989).

Numerical double porosity models

Barker (1985*c*) gave a general mathematical formulation of the double porosity and QSS models as a partial integro-differential equation. The various numerical models of double-porosity behaviour that have been developed can be regarded as numerical formulations of that equation. Bibby (1981) presented a model with parallel fractures (effectively a double-porosity model with slab-shaped matrix blocks) which combined a finite-element fracture formulation with an analytical solution for the matrix diffusion. Preuss and Narasimhan (1985) introduced a multiple interacting continua (MINC) model, based on the double-porosity medium approach, where the matrix material is divided into layers of various distances from the fracture/matrix block surface. This conceptual model was extended by Neretnieks and Rasmuson (1984) to model mixtures of blocks of various shapes and sizes. They added contributions from all blocks, of whatever shape and size, into an equivalent single block, referred to as a pseudobody.

NAMSOL is a very general transport code developed at Harwell which includes matrix diffusion (Cherrill *et al.* 1987). It uses the approach introduced by Huyakorn *et al.* (1983) where each fracture node is connected to a set of nodes extending into the matrix blocks, which may be either planar or spherical.

Multiphase models

Multiphase flow modelling of double porosity formations is very much in its infancy in hydrogeology. However, it is a problem which by necessity has been addressed in studies of petroleum reservoirs. The problems studied involve the processes of gas/oil drainage and water/oil imbibition (which are analogous to the hydrogeological processes of unsaturated flow and pollution by immiscible fluids). A recent paper by Rossen and Shen (1989) provides a brief review of the modelling literature, while van Golf-Racht (1982) and Torsaeter *et al.* (1987) provided descriptions of the processes involved. Dykhuizen (1987) summarized the modelling of transport through unsaturated fractured rock.

Analytical techniques

The transport equations are often linear: the sum of two solutions is also a solution. Under these conditions the solution for a homogeneous system with simple boundary conditions can be found in the form of a Laplace transform (of the time-dependent head, concentration, or temperature). Inversion of the transforms, to reveal solutions in terms of time rather than a transform parameter, is often very difficult (in many cases impractical), and the results very complex (several erroneous results have appeared in the literature). Also, these results normally involve singular integrals which are numerically difficult and expensive to evaluate. An alternative technique advocated by Barker (1982, 1985a) is to evaluate the time-dependent solution directly from the transform by numerical inversion. The inversion algorithm described by Talbot (1979), was found to be particularly efficient and accurate. This approach has been fairly widely adopted; for example, it pro-

vides the basis of the NAM1D program (Hodgkinson *et al.* 1984) for one-dimensional transport in fractured rock which includes radioactive chain decays.

Laplace-transform solutions for solute transport in a set of parallel fractures were given by Barker (1982): a more comprehensive set of solutions was given by Lever *et al.* (1983), who provided a careful account of the various definitions of the diffusion coefficient and the parameters determining retardation. Solutions for pumping tests in fractured rock can be found in Moench (1984) and Barker (1985a).

Analytical solutions play an important role in that they permit a better understanding of the effects of uncertainties both in parameter values and the form of governing equations. When Monte Carlo methods are necessary, many thousands of realizations are readily computed for analytical solutions (e.g. Chilton *et al.* 1990).

International programmes

Because of the particularly stringent controls necessary for the safe disposal of radioactive waste, a number of international exercises have been undertaken to test the validity of codes for repository simulations. The reports of these programmes provide many examples of the use of fractured-rock transport models.

Starting in 1981, three projects: INTRACOIN, HYDROCOIN, and INTRAVAL, were set up by the Swedish Nuclear Power Inspectorate (SKI) to improve the understanding of the reliability of groundwater modelling strategies for radioactive waste disposal safety assessments. The first two were divided into three levels: verification, validation, and application. To give some idea of the scale of the exercise, twenty organizations from eleven countries took part in the verification phase of HYDROCOIN: lessons learnt from the project were reported by Cole *et al.* (1987).

INTRAVAL was initiated in October 1987 as a successor to INTRACOIN and HYDROCOIN, with the aim of considering the problems of validating geosphere models using data from experiments in the field and the laboratory as well as from natural analogues in a systematic manner. In a related undertaking, the Commission of Euro-

pean Communities established the Natural Analogues Working Group, the aim of which is to
bring together earth scientists and modellers to
study natural geological systems as analogues of
waste repositories.

Field and laboratory investigations

Fractured formations are generally highly heterogeneous and anisotropic, data requirements are
rather different from those of granular formations.

Field laboratories

Many field laboratories have now been established
for the *in situ* study of transport in fractured rock.
Perhaps the best known is the underground
research facility at Stripa; it is situated in a granite
body adjacent to a leptite which was mined for
iron ore until 1976. It is used for research into
nuclear waste storage: this began in 1977, and the
International Stripa Project was initiated in 1980.
Similar facilities exist in the Fanay-Augeres uranium mine in France and the Grimsel rock laboratory in central Switzerland. Atomic Energy of
Canada Ltd have an underground research laboratory in a large granitic pluton in Manitoba.

An experimental hot dry rock system was established in 1980 in the Carnmenellis granite at Rosemanowes in Cornwall, as part of the Department
of Energy's Renewable Energy Research and
Development Programme; it is described in Chapter 17. Observations of pressure and temperature
changes as well as the results of various tracer tests
for the period 1983–1986 are summarized by the
Camborne School of Mines (1986). A number of
models have been employed in the analysis of the
system, being generally in the form of a small set
of parallel fractures, conceptually similar to the
channel models described earlier.

Tracer tests

In contrast to hydraulic tests, which monitor the
diffusive transmission of pressure pulses, tracer
tests monitor the actual movement of the fluid,
which is governed by a convective equation. The

importance of tracer tests in predicting transport
was emphasized by Horne (1989) who compiled
the results of a large number of tracer tests on
fractured geothermal reservoirs. He concluded
that transport rates could not be predicted without
the results of tracer tests.

The tracer can take many forms: chemicals,
isotopes, fluorescent dyes, particles, bacterio-
phages, (Skilton 1987; Skilton and Wheeler 1989)
heat, or even fresh water. Different tracers tend to
behave differently, for example larger tracers diffuse more slowly from the transient fracture water
into the stagnant water of microfissures or the
pores of the rock matrix. Analysis of these differences can aid the understanding of a fracture
system.

A significant problem in the interpretation of
the results of tracer tests is that several loss and
retardation mechanisms are normally active
(adsorption, chemical reaction, ion exchange, and
diffusion into various regions of stagnant water).
These phenomena are described by similar mathematical terms in the transport equations and are
therefore difficult to distinguish. This leads to a
lack of uniqueness both in the parameter sets
obtained and the form of the transport model:
successful calibration of a model must not be
confused with validation.

When the tracer is injected as a short pulse the
breakthrough at any observation point has a very
characteristic form in fractured rock (Fig. 13.4):
after a relatively sharp rise, the concentration/time
curve usually exhibits a long tail. The tail is interpreted as being due to relatively slow paths and/or

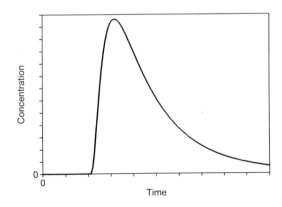

Fig. 13.4. Typical form of a breakthrough curve
for a pulse-injection tracer test in fractured rock.

retardation in stagnant water. Both channel models and double porosity models have enough free parameters to fit such a curve, so extra information must be obtained to avoid the lack of uniqueness in the interpretation. The theory of tracer tests in double porosity systems was reviewed by Maloszewski and Zuber (1985), who have made very significant contributions to the field.

Black and Kipp (1983) described a tracer experiment on an irrigated plot on the Lower Chalk. A quasi-steady-state double-porosity model fitted the data well but revealed a large fracture porosity. Wellings (1984) described a similar tracer test on unsaturated Upper Chalk and concluded that the flow was predominantly through the matrix pores. Price *et al.* (1989) presented a study of a potential pollution problem in the Chalk using fluorescent dye and bacteriophage tracers: speeds of movement in excess of 1 km d^{-1} were observed.

Radon is a potentially very useful tracer in granitic rocks (Chapter 15). It is produced within the rock by direct alpha-recoil and diffuses along grain boundaries or microfractures into the fractures. Andrews *et al.* (1986) used laboratory measurements in conjunction with a simple flow model to estimate the heat-transfer area of the Rosemanowes hot dry rock system. Many other tracer tests for that system are described in CSM (1986, 1988). The possibility of using colloids as tracers in the system was considered by Stelfox (1989).

Hodgkinson and Lever (1983) successfully analysed a field tracer test, of strontium and iodine, (as part of the INTRACOIN study) using a single fracture model with matrix diffusion and kinetic sorption. They concluded that matrix diffusion is the dominant dispersion mechanism. The diffusion coefficient they obtained was several orders of magnitude higher than the laboratory values obtained by Bradbury *et al.* (1982) and Skagius and Neretnieks (1982) for intact samples of granitic rock. This difference was taken to be the result of alteration of the rock in the vicinity of the natural fracture. On the other hand, they obtained a value for the diffusion of strontium which was seven hundred times larger than a laboratory value obtained by Skagius *et al.* (1982). But in this case the difference was probably due to the introduction of cracks and microfissures as a result of crushing of the laboratory samples.

With large cores quite sophisticated laboratory investigations are possible. For example Neretnieks *et al.* (1982) performed a tracer test on a 30 cm long, 20 cm diameter core from the Stripa mine. Both sorbing and non-sorbing tracers were used, and the results indicated that there was substantial diffusion into and sorption within the intercrystalline microfissures of the rock matrix.

Hydraulic tests

Standard forms of pumping test have been adapted, both in terms of field technique and data analysis, to the fractured rock environment.

A straddle packer system developed for testing low-permeability formations (Holmes 1981), has been used to measure the hydraulic conductivity in three deep boreholes in the Strath Halladale granite of Caithness. Less than 20 per cent of the tested sections carried 80 per cent of the water, and there was a significant decrease in permeability with depth.

An extensive research programme on the hydraulic properties of single fractures in Cornish granite was described by Heath and Durrance (1984) and Bourke *et al.* (1985). Constant head tests between packers were carried out on every metre of a 700 metre borehole, and analysed (Hodgkinson 1984) using a model similar to that developed by Barker (1981*a*). This study revealed an approximately log-normal distribution of fracture transmissivities, but also that only 10 per cent of the fractures either recorded by the seisviewer or seen in the core had a detectable flow rate: this was taken as strong evidence for channelling.

Price *et al.* (1982) described double-packer injection tests in the Chalk of Hampshire and the Permian Penrith Sandstone of Cumbria. Data from intervals containing fissures were analysed using a specially developed formula (Barker 1981*a*).

Using an analytical model, Barker and Black (1983) showed that a standard slug-test analysis in a double-porosity fractured rock will always tend to overestimate the transmissivity values. They also presented graphs for estimating the radius of influence of the test. Black (1985) gave a more practical discussion of the problems of interpreting slug tests in fractured rock.

Cross-hole packer tests are particularly applicable to fractured media. Hsieh and Neuman (1985*a*, 1985*b*) described a method for determining the three-dimensional hydraulic conductivity tensor. Black *et al.* (1986) described results of sinusoidal cross-hole tests in the Stripa mine: this technique was introduced by Black and Kipp (1981), who discussed its advantages and limitations.

Constant rate tests in double-porosity formations exhibit a particularly important characteristic behaviour (Fig. 13.5). At early times (period I in the figure) the drawdown depends only on the fracture system until leakage from the matrix porosity decreases the rate of increase of drawdown, during what is sometimes referred to as the interporosity-flow period (period II). Then, as the matrix blocks approach hydraulic equilibrium with the fractures, the rate of leakage from them decreases and the rate of drawdown increases again (period III). When the results of such a test are plotted on a semi-log scale, as in Fig. 13.5, the three characteristic slopes are inversely proportional to the aquifer (fracture) transmissivity at early times, twice that value at intermediate times, and that value again at long times. Such behaviour can easily be misinterpreted as being due to barriers and/or recharge boundaries. Often the first change of slope will be at such an early time that it will be missed; it is also masked by the complicating effects of wellbore storage and formation damage near the well. The behaviour depicted in Fig. 13.5 is not unique to fractured formations, it is also characteristic of layering with significant permeability contrasts between the layers (Barker 1981*b*; Gringarten 1984).

In most studies of hydrogeological systems the geometry of the system is well known. In fractured systems this is far from being the case, yet in most investigations the geometry must be specified a priori. When performing pumping tests the data analysis will conventionally be based on a model of one, two, or three dimensional flow. In fractured rock the problem is not simply that the correct dimension is not obvious, but also that a single dimension may not be adequate or that a non-integer dimension may be appropriate! Barker (1988) presented a model of hydraulic tests for an arbitrary dimension, thus generalizing many well-known solutions such as the Theis and Thiem equations. When applied to sinusoidal cross-hole tests in the Stripa mine (Noy *et al.* 1988) the model revealed dimensions in the range 1.2–2.2. This new model clearly requires further validation, and will only be applicable under certain conditions (relating to fracture density and connectivity). It does, however, appear to have a connection with the fractal description of fracture densities.

Geophysical techniques

Surface geophysical methods can help identify the position and orientation of major fracture systems: vertical seismic profiling (Stewart *et al.* 1981) is being developed into a particularly effective tool. Standard borehole geophysical methods can provide estimates of fracture densities (somewhat more reliably than with cores) but Nelson (1985) warned of non-uniqueness in the interpretation of well logs. An acoustic televiewer can detect fracture apertures as small as 1 mm, and indicate their orientations. Down-hole cameras can provide similar information. The high sensitivity of heat-pulse and electromagnetic flowmeters allows the measurement of the vertical movement of water in boreholes under natural as well as pumping conditions: contributions to the flow of individual fractures may thus be determined.

Radar reflection techniques and seismic and radar tomography, which have improved greatly in recent years, have begun to provide valuable spatial information on fractures (Olsson *et al.* 1987; Black 1987, and references therein).

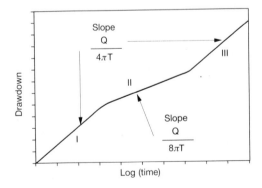

Fig. 13.5. Characteristic form of the drawdown curve for a constant-rate pumping test in a double porosity system.

Isotope techniques

Neretnieks (1981) pointed out that in double porosity systems the residence time of flowing water can differ (by several orders of magnitude) from the residence time of diffusing species (as would normally be determined from age dating with ^{14}C or ^3H). The implications for the dating of groundwater in the Chalk and Pleistocene sands of Suffolk were considered briefly as part of a study by Bath *et al.* (1985): they concluded that the apparent age probably reflects the average residence time in the fissures and matrix.

For numerous case studies and detailed technical information the reader is directed towards the proceedings of a specialists' meeting on the topic: IAEA (1989).

Core testing

During drilling operations the core material, if properly handled, can provide information on the density and orientation of fractures. Laboratory studies can reveal further information on the roughness and surface properties (often affected by mineral precipitation). The porosity and permeability of the rock matrix are, as always, important but sometimes the adsoprtion and diffusion characteristics need also to be determined.

Fracture porosity has been measured using the techniques of computerized tomographic scanning and fluorescent epoxy impregnation (Bergosh and Lord 1987).

Measurements of diffusion coefficients

The importance of matrix diffusion in the retardation of solute migration was emphasized earlier. There is a clear need for modellers to know the values of diffusion coefficients of various species in various types of rocks. Many groups have now made such measurements. Typically a saturated disc of the rock is placed between two water-filled chambers into one of which the solute is introduced. Analysis of changes in concentration in the chambers with time are used to deduce the diffusion coefficient.

There is significant room for confusion over the meaning of the term diffusion coefficient. Brief clarification is given here for the case of diffusion without adsorption or surface diffusion, and when the concentration is low, so diffusion is independent of concentration. Lever *et al.* (1983) discuss diffusion coefficients under more general conditions.

Fick's first law applied to diffusion of a solute in water (or any other solvent) relates the diffusive flux, F, to the concentration gradient. For one-dimensional transport in the x direction:

$$F = - D_T \frac{\partial c}{\partial x}$$

where D_T is termed the tracer (or free-water) diffusion coefficient (the term self-diffusion coefficient applies only to diffusion of molecules within identical molecules).

The above equation applies to a saturated porous medium but a smaller diffusion coefficient, D_E, must be used because of the impediment to transport caused by the matrix. The terms effective and intrinsic are used to describe this coefficient.

Yet another value, the apparent diffusion coefficient, D_A, arises when time-dependent diffusion in a porous medium is described by Fick's second law:

$$\frac{\partial c}{\partial t} = D_A \frac{\partial^2 c}{\partial x^2}$$

These diffusion coefficients are related by $D_A = D_E/\phi$ and $D_E = D_T \Psi$ where ϕ is the porosity and Ψ is the diffusibility. (The possibility of dead-end pores which contribute to storage but not the diffusion surface (see for example Lever *et al.* 1985) has been ignored.) The diffusibility is taken to be an intrinsic property of the porous medium (rather like permeability), and is sometimes expressed as $\Psi = \phi \delta / \tau^2$, where δ is the constrictivity and τ is the tortuosity.

Hill (1984) reported measurements of the effective diffusion coefficients of nitrate, chloride, sulphate, and water (labelled using tritium) in samples of both cracked and uncracked chalk. The smallest values were for sulphate (0.28×10^{-10}–1.47×10^{-10} m^2 s^{-1}); similar values were found for nitrate (0.53×10^{-10}–3.2×10^{-10} m^2 s^{-1}) and chloride (0.52×10^{-10}–3.23×10^{-10} m^2 s^{-1}), and tritiated water gave slightly larger values (0.6×10^{-10}–3.51×10^{-10} m^2 s^{-1}).

Numerous diffusion experiments have been carried out at the Harwell Laboratory (Bradbury *et al.* 1982). Short-time behaviour in these experiments indicated non-Fickian behaviour, but a dead-end pore model (Lever *et al.* 1985) gave a reasonable fit to all of the data. Bradbury and Green (1985) report diffusion coefficients (mostly for iodide) and related parameters for four United Kingdom granites: Ossian, Scottish Lowland, Skene Complex, and the Cornish Carnmenellis. Lever *et al.* (1985) measured the effective diffusion coefficient of potassium iodide in Darley Dale sandstone (in addition to that of some granites), and obtained a value of 2.8×10^{-11} m^2 s^{-1}. Feenstra *et al.* (1984) measured effective diffusion coefficients of chloride in 11 core samples of sandstone and obtained values in the range 3.4×10^{-12}–3.2×10^{-11} m^2 s^{-1}.

Measurements of the diffusion of radon in various rock types are summarized by Andrews *et al.* (1986). There have been numerous measurements of the diffusion coefficients for clays because of their importance as barriers to migration in waste repositories.

Conclusions

A complete characterization of a fractured rock system includes the dimensions of the system—the fracture positions, lengths, widths, and apertures (and spatial variations of those apertures)—and the porosity and permeability of the matrix. The collection of such information is neither feasible nor economic within the context of most groundwater investigations. Even within the context of feasibility studies of radioactive waste disposal, where financial resources are relatively large, only limited success can be claimed. Without detailed characterization of fractured-rock systems, uncertainty in predictions is a fact that will have to be endured for the foreseeable future. Further effort must be made to develop geophysical, tracer, and pumping-test techniques, both individually and as combined tools, to obtain a reasonable body of information for model formulation and calibration. But even then predictions of behaviour of the system must be formulated and presented in a probabilistic manner.

Much effort has been put into the development of a variety of mathematical models to deal with various types of fractured rock (differing particularly in terms of the fracture topology). The calibration and validation of all but the simplest of these models poses a formidable task, with great demands on hydrogeologists and geophysicists to provide adequate data. So, although the models often have the potential to provide useful predictions, they can only be used to investigate a variety of possible scenarios with parameters being specified in the form of distributions of values.

Those working outside the radioactive waste field should take note of the work carried out in this area, both for the findings reported and to be aware of the enormous effort required to understand and characterize fractured rock systems. Also, the successful international collaborations in the radwaste field might be emulated in relation to equally important world-wide problems such as the pollution of groundwaters by pesticides and organic solvents.

Acknowledgements

This chapter is published with the approval of the Director, British Geological Survey (NERC).

References

Abelin, H. (1986). Migration in a single fracture: An *in situ* experiment in a natural fracture. Ph.D. thesis. Royal Institute of Technology, Stockholm.

Abelin, H., Neretnieks, I., Tunbrant, S., and Moreno, L. (1985). *Final report on the migration in a single fracture: experimental results and evaluation*. KBS Technical Report TR 85–03. Swedish Nuclear Fuel and Waste Management Co., Stockholm.

Andersson, J., Shapiro, A. M., and Bear, J. (1984). A stochastic model of fractured rock conditioned by measured information. *Water Resour. Res.*, **20**, 79–88.

Andrews, J. N., Hussain, N., Batchelor, A. S., and Kwakwa, K. (1986). ^{222}Rn solution in the circulating fluids in a 'hot dry rock' geothermal reservoir. *App. Geochem.*, **1**, 647–57.

Avogadro, A. and de Marsily, G. (1984). The role of colloids in nuclear waste disposal. *Proc. Mat. Res. Soc. Symp.*, **26**, 495–505.

Baca, R. G., Arnett, R. C., and Langford, D. W. (1984). Modelling fluid flow in fractured-porous rock masses by finite-element techniques. *Int. J. Num. Meth. in Fluids*, **4**, 337–48.

Barenblatt, G. I., Zheltov, Y. P., and Kochina, I. N. (1960). Basic concepts in the theory of seepage of homogenous liquids in fissured rocks. *J. Appl. Math. Mech.* (English translation), **24**, 852–64.

Barker, J. A. (1981*a*). A formula for estimating fissure transmissivities from steady-state injection-test data. *J. Hydrol.*, **52**, 337–46.

Barker, J. A. (1981*b*). A *multilayered aquifer model for the Marchwood pumping test*. Technical Report, WD/ST/81/14. British Geological Survey, Keyworth.

Barker, J. A. (1982). Laplace transform solutions for solute transport in fissured aquifers. *Advan. Water Res.*, **5**, 98–104.

Barker, J. A. (1985*a*). Generalized well-function evaluation for homogeneous and fissured aquifers. *J. Hydrol.*, **76**, 143–154.

Barker, J. A. (1985*b*). Block-geometry functions characterizing transport in densely fissured media. *J. Hydrol.*, **77**, 263–79.

Barker, J. A. (1985*c*). Modelling the effects of matrix diffusion on transport in densely fissured media. In *Hydrogeology in the service of man*, pp. 250–69. Memoirs 18th Congress Int. Assoc. of Hydrogeo., Cambridge.

Barker, J. A. (1988). A generalized radial flow model for hydraulic tests in fractured rock. *Water Resour. Res.*, **24**, 1796–804.

Barker, J. A. and Black, J. H. (1983). Slug tests in fissured aquifers. *Water Resour. Res.*, **19**, 1558–64.

Barton, C. C. and Larsen, E. (1985). Fractal geometry of two-dimensional fracture networks in Yucca Mountain, Southwest Nevada. *Proc. Int. Symp. on Fundamentals of rock joints*, (ed. O. Stephensson), pp. 77–84, Bjorkliden, Lapland, Sweden.

Bath, A. H., Downing, R. A., and Barker, J. A. (1985). *The age of groundwaters in the Chalk and Pleistocene Sands of north-east Suffolk*. Technical Report, WD/ST/85/1. British Geological Survey.

Bergosh, J. L. and Lord, G. D. (1987). *New developments in the analysis of cores from naturally fractured reservoirs*. SPE Ann. Tech. Conf. SPE 16805, Dallas.

Bibby, R. (1981). Mass transport studies of solutes in dual-porosity media. *Water Resour. Res.*, **17**, 1075–81.

Black, J. H. (1985). The interpretation of slug tests in fissured rocks. *Q. J. Eng. Geol.*, **18**, 161–71.

Black, J. H. (1987). Field measurement, model prediction and model assessment: Their role in phase III of the Stripa Project. *Proc. GEOVAL-87*, **12**, pp. 457–70. Swedish Nuclear Power Inspectorate, Stockholm.

Black, J. H., Barker, J. A., and Noy, D. J. (1986). *The method, theory and analysis of crosshole sinusoidal pressure tests in fissured rock*. Stripa Project Report TR 86–03, Swedish Nuclear Fuel and Waste Management Co., Stockholm.

Black, J. H. and Kipp, Jr, K. L. (1981). Determination of hydrogeological parameters using sinusoidal pressure tests: a theoretical appraisal. *Water Resour. Res.*, **17**, 686–92.

Black, J. H. and Kipp, Jr, K. L. (1983). Movement of tracers through dual-porosity media—experiments and modelling in the Cretaceous Chalk. *J. Hydrol.*, **62**, 287–312.

Bles, J. L. and Feuga, B. (1986). *The fracture of rocks*. Elsevier, New York.

Bonano, E. J. and Bayeler, W. E. (1985). Transport and capture of colloidal particles in single fractures. In *Scientific basis for nuclear waste management*, No. VIII (ed. C. M. Jantzen *et al.*), pp. 385–92. Proc. Mat. Res. Soc. Sym.

Bourke, P. J., Durrance, E. M., Hodgkinson, D. P., and Heath, M. J. (1985). *Fracture hydrology relevant to radionuclide transport*. Report AERE-R 11414 of the Atomic Energy Research Establishment, Harwell.

Bradbury, M. H. and Green, A. (1985). Measurement of important parameters determining aqueous phase diffusion rates through crystalline rock matrices. *J. Hydrol.*, **82**, 39–55.

Bradbury, M. H. and Green, A. (1986). Investigations into the factors influencing long range diffusion rates and pore space accessibility at depth in granite. *J. Hydrol.*, **89**, 123–39.

Bradbury, M. H., Lever, D. A., and Kinsey, D. V. (1982). Aqueous phase diffusion in crystalline rocks. *Scientific basis for radioactive waste management*, (ed. W. Lutse) **V**, pp. 569–78, Elsevier, Berlin.

Brown, S. R. (1989). Transport of fluid and electric current through a single fracture. *J. Geophys. Res.*, **94**, 9429–38.

Cherrill, T. P., Herbert, A. W., and Jackson, C. P. (1987). *The extension of NAMSOL to model the effect of rock-matrix diffusion*. Report AERE-R 12231 of the Atomic Energy Research Establishment, Harwell.

Chiles, J-P. (1989). Three-dimensional geometric modelling of a fracture network. *Proc. conf. on Geostatistical, sensitivity, and uncertainty methods for ground-water flow and radionuclide transport modeling*, (ed. B. E. Buxton) pp. 361–85. Battelle Press, Columbus.

Chilton, P. J., Lawrence, A. R., and Barker, J. A. (1990). Chlorinated solvents in chalk aquifers: Some preliminary observations on behaviour and transport. In *Chalk*, pp. 605–10, *Proc. Int. Chalk Symp.* (Brighton, Sussex, September 1989). Thomas Telford, London.

Coats, K. H. and Smith, B. D. (1964). Dead-end pore volume and dispersion in porous media. *Soc. Petrol. Eng. J.*, **4**, 73–84.

Cole, C. R., Nicolson, T. J., Davis, P., and McCartin, T. J. (1987). Lessons learned from the HYDROCOIN experience. *Proc. GEOVAL-87*, **1**, pp. 269–85. Swedish Nuclear Power Inspectorate, Stockholm.

CSM (Camborne School of Mines). (1986). *Geothermal energy project: phase 2B tracer results*. Volume 2. Internal report 2B-40 of the Camborne School of Mines, Cornwall.

CSM (Camborne School of Mines). (1988). *Camborne geothermal energy project circulation results*. Contractors report: ETSU G 137–P5. Energy Technology Support Unit, Department of Energy.

Dykhuizen, R. C. (1987). Transport of solutes through unsaturated fractured media. *Water Resour. Res.*, **21**, 1531–39.

Feenstra, S., Cherry, J. A., Sudicky, E. A., and Haq, Z. (1984). Matrix diffusion effects on contaminant migration from an injection well in fractured sandstone. *Ground Water*, **22**, 307–16.

Foster, S. S. D. (1975). The Chalk groundwater tritium anomaly—a possible explanation. *J. Hydrol.*, **25**, 159–65.

Gringarten, A. C. (1984). Interpretation of tests in fissured and multilayered reservoirs with double-porosity behaviour: Theory and practice. *J. Petrol. Tech.*, **36**, 549–64.

Grindrod, P. (1989). *Colloid–nuclide migration in fractured rock: Mathematical model specification*. Intera-Exploration Consultants Limited, I2145–1, Draft Report Prepared for the Project 90 Workshop, June 1989.

Grisak, G. E. and Pickens, J. F. (1980). Solute transport through fractured media. *Water Resour. Res.*, **16**, 719–30.

Heath, M. J. and Durrance, E. M. (1984). *Radionuclide migration in fractured rock: hydrological investigations at an experimental site in the Carnmenellis Granite, Cornwall*. Report AERE-R 11402 of the Atomic Energy Research Establishment, Harwell.

Herbert, A. W., Hodgkinson, D. P., and Lever, D. A. (1986). Mathematical modelling of radionuclide migration in groundwater. *Q. J. Eng. Geol.*, **19**, 109–20.

Hill, D. (1984). Diffusion coefficients of nitrate, chloride, sulphate and water in cracked and uncracked Chalk. *J. Soil Sci.*, **35**, 27–33.

Hodgkinson, D. P. (1984). *Analysis of steady-state hydraulic tests in fractured rock*. Report AERE-R 11287 of the Atomic Energy Research Establishment, Harwell.

Hodgkinson, D. P. and Lever, D. A. (1983). Interpretation of a field experiment on the transport of sorbed and non-sorbed tracers through a fracture in crystalline rock. *Radioactive Waste Management and the Nuclear Fuel Cycle*, **4**, 129–58.

Hodgkinson, D. P., Lever, D. A., and England, T. H. (1984). Mathematical modelling of radionuclide transport through fractured rock using numerical inversion of Laplace transforms: Application to INTRACOIN Level 2. *Ann. Nucl. Energy*, **11**, 111–22.

Holmes, D. C. (1981). *Hydraulic testing in deep boreholes at Altnabreac: development of testing system and initial results*. Report ENPU 81–4. British Geological Survey, Keyworth.

Horne, R. N. (1989). Uncertainty in forecasting breakthrough of fluid transported through fractures. *Proc. conf. on Geostatistical, sensitivity, and uncertainty methods for ground-water flow and radionuclide transport modeling*, (ed. B. E. Buxton), pp. 261–74. Battelle Press, Columbus.

Hsieh, P. A. and Neuman, S. P. (1985a). Field determination of the three-dimensional hydraulic conductivity tensor of anisotropic media, 1. Theory. *Water Resour. Res.*, **21**, 1655–65.

Hsieh, P. A. and Neuman, S. P. (1985b). Field determination of the three-dimensional hydraulic conductivity tensor of anisotropic media, 2. Methodology and application to fractured rock. *Water Resour. Res.*, **21**, 1667–76.

Hull, L. C. and Koslow, K. N. (1986). Streamline routing through fracture junctions. *Water Resour. Res.*, **22**, 1731–4.

Huyakorn, P. S., Lester, B. H., and Mercer, J. W. (1983). An efficient finite element technique for modelling transport in fractured porous media I. Single species transport. *Water Resour. Res.*, **19**, 841–54.

IAEA (1989). *Isotope techniques in the study of the hydrology of fractured and fissured rocks*. Proc. of an advisory group meeting, International Atomic Energy Agency, Vienna, November 1986.

Jacquin, C. G. and Adler, P. M. (1987). Fractal porous media II: Geometry of porous geological structures. *Transport in Porous Media*, **2**, 571–96.

Khaleel, R. (1989). Scale dependence of continuum models for fractured basalts. *Water Resour. Res.*,

25, 1847–55.

Lever, D. A., Bradbury, M. H., and Hemingway, S. J. (1983). Modelling the effect of diffusion into the rock matrix on radionuclide migration. *Progress in Nuclear Energy*, **12**, 85–117.

Lever, D. A., Bradbury, M. H., and Hemingway, S. J. (1985). The effect of dead-end porosity on rock-matrix diffusion. *J. Hydrol.*, **80**, 45–76.

Long, J. C. S. and Billaux, D. M. (1987). From field data to fracture network modelling: an example incorporating spatial structure. *Water Resour. Res.*, **23**, 1201–16.

Long, J. C. S., Remer, J. S., Wilson, C. R., and Witherspoon, P. A. (1982). Porous media equivalents for networks of discontinuous fractures. *Water Resour. Res.*, **18**, 645–58.

Long, J. C. S. and Witherspoon, P. A. (1985). A relationship of the degree of interconnection to permeability of fractured networks. *J. Geophys. Res.*, **90**, 3087–98.

Maloszewski, P. and Zuber, A. (1985). On the theory of tracer experiments in fissured rocks with a porous matrix. *J. Hydrol.*, **79**, 333–58.

Miller, J. D. and Clemo, T. M. (1988). Dual permeability modelling of flow in a fractured geothermal reservoir. *Proc. 13th annual workshop on geothermal reservoir engineering*, Stanford University, California.

Moench, A. F. (1984). Double porosity model for a fissured groundwater reservoir with fracture skin. *Water Resour. Res.*, **20**, 831–46.

Nelson, R. A. (1985). Geologic analysis of naturally fractured reservoirs. *Contributions in Petroleum Geology and Engineering*, Vol. 1. Gulf Publishing Company, Houston, Texas.

Nelson, R. A. (1987). Fractured reservoirs: Turning knowledge into practice. *J. Petrol. Tech.*, **39**, 407–14.

Neretnieks, I. (1980). Diffusion in the rock matrix: An important factor in radionuclide retardation? *J. Geophys. Res.*, **85**, 4379–97.

Neretnieks, I. (1981). Age dating of groundwater in fissured rock: Influence of water volume in micropores. *Water Resour. Res.*, **17**, 421–2.

Neretnieks, I. (1985). Transport in fractured rocks. In *Proceedings, Memoirs of the 17th Int. Cong. of IAH*, Vol. 17, pp. 301–18, Int. Assoc. Hydrogeol., Tucson, Arizona.

Neretnieks, I. and Rasmuson, A. (1984). An approach to modelling radionuclide migration in a medium with strongly varying velocity and block sizes along the flow path. *Water Resour. Res.*, **20**, 1823–36.

Neretnieks, I., Eriksen, T., and Tahtinen, P. (1982).

Tracer movement in a single fissure in granitic rock: some experimental results and their interpretation. *Water Resour. Res.*, **18**, 849–58.

Noy, D. J., (1985). *Computer codes for three dimensional mass transport with non-linear sorption.* Technical Report, FLPU-85-4. British Geological Survey, Keyworth.

Noy, D. J., Barker, J. A., Black, J. H., and Holmes D. C. (1988). *Crosshole investigations: implementation and fractional dimension interpretations of sinusoidal tests.* Stripa Project Report TR 88–01, Swedish Nuclear Fuel and Waste Management Co., Stockholm.

Oakes, D. B. (1982). Nitrate pollution of groundwater resources—mechanisms. In *Non-point pollution of municipal water supply sources: issues of applications and control*, IIASA Collaboration Proceedings Series CP-82-S4.

Oakes, D. B. (1989). The impact of agricultural practices on groundwater nitrate concentrations. In *World Water* 1989, 45–9. Thomas Telford, London.

Olsson, O., Falk, L., Forslund, O., Lundmark, L., and Sandberg, E. (1987). *Crosshole investigations— results from borehole radar investigations.* Stripa Project Report TR 87–11, Swedish Nuclear Fuel and Waste Management Co., Stockholm.

O'Neill, K. (1977). The transient three-dimensional transport of liquid and heat in fractured porous media. Ph.D. thesis. Princeton University.

Pankow, J. F., Johnson, J. P., Hewetson, J. P., and Cherry, J. A. (1986). An evaluation of contamination migration patterns at two waste disposal sites on fractured porous media in terms of the equivalent porous medium (EPM) model. *J. Cont. Hydrol.*, **1**, 65–76.

Preuss, K. and Narasimhan, T. N. (1985). A practical method for modelling fluid and heat flow in fractured porous media. *Soc. Pet. Eng. J.*, **25**, 14–26.

Price, M., Atkinson, T. C., Wheeler, D., Barker, J. A., and Monkhouse, R. A. (1989). *Highway drainage to the Chalk aquifer: the movement of groundwater in the Chalk aquifer near Brickett Wood, Hertfordshire, and its possible pollution by drainage from the M25 motorway.* Technical Report, WD/89/3, British Geological Survey, Keyworth.

Price, M., Morris, B. L., and Robertson, A. S. (1982). A study of intergranular and fissure permeability in Chalk and Permian aquifers, using double-packer injection testing. *J. Hydrol.*, **54**, 401–23.

Pyrak-Nolte, L. J., Myer, L. R., and Cook, N. G. W. (1987). Hydraulic and mechanical properties of natural fractures in low permeability rocks. *Proc. 6th Int. Congr. on Rock Mechanics*, pp. 225–31. Montreal.

Rasmuson, A., and Neretnieks, I. (1983). *Surface migration in sorption processes*. KBS Technical Report 83–37, Swedish Nuclear Fuel and Waste Management Co., Stockholm.

Rasmussen, T. C., Yeh, T. C. J., and Evans, D. D. (1989). Effect of variable fracture permeability/matrix permeability ratios on three-dimensional fractured rock hydraulic conductivity. Proc. conf. on Geostatistical, sensitivity, and uncertainty methods for ground-water flow and radionuclide transport modeling, (ed. B. E. Buxton) pp. 337–358. Battelle Press, Columbus.

Raven, K. G., Novakowski, K. S., and Lapcevic, P. A. (1988). Interpretation of field tracer test of a single fracture using a transient solute storage model. *Water Resour. Res.*, **24**, 2019–32.

Robinson, P. C. (1982). *NAMNET—Network flow program*. Report AERE-R 10510 of the Atomic Energy Research Establishment, Harwell.

Robinson, P. C. (1983). Connectivity of fracture systems—a percolation theory approach. *J. Phys. A: Math. Gen.*, **16**, 605–14.

Robinson, P. C., (1984). Numerical calculations of critical densities for lines and planes. *J. Phys. A: Math. Gen.*, **17**, 2823–30.

Rossen, R. H. and Shen, E. I. C (1989). Simulation of gas/oil drainage and water/oil imbibition in naturally fractured reservoirs. *SPE Reservoir Engineering*, **4**, 464–70.

Sagar, B. and Runchal, A. (1982). Permeability of fractured rock: Effect of fracture size and data analysis. *Water Resour. Res.*, **18**, 266–74.

Schwille, F. (1988). *Dense chlorinated solvents in porous and fractured media*, (trans. by J. Pankow). Lewis Publishers, Chelsea, MI.

Skagius, C. and Neretnieks, I. (1982). Diffusion in crystalline rocks. *Scientific basis for radioactive waste management*, **V**, (ed. W. Lutse) pp. 509–18. Elsevier, Berlin.

Skagius, C., Svedberg, G., and Neretnieks, I. (1982). A study of strontium and caesium sorption on granite. *Nuclear Tech.*, **59**, 302–13.

Skilton, H. E. (1987). Bacteriophage as models of pathogenic virus behaviour in groundwater. Ph.D. thesis. University of Surrey.

Skilton H. E. and Wheeler, D. (1989). The application of bacteriophage as tracers of chalk aquifer systems. *J. Applied Bacteriology*, **66**, 549–57.

Smith, L. and Schwartz F. W. (1984). An analysis of the influence of fracture geometry on mass transport in fractured media. *Water Resour. Res.*, **20**, 1241–52.

Stelfox, L. (1989). A feasibility study: To determine the value of colloids as tracers in a hot dry rock geothermal reservoir. M.Sc. thesis. University College, University of London.

Stewart, R. R., Turpening, R. M., and Tolsoz, M. N. (1981). Study of subsurface fracture zone by vertical seismic profiling. *Geophys. Res. Lett.*, **8**, 1132–5.

Talbot A. (1979). Accurate numerical inversion of Laplace transforms. *J. Inst. Math. Appl.*, **23**, 97–120.

Torsaeter, O., Kleppe, J., and van Golf-Racht, T. D. (1987). Multiphase flow in fractured rock. In *Advances in transport in porous media*, (ed. J. Bear and M. Y. Corapcioglu). NATO ISI Series, Martinus Nijhoff, Dordrecht.

Troisi, S., Vurro, M., and Castellano, L. (1989). Aspects of hydrodynamics in fissured systems. In *Isotope techniques in the study of the hydrology of fractured and fissured rocks*. Proc. of an advisory group meeting, 17–27. International Atomic Energy Agency, Vienna, November 1986.

Tsang, Y. W. (1984). The effect of tortuosity of liquid flow through a single fissure. *Water Resour. Res.*, **20**, 1209–15.

Tsang, Y. W. and Tsang, C. F. (1987). Channel model of flow through fractured media. *Water Resour. Res.*, **23**, 467–79.

van Golf-Racht, T. D. (1982). *Fundamentals of fractured reservoir engineering*. Elsevier, Amsterdam.

Wellings, S. R. (1984). Recharge in the Upper Chalk aquifer at a site in Hampshire, England, 2. Solute movement. *J. Hydrol.*, **69**, 275–85.

Witherspoon, P. A., Wang, J. S., Iwai, K., and Gale, J. E. (1980). Validity of the cubic law for fluid flow in deformed rock fractures. *Water Resour. Res.*, **16**, 1016–24.

Young, C. P., Oakes, D. B., and Wilkinson, W. B. (1976). Prediction of future nitrate concentrations in ground water. *Ground Water*, **14**, 426–38.

14. Statistical methods of characterizing hydrogeological parameters

R. Mackay and P. E. O'Connell

Introduction

During the 1980s, one of the dominant themes in hydrogeological research was subsurface flow and transport modelling. This research has focused largely on the problem of developing better predictive models of flow and tranport phenomena in both the saturated and unsaturated zones. A major component of this research has been to gain an understanding of the cause of discrepancies between model predictions and observed behaviour. These discrepancies or errors arise from a variety of sources including:

(1) the description of the physical processes governing flow and transport adopted in the model;

(2) the translation of field data into appropriate values for the coefficients of the model (e.g. hydraulic conductivity, porosity, and dispersivity);

(3) the solution method employed.

Methods for quantifying and minimizing the errors arising from each of these sources are actively being sought. The cumulative result of all errors in the modelling process is the difference between the true response of the system under study and the simulated response produced by the model. Uncertainty is the inability a priori to calculate this difference. Over the last ten years, the literature on each of the known sources of error and, more importantly, on the uncertainty that is introduced into the model results has expanded rapidly.

Geological strata are naturally heterogeneous. This heterogeneity is observable at all scales ranging from the microscopic scale at which individual rock grains are identified through to the megascopic scale of entire formations. Moreover, the patterns of fluid movement through the voids of the media are heterogeneous. In general, models of flow and transport do not attempt to describe completely the flow patterns through the media but instead the average (or mean) behaviour of the flow system is simulated using simplified representations of the hydrogeological characteristics of the media. Bear (1972) showed that it is possible to model the movement of fluids in geological media using smoothed representations of the hydraulic parameters of the governing equations of flow by averaging over a sufficiently large volume of the media (the representative elementary volume (REV)). This concept of an equivalent system of smoothly varying parameter distributions that is representative of the averaged conditions in the actual geological media has underpinned the development of almost all the flow and transport models formulated to date.

The first models of groundwater flow were deterministic. Such models represent, in some sense, a 'best' estimate of the actual conditions found in the field. Generally, these models were found to be adequate for the purpose for which they were developed. Errors in the model results were neither so large nor their consequences so severe as to demand a major research effort aimed at understanding or eliminating them. However, early applications of deterministic solute transport models generated far less satisfactory results. Significant errors between the model results and the field data were observed which led to questions being asked about the potential of deterministic models for predicting transport in geological media. One result of this debate has been a move away from deterministic models towards the application of stochastic models.

Stochastic modelling techniques explicitly include for the uncertainty in the parameters of the model and produce predictions which reflect this uncertainty. Since the major source of error in model predictions arises from the lack of knowledge of the hydrogeological properties of the media, stochastic modelling provides a means of quantifying the uncertainty of predictions as a function of the available hydrogeological information. In general, stochastic models require as input statistical information describing the variability of the parameters of the model at any point in the space/time domain.

In this review attention is focused on statistical methods for characterizing the heterogeneity of hydrogeological parameters and on those areas of hydrogeological research to which they are being applied. In the United Kingdom, the radioactive waste disposal research programmes established in the early 1980s generated considerable interest in media characterization. Given the nature of the potential host geological environments for buried repositories it was quickly recognized that uncertainties in model parameterization could not be ignored. As a result much of the research described here has been supported largely by the radioactive waste disposal research programmes. In addition, reference is made to research sponsored by the petroleum industry on problems which are identical, in essence, to those faced by the waste disposal industries. However, the review of contributions from the oil industry is not intended to be exhaustive.

The next section briefly introduces the various research topics in stochastic hydrogeology that are presented. The United Kingdom contributions to this field of research are placed in context within the overall international research effort. More detailed reviews of each of the main areas of research are then presented and, finally some of the more pressing problems in this research field are identified and future prospects for stochastic methods of hydrogeological research and application are considered.

Research topics in stochastic hydrogeology

The term 'stochastic hydrogeology' encompasses various stochastic modelling approaches and statistical estimation methods for characterizing hydrogeological parameters. Research in stochastic hydrogeology falls into a number of distinct but connected categories. Figure 14.1 provides a summary of these categories in the form of a flow chart which indicates the principal interactions between subject areas.

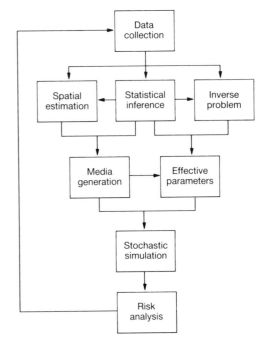

Fig. 14.1. Research areas in stochastic hydrogeology.

In this section, as well as brief descriptions of the main research areas, some definitions of the terminology used throughout this chapter are provided. Two terms, used frequently in the following discussions, warrant immediate clarification. These are 'distribution' and 'simulation'.

The term distribution on its own is only used where reference is made to a spatial distribution such as the distribution of hydraulic conductivity within a geological formation. Where it is used in a statistical sense, for example, probability distribution, then it is explicitly referenced as such.

The term simulation is frequently used in two senses in the literature: first, to describe the mathematical or numerical generation of a parameter distribution, such as hydraulic conductivity, over space and, second, to describe the application of a model for prediction of flow and transport. To avoid any confusion in this chapter, the term simulation is employed only in connection with the application of flow and transport models. The term generation is used to describe the simulation of parameter distributions.

Statistical inference and spatial estimation

Measurements of a hydrogeological parameter of interest are needed to identify its variation within a region. Hard data, i.e. direct measurements of the parameter value, are normally limited to a few sparsely distributed locations. However, additional soft data are frequently available. Soft data include hard measurements of related parameters or measurements using low resolution techniques such as surface geophysics. Interpretations of the nature of the geological environment hypothesized by geologists or hydrogeologists may also be regarded as soft data. Statistical approaches to the problem of identifying parameter variation within a region from the available data, have been developed based on the theory of regionalized variables. In this approach, the media parameters are treated as random functions (de Marsily 1986) on which a limited number of observations are available over the domain of interest.

Various assumptions need to be invoked about the behaviour of a random function in space. A random function is defined by an infinite number (the ensemble) of realizations. Thus, the probability law governing the behaviour of the random function should, in principle, be defined from the ensemble. However, in reality only one realization exists. To overcome this problem, an assumption of ergodicity is made which implies that the single available realization behaves in space with the same probability law as the ensemble of possible realizations. In general, the full form of the probability law of the random function cannot be inferred from the available data. However, additional assumptions permit a probability model (a reduced form of the probability law) of the random

function to be established from the available data. The probability model usually comprises at least two components: a univariate probability distribution describing point value variability over the ensemble of realizations and a covariance function describing the spatial coherence of the point values. The spatial coherence or correlation expresses the observation that measurements at two points are more likely to be similar if the points are close together than if they are far apart.

Various techniques are being developed to use both hard and soft data sources for inferring the probability model describing parameter heterogeneity. This process is known as statistical inference. The term 'geostatistics' has been coined to describe the random function approach to quantifying the variability of spatially heterogeneous parameters and the process of inferring the probability model describing their behaviour. Methods of statistical inference are reviewed later in the section dealing with estimating parameters for hydrogeological models.

Using the probability model and the observed data it is possible to estimate at any point over the domain the most likely value for the parameter. The definition of most likely depends on the assumptions used to perform the estimation. One estimation procedure that is based on the geostatistical approach to parameter identification, Kriging, is a 'best linear unbiased' estimator. The advantages of this method over traditional interpolation schemes such as hand contouring, least squares methods, and mathematical surface fitting methods are twofold: the estimation procedure returns the value of the parameter at the measurement points and an estimate of the possible error in the magnitude of the parameter at any point in space is produced. The application of the various Kriging methods for the estimation of hydrogeological parameters is reviewed below.

The inverse problem

A major problem in modelling flow and transport is the identification of the parameters of the governing equations. The calculation of the parameters from information on the dependent variables, head and concentration, given known geometry and boundary conditions, is known as the inverse problem. Two approaches to the inverse problem are found in the literature, the direct and the indirect approach. The first treats the parameters as the dependent variables of the governing equations and solves for their distribution. The second leaves the heads or concentrations as the dependent variables and iteratively updates the parameter values according to some procedure for minimizing the error between observed and simulated dependent variable distributions. The robustness of inverse techniques for parameter estimation depends on the techniques used for minimization of the error, the level of data availability, and the degree of non-linearity of the problem being solved. In the review of inverse methods presented later only indirect methods are discussed, since they are usually formulated within a statistical framework.

Media generation

The ability to generate, either mathematically or numerically, a realization or realizations of the hydrogeological parameter distributions within a volume of geological media is central to the problem of modelling subsurface flow and transport phenomena. Of particular interest are the spatial characteristics of the hydraulic and hydrogeochemical parameters of the media, notably hydraulic conductivity and dispersivity. Values assigned to these parameters are not true point values but represent averages over a given volume of the media. In order to generate a heterogeneous parameter distribution it is necessary to define the scale of averaging of the parameters. Different field techniques (e.g. pumping tests and core measurements) used to collect data on these parameters adopt different scales of averaging. Relating the results from these different scales of averaging is one of the key issues facing hydrogeological research.

In studies of the impact of heterogeneity on flow and transport simulation and on the scaling of parameters over different averaging volumes, the small scale variations of the parameters need to be generated. In general, insufficient data are available to define the exact form of these variations and statistical methods are used to characterize this heterogeneity. Thus, media generation is nor-

mally carried out using the stochastic approach in which the spatial structure of the media is characterized through a random function. If the probability model governing the behaviour of the random function in space can be defined, then a large number of equally probable realizations of the parameter distribution over the domain may be generated. This approach to media generation requires the solution of two problems;

(1) the determination of the probability model that describes the structure of variations of the parameter over the domain of interest based on the information available from measurements of the parameter;

(2) the generation of parameter values at a defined set of discrete points in space which retain the properties of the media defined by the probability model (unconditional generation) but which can also retain the measurement values in the distribution (conditional generation).

Various techniques have been developed for generating geological media parameters using the stochastic approach. These are reviewed in detail in the section on stochastic simulation models. The nature of the heterogeneity of geological media is such that the choice of appropriate probability models and methods for generating media realizations are still the subject of a large research effort.

Modelling flow and transport and effective parameters

The general equation for flow in a porous medium is:

$$\frac{\partial}{\partial x_i}\left[\frac{\varrho k_{ij}}{\mu}\left(\frac{\partial p}{\partial x_i} + \varrho g \frac{\partial z}{\partial x_j}\right)\right]$$

$$+ \varrho_s w_s = \varrho a \frac{\partial p}{\partial t} + \varrho_0 \varepsilon \beta \frac{\partial \mu}{\partial t} \quad (1)$$

whilst the general equation for single species solute transport in a porous medium can be written as:

$$\varepsilon \frac{\partial c}{\partial t} = \frac{\partial}{\partial x_i}\left[\varepsilon\left(cv_i - D_{ij}\frac{\partial c}{\partial x_j}\right)\right] - c_s w_s \quad (2)$$

where

ϱ is the fluid density $[ML^{-3}]$

μ is the dynamic viscosity $[ML^{-1}T^{-1}]$

k_{ij} is the intrinsic permeability $[L^2]$

p is the fluid pressure $[ML^{-1}T^{-2}]$

g is the acceleration due to gravity $[LT^{-2}]$

z is the elevation above datum $[L]$

ϱ_s is the fluid source/sink density $[ML^{-3}]$

w_s is the fluid source inflow $(+)$ rate or outflow $(-)$ rate $[T^{-1}]$

α is the compressibility coefficient of the medium $[LM^{-1}T^2]$

ϱ_0 is the fluid density at reference pressure/temperature/concentration. $[ML^{-3}]$

ε is the effective porosity

β is the compressibility coefficient of the fluid $[LM^{-}T^2]$

x_i are the coordinates $[L]$

t is time $[T]$

c is concentration of the solute $[ML^{-3}]$

v_i are the components of the Darcian velocity vector $[LT^{-1}]$

D_{ij} are the dispersion coefficients $[L^2T^{-1}]$

c_s is the concentration of the source/sink solute $[ML^{-3}]$

These equations may be solved for any problem subject to appropriate initial and boundary conditions as long as knowledge of the variation of the coefficients (the parameters of the model) over space and time is available. Both deterministic and stochastic approaches may be used for the solution of the equations.

In the deterministic (traditional) approach, the media parameters are assumed to vary in space and time in a manner defined by the modeller on the basis of a conceptual (simplified) model of the system. Using appropriate solution techniques (numerical or analytical) a 'best model estimate' of the true conditions is obtained. To address the problem of possible error, sensitivity analyses can be carried out to examine the response of the model to variations in the form or magnitude of the input parameters. Calibration may also be undertaken to refine the conceptual model and the parameter distributions.

In the stochastic approach, as described previously, each media parameter is treated as a random function. The realizations of the parameters of the model, generated from the probability law of the random functions and conditioned on the available data, then form equally likely representations of the media parameters, and may form the basis of a Monte Carlo experiment. Repetitive solution of the same problem using different realizations of the input parameters allows an estimate of the probability distribution of selected output variables (e.g. head at a point, x) to be made. These probability distributions represent an estimate of the uncertainty in the output of the model resulting from the uncertainty in the input parameters. This approach is referred to here as stochastic simulation.

An alternative approach to Monte Carlo simulation involves the rewriting of equations (1) and (2) in the form of stochastic partial differential equations. In this case perturbations around the expected values for each of the parameters are introduced into the equations and the resulting stochastic partial differential equations solved either analytically or numerically. Again the results of this approach give the probability distribution of the output variables resulting from the input uncertainty defined in the perturbations of the parameters.

In either case, the aim of the stochastic approach is to obtain the probability distribution of the output variables over the set of possible realizations of the input parameters such that statements about uncertainty and risk can be elicited. For simple flow and transport systems and for small perturbations of the input parameters, analytical methods can be used; otherwise, Monte Carlo simulation must be employed.

In the development of the conceptual model for both deterministic and stochastic modelling exercises, local spatial variability below a specified scale of the parameter values is avoided by implicit averaging of the governing equations. In the case of numerical solutions to the governing equations the parameters are usually assumed to be constant or piecewise linear over the grid blocks or elements of the mesh employed to discretize the domain. These averaged parameters are referred to as 'effective parameters'. The validity of effective parameter models for representing flow and transport characteristics in heterogeneous media is an important problem in both stochastic and deterministic hydrogeological modelling. Research into the use and misuse of effective parameter models is reviewed later.

United Kingdom contributions in perspective

The problems of stochastic hydrogeology and media characterization in flow and transport models have been addressed by many researchers world-wide over the last two decades. Much of the early work was related to understanding flow through heterogeneous media but recently more attention has focused on the problems of contaminant transport. Although stochastic simulation of flow through heterogeneous porous media was undertaken as early as 1960 by Warren and Price (1961), it was not until the work of Freeze (1975) on effective parameterization of one-dimensional heterogeneous saturated flow systems that stochastic hydrogeology really attracted significant research interest. Since then, the number of papers published in this field has increased every year. In spite of this growth, stochastic hydrogeology is still a young science. Much of the increased research activity has been facilitated by the increases in computer power.

A complete appraisal of all international contributions to this subject over the last ten years is beyond the scope of this review. To put the UK research programme in context, some of the contributions from outside the United Kingdom that have had a significant influence on the development of the subject are also considered here.

Media generation, statistical inference, and spatial estimation

Matheron and co-workers at the Ecole des Mines in Paris were largely responsible for the development of geostatistical techniques and their application to problems of spatial estimation and geological media generation. The theory of regionalized variables and the spatial estimation technique, Kriging, developed by Matheron were applied initially to problems of ore estimation in the mining industry and only entered into use in hydrogeological studies in the late 1970s. Geosta-

tistical methods for generating heterogeneous transmissivity fields over an aquifer domain were successfully employed to investigate uncertainties in the two-dimensional modelling of a regional aquifer (Delhomme 1979).

Whilst geostatistical techniques found early favour because of their tractability for statistical inference and estimation, the restriction to Gaussian models with geometric spatial correlation functions has been criticized for producing unrealistic spatial patterns of heterogeneity. Various techniques of media generation are now being researched including fractals (Hewett 1986) and Markov random fields in addition to extensions of the traditional geostatistical methods. These approaches potentially permit better representations of the heterogeneity observed in geological media parameters.

Extensions to geostatistical methods have derived from the Ecole des Mines, Paris (Guerillot *et al.* 1990) as well as from Journel and his group in the USA (Journel 1989). Journel's methodology specifically aims to incorporate subjective information to supplement hard data in generating media realizations. Journel employs indicator methods to obtain non-Gaussian models of parameter variability.

Stochastic simulation, effective parameters, and the inverse problem

Since Freeze's early work, many researchers have tackled the problem of the validity of, or more importantly, the range of applicability of effective parameter models. Smith and Freeze (1979*a,b*), Smith and Schwartz (1980), Delhomme (1979) and El-Kadi and Brutsaert (1985) are amongst the best known contributions. They each describe the results of Monte Carlo experiments with numerical models aimed at evaluating the relationship between effective parameters and the underlying spatial variability of the media. In general, the experiments are restricted to one or two-dimensional problems. Of greater importance, in terms of their impact on this field of research, are the results of analytical approaches employed to solve the stochastic partial differential equations of flow and solute transport. Two analytical approaches are found in the literature. The first is based on the solution of the perturbed form of the partial differential equations. The most prominent use of this technique is the research conducted by Gelhar and co-workers (Gelhar 1984, 1986). The second approach is the so called 'embedding matrix' method of Dagan (1979, 1981, 1982*a*). Both methods have advantages and disadvantages but esentially they permit the most concise analysis of the validity of effective parameter models and their range of application yet devised. However, the applicability of the results to real geological media is constrained by the restrictive conditions on spatial variability imposed to permit an analytical treatment of the problem.

The contributions of Neuman and his co-workers at the University of Arizona must also be men-

tioned, particularly in relation to an inverse method which is consistent with the stochastic approach to groundwater modelling (Neuman and Yakowitz 1979; Neuman 1980; Clifton and Neuman 1982; Carrera and Neuman 1986*a*,*b*,*c*).

United Kingdom research

In spite of (or, perhaps, because of) the high profile international activity of groups in the United States of America, Canada, Israel, and France over the last fifteen years, research in stochastic hydrogeology has not been prominent in the United Kingdom. Nevertheless, a growing body of expertise in this field has been developing during the last ten years focused on a subset of the areas identified in Fig.14.1. This subset comprises media generation, effective parameter studies (including field data collection) and stochastic simulation (including risk analysis studies). Table 14.1 summarizes the UK contributions to research in stochastic hydrogeology in terms of the principal contributors and their published research areas. The research topics assigned to each group reflect the published or available literature and do not necessarily reflect the full range of research undertaken by the groups in the field of stochastic hydrogeology.

Table 14.1 UK contributors to research in stochastic hydrogeology in the 1980s

Organisation	Research topic
HMIP assessment team* (Department of the Environment)	Radionuclide transport in geological media including: risk analysis; uncertainty and bias; stochastic simulation.
UKAEA Technology (Harwell)	Fracture network modelling
UKAEA Technology (Winfrith)	Media generation Effective parameters
British Geological Survey	Data collection Effective parameters
University of Newcastle upon Tyne	Data collection Media generation Effective parameters
University of Strathclyde	Media generation
University of Lancaster	Stochastic simulation Effective parameters
GEOSTOKOS	Spatial estimation Stochastic simulation
BP Research Centre	Effective parameters

* Comprises a number of engineering and scientific consultancy groups.

Parameter estimation for hydrogeological models

Spatial estimation: Kriging

Measurements of hydrogeological properties are invariably sparsely and irregularly distributed in space, while a numerical model of groundwater flow and transport requires variables such as head or hydraulic conductivity to be specified at the model grid points. The traditional hydrogeological approach to this problem is to produce a map or surface of the property of interest based on the available measurements and any other relevant information; the a priori knowledge of an experienced hydrogeologist is a useful ingredient of the mapping exercise. However, a disadvantage of this subjective approach is that, from the same data, different hydrogeologists will form different interpretations and produce different maps, implying a degree of uncertainty in estimating the property of interest where no measurements are available.

The technique of Kriging has been developed in response to the need for an objective procedure for geological mapping and for estimating the extent of ore bodies. Named after the South African mining engineer who pioneered the use of the technique, Kriging owes a great deal to the contributions of Matheron and his co-workers at the Ecoles des Mines. They have provided the theoretical foundations and practical framework for using Kriging and a whole range of related techniques which now constitute the field of geostatistics. During the past decade, these techniques have found widespread application in such fields as hydrology, hydrogeology, and soil science; an excellent treatment of the subject in a hydrogeological context is given by de Marsily (1986).

Delhomme (1979) noting the evolution at that time of the new subject of 'stochastic hydrogeology', demonstrated the potential of Kriging in mapping the transmissivity field of an aquifer, and in quantifying the uncertainty in the mapped estimates. The starting point for Kriging is to regard the parameters of interest as a random function, while a set of observations on the parameter over the spatial domain in question is treated as a realization of the random function, or 'regionalized variable' in geostatistical terminology. As already observed, the identification of the probability model governing the behaviour of the random function in space is central to the application of the geostatistical approach. Here, the concept of stationarity is invoked to distinguish between varying degrees of heterogeneity of the random function in space. A hierarchy of hypotheses concerning the stationarity of the random function may be invoked as follows: the object is to identify the simplest hypothesis which is consistent with the behaviour of the observed realization (Delhomme 1979). Only the concepts will be outlined here; the mathematical details can be found elsewhere (e.g. de Marsily 1986):

1. The expected or average value of the random function $Z(x)$ at every location in space x is assumed to satisfy

$$E[Z(x)] = m \qquad (3)$$

where m is a constant independent of x, and

$$C(x,x') = E[(Z(x) - m)(Z(x') - m)] \qquad (4)$$
$$= C(h)$$

where x and x' are the coordinates of points separated by a distance h and $C(h)$ is the covariance. The variance $C(0)$ is assumed to be finite under this hypothesis which corresponds to weak stationarity for the random function. In essence, these assumptions imply the absence of spatial drifts in the behaviour of the parameter.

2. In many cases, the hypothesis of second order stationarity does not hold: for example, the variance $C(0)$ may be observed to increase with the area under consideration. A less restrictive hypothesis than (1), called the 'intrinsic hypothesis', may then be invoked: this also has the advantage of not requiring the estimation of the unknown constant m in equation (3). Here, attention is focused on the increments $[Z(x) - Z(x')]$, and the random function is then characterized by the variogram

$$\gamma(h) = 1/2 \, E \, [(Z(x) - Z(x'))^2] \qquad (5)$$

which may be estimated by

$$\hat{\gamma}(h) = \frac{1}{n_h} \sum (Z_i - Z_j)^2 \qquad (6)$$

where n_h is the number of pairs of observations (Z_i, Z_j) corresponding to a distance h. The practicalities of estimating the variogram are discussed by de Marsily (1986).

From the sample variogram estimated by equation (6), the form of the underlying 'true' variogram has to be inferred. Various models for the variogram have been proposed, of which the spherical model given by

$$\gamma(h) = \omega \left[3/2 \, \frac{|h|}{a} - 1/2 \left(\frac{|h|}{a} \right)^3 \right] \quad h < a$$
$$= \omega \qquad\qquad\qquad\qquad h \geqslant a \qquad (7)$$

is a typical example. The process of inferring the variogram model and its parameters from the sample variogram is normally subjective and based on graphical analysis, rather than on formal statistical inference. Once identified, the variogram constitutes the probability model governing the behaviour of the random function.

3. In the presence of a simple large-scale drift or trend the random function $Z(x)$ may be expressed as

$$Z(x) = m(x) + Y(x) \qquad (8)$$

where $m(x)$ represents the drift; an estimate of the latter can be obtained, for example, by least squares and subtracted from $Z(x)$ to give the residuals $Y(x)$. However, the variogram inferred from these residuals is known to be biased, although the bias can be neglected for small values of (h).

4. In the presence of a complex drift, such as might be observed for hydraulic heads in a hilly domain, the intrinsic hypothesis must be generalized. This is achieved by assuming that the drift can be described locally by an nth degree polynomial; in this case, the behaviour of the random function is described by a generalized covariance function $K(h)$. The version of Kriging corresponding to this hypothesis is known as Universal Kriging.

Kriging equations corresponding to the above hypothesis are available in de Marsily (1986); only a brief summary of the equations for case (2) is provided here. Given a set of measurements $Z(x_i) = Z_i$ at locations x_i, an estimate of Z_0 at an unmeasured point is formed as

$$Z_0^* = \sum_{i=1}^{n} \lambda_0^i Z_i \qquad (9)$$

where n is the number of observations. A set of Kriging equations is derived by minimizing the quantity $E(Z_0 - Z_0^*)^2$ i.e. the mean square error of the estimate:

$$\sum_{j} \lambda_0^j \gamma(x_i - x_j) + \mu = \gamma(x_i - x_0)$$
$$i = 1, \ldots, n \quad (10)$$
$$\sum_{i} \lambda_0^i = 1$$

The quantity $\gamma(x_i - x_j)$ is the variogram ordinate corresponding to a separation distance $(x_i - x_j)$, and μ is a Lagrange multiplier. The system of linear equations is solved to give the weights λ_0^i; the constraint $\sum \lambda_0^i = 1$ ensures that the estimate is unbiased. Note that the weights change as a function of the location of x_0. The estimation variance of Z_0^* is given by

$$E(Z_0^* - Z_0)^2 = \sum_{i} \lambda_0^i \gamma(x_i - x_0) + \mu. \qquad (11)$$

Consequently, not only can a grid of Z_0^* values over the domain of interest be calculated, but the variance of Z_0^* can also be mapped, thus allowing areas of greatest uncertainty in the estimation to be identified. Kriging also has the following useful attributes:

(1) it is an exact interpolator, i.e. it returns the observed values at measurement points;

(2) it is a best linear unbiased estimator;

(3) the Kriging equations do not involve the measured values;

(4) measurements which represent averages over some specified area (e.g. block averages) can be accommodated;

(5) measurement errors can also be accounted for.

Although an extensive literature on the theoretical and applied aspects of Kriging and associated geostatistical techniques has emerged over the past decade, contributions from the UK have been relatively few. Clark of Geostokos Ltd, in collaboration with American research workers (Clark *et al.* 1989), developed a novel approach to Co-Kriging termed Multivariate Universal Co-Kriging (MUCK). Co-Kriging is an extension of Kriging attributed to Matheron (1971) in which additional related variables can be used in the Kriging framework with the aim of reducing the estimation variance of Z_0^*. Clark *et al.* (1989) discussed a number of difficulties associated with the use of Matheron's definition of the cross-variogram

$$\gamma_{WZ}(h) = \frac{1}{2} E[(Z_i - Z_j)(W_i - W_j)] \qquad (12)$$

which measures the degree of association between two variables Z_i and W_i. Such difficulties are primarily attributed to the fact that $\gamma_{wz}(h)$ can be negative when W and Z are inversely correlated. Since the variogram can then decrease as distance increases, this can result in some very strange Kriging weightings for large distances and possibly negative estimation variances. As an alternative to equation (12), Clark *et al.* (1989) proposed the use of

$$\gamma_{wz}(h) = \tfrac{1}{2}E\{Z_i - W_j\}^2 \qquad (13)$$

which, apart from avoiding the negative behaviour in $\gamma_{wz}(h)$ associated with equation (12), has the advantage of being able to accommodate all potential pairs of values, irrespective of whether or not they are available at the same location. The equations for MUCK which lead to optimal estimates of all the variables simultaneously in the presence of significant drift were then formulated.

A case study of the application of MUCK was described involving piezometric data from two deep brine aquifers, the Pennsylvanian and the Wolfcamp, in northern Texas. Maps of the MUCK estimation variances show that, in areas where Pennsylvanian samples are scarce, the Wolfcamp samples significantly reduce the estimation variances, and vice versa. The average ratio of MUCK to Universal Kriging estimates for the Pennsylvanian aquifer was found to be 1.15 and, for the Wolfcamp aquifer, 1.09, indicating a useful reduction in the estimation variance through the inclusion of additional information.

In a case study involving mass transport modelling of the saturated zone for an industrial waste disposal site at Villa Farm, near Coventry (see Chapter 8), Universal Kriging was employed to interpolate a number of aquifer parameters (groundwater levels, ground surface, top, and base of aquifer) onto the grid points of a finite element model (Mackay *et al.* 1986). In order to deal with the problem of estimating the variogram in the presence of drift, the increments of pairs of data points were divided into subsets defined by their vector orientation in space. Variograms for the different subsets were then used to estimate the known values at the data points but omitting each data point in turn from the Kriging data set. The resultant errors, standardized by the theoretical

standard deviations of the Kriged estimates, should have zero mean, unit standard deviation, be normally distributed, and spatially uncorrelated. The variogram which best satisfied these criteria was adopted for use, and Kriged estimates of the various aquifer parameters were then generated at the model grid points. These 'surfaces' were then used in an iterative model calibration procedure whereby the permeability estimates fed into the model were adjusted successively in response to the derived head distribution for a number of plausible boundary conditions. A range of possible flow directions consistent with the available data were derived, indicating that further data would be needed to reduce this uncertainty.

The inverse problem

The inverse problem has long been recognized as one of the most challenging problems associated with the implementation of groundwater flow and transport models. In the context of groundwater flow modelling, the main difficulties arise for two reasons (Neuman and Yakowitz 1979).

1. The transmissivities are dependent on the derivatives of the heads and not on the heads themselves. Small errors in the head values may therefore introduce large fluctuations in the values of the calculated transmissivities.

2. Even in the case where the head data are precise, the solution to the inverse problem may be non-unique; this may be caused by insufficient information about lateral flow rates or by the lack of sufficiently large hydraulic gradients in some parts of the aquifer.

In the second section, a distinction was drawn between direct and indirect approaches to the inverse problem. Much of the research carried out in the past decade has been based on a statistical approach to the inverse problem, and has been carried out primarily by American research groups. However, a brief résumé will be given of the major contributions to the literature from these sources, since the authors are not aware of any published material on the subject from the UK in this period.

A very useful summary of recent research on the inverse problem has been given by Peck *et al.* (1988) who pointed out that the indirect parameter estimation technique most widely used is the method of trial and error calibration. The object of this approach is to find the set of model parameters (e.g. transmissivities at the nodes of the numerical model) which minimizes some function of the differences between observed and simulated hydraulic heads. The groundwater flow model is run repeatedly until the desired agreement between the observed and simulated heads is obtained. This process allows the accumulated knowledge/experience of the modeller, and his/her understanding of the hydrogeology of the aquifer in question, to be brought to bear on the parameter estimation problem.

The laborious nature of the trial and error approach is doubtless a disadvantage; none the less, it is a process which should result in a physically plausible calibrated model, largely through the subjective information which is introduced into the estimation problem. A further disadvantage is that no estimates of the standard errors of the estimated parameters are available.

Most of the notable contributions on statistical approaches to the inverse problem have sought to develop a framework where prior information can be combined with the observed data within the estimation framework. Three statistical approaches were summarized by Peck *et al.* (1988): weighted least squares, Bayesian, and maximum likelihood estimation. A model of groundwater flow in an aquifer can be written as

$$h^* = g(p) + \varepsilon_h \qquad (14)$$

where

h^* is a vector of head measurements
p is a vector of unknown parameters
$g(p)$ is the modelled hydraulic head, a non-linear function of p
ε_h is a vector of hydraulic head measurements.

The vector of prior flow parameter measurements, p^*, can be represented by

$$p^* = p + \varepsilon_p \qquad (15)$$

where ε_p is a vector of parameter errors.

The weighted least squares estimator proposed by Neuman and Yakowitz (1979) yields parameter estimates that minimize a modified sum of squared errors

$$F = F_h + \lambda F_p \qquad (16)$$

where

$$F_h = [h^* - g(\hat{p})]^T V_h^{-1} [h^* - g(\hat{p})]$$
$$F_p = [p^* - \hat{p}]^T V_p^{-1} [p^* - \hat{p}] \qquad (17)$$

and $\hat{p}$ represents the vector of parameter estimates. The factor λ allows for the relative weighting of head measurements and prior information. The weighting matrices V_h and V_p are chosen on the basis of experience, and allow more emphasis to be placed on measurements with higher reliability. A special case is where V_h and V_p represent the covariance matrices of the heads and prior parameter estimates which correspond to the weighted least squares approach. Applications of this approach were described by Cooley (1982, 1983) and Cooley *et al.* (1986). Cooley (1982) has also presented an expression for calculating the covariance matrix of the parameter estimates; however, this is based on the assumption of a linear relationship between the model parameters and hydraulic heads which can lead to underestimates of the variances of the parameter.

In the Bayesian approach, the unknown parameters, p, are assumed to be random variables; a priori information on the parameters is combined with the likelihood function, which accounts for the information in the head data, to yield the a posteriori distribution of p. The estimate of p is then obtained by minimizing the conditional Bayes risk which involves integrating a loss function with the a posteriori distribution. However, Peck *et al.* (1988) pointed out that a rigorous application of Bayesian estimation is not possible, since h is a nonlinear function of p. A comprehensive development of the Bayesian approach was given by Neuman and Yakowitz (1979) and an application in which Kriging was used to obtain the prior information was described by Clifton and Neuman (1982).

The maximum likelihood (ML) approach to the inverse problem assumes that the parameters are fixed but unknown quantities; prior information consists of 'noisy' parameter estimates. With the ML approach, a likelihood function is defined as the probability density function of h as a function of p; ML estimates of p are those which maximize the probability of obtaining the observations h^*. Carrera and Neuman (1986a,b,c) gave a detailed development of the ML approach and presented applications to hypothetical and field data.

Peck *et al.* (1988) observed that weighted least squares is a special case of Bayesian and ML estimation; however, the latter two approaches are based on fundamentally different philosophies. Moreover, the Bayesian approach requires the assumption of a linear relationship between heads and parameters, while the ML approach does not require this assumption.

Peck *et al.* (1988) reviewed a number of geostatistical approaches to parameter estimation. Some of these approaches involve straightforward Kriging or Co-Kriging of the data (e.g. transmissivities: Delhomme 1979) without any use of the groundwater model itself in solving the inverse problem. Kitanidis and Vomvoris (1983) described a geostatistical approach which accounts for the relationship between hydraulic head and transmissivity by linearizing the groundwater flow equation. This assumption is a limitation which was also noted for the Bayesian approach.

Attempts to solve the inverse problem for contaminant transport problems have been relatively few to date. Wagner and Gorelick (1986) used a weighted nonlinear least squares approach to estimate transport parameters; they used a similar approach to estimate simultaneously both flow and transport parameters (Wagner and Gorelick 1987).

Stochastic simulation models of hydrogeological media

Porous media

The need to quantify the uncertainty in predictions of the travel times of hazardous contaminants has focused attention on the need for stochastic models of the media properties. Observations of exposures of sand and gravel deposits show, for example, that properties such as grain size vary enormously, even in individual formations. This variability results in correspondingly large variations in flow parameters such as hydraulic conductivity; for example, data based on small laboratory cores taken from boreholes show that the standard deviation of the natural logarithm of hydraulic conductivity can vary from 0.2 to as much as 5 in various types of natural deposits. However, this natural intraformational heterogeneity exhibits a degree of spatial structure or coherence in that observations on hydraulic conductivity, for example, at spatially adjacent points will reflect the layered or lenticular structure of the formation. Gelhar (1987) discussed three possible approaches to incorporating this natural heterogeneity into flow and transport models.

1. Treat formations, layers, or zones within the aquifer as being homogeneous and construct a model based on discretization which reflects this zoning or layering. However, this classical deterministic approach ignores the spatial variability of flow parameters within each formation, and cannot provide a quantitative measure of the error introduced by this assumption.

2. Represent the actual detailed heterogeneity of the aquifer in a complex three-dimensional numerical model. However, this approach is precluded by the need for extremely detailed measurements of the three-dimensional spatial distribution of hydraulic properties, and by the excessive computational demands of such an approach.

3. Characterize the spatial variability through a stochastic model which is defined by a relatively small number of statistical parameters; techniques of statistical inference can then be employed to estimate these parameters from a limited set of observations. Such a model cannot, of course, be used to predict, in a deterministic sense, what the hydraulic properties are at points where no measurements are available; rather, by carrying out Monte Carlo sampling experiments with the stochastic model, many possible realizations of the distribution of hydraulic properties in space can be generated. Each realization, when coupled with the deterministic flow and transport equations in a numerical model, provides a possible scenario of contaminant migration, described for example, through a concentration-time curve at a point of interest. The set of realizations provides a set of

such curves; the mean concentration curve provides an estimate of the average travel time, and the variance about the mean curve provides an estimate of the uncertainty in the mean estimate associated with the lack of knowledge of the actual spatial distribution of media properties.

The last of these approaches, which constitutes the stochastic approach to modelling groundwater flow and transport in porous media, has evolved from the seminal papers of Freeze (1975), Smith and Freeze (1979 *a,b*), and Delhomme (1979). Freeze (1975) assumed that the hydraulic conductivity values in adjacent blocks of a one-dimensional discretized flow domain were lognormally distributed and statistically independent; probability distributions of hydraulic head were then generated through Monte Carlo experiments and interpreted as a function of the parameters of the lognormal distribution and for steady and unsteady flow regimes. A more realistic, but still idealized approach to characterizing the spatial structure of hydraulic conductivity was subsequently adopted for idealized one-dimensional (Smith and Freeze 1979*a*) and two-dimensional (Smith and Freeze 1979*b*) steady flow regimes by using a nearest neighbour stochastic model with an autoregressive structure to generate spatially correlated block values of hydraulic conductivity which were again assumed to be lognormally distributed. Probability distributions of hydraulic head corresponding to both isotropic and anisotropic fields of hydraulic conductivity were generated by Smith and Freeze (1979*b*) and interpreted as a function of selected values of the parameters of the stochastic model. Delhomme (1979) described a geostatistical approach to media generation and demonstrated its application through a case study of an aquifer in northern France. The stochastic model was again based on the appropriate hypothesis for the random function as outlined in the previous section. Two methods of generating media realizations were described by Delhomme:

(1) *unconditional realizations* where the sampling at the various points in the spatial domain does not take account of any known (observed) values at some of the points;

(2) *conditional realizations* where the known

values of a media parameter at a sub-set of points within the spatial domain are reproduced exactly in each realization. Random sampling is thus restricted to those points where no data exist.

In the geostatistical literature, the preferred algorithm for generating unconditional media realizations is the turning bands method (TBM) developed by Matheron and his co-workers (e.g. Journel and Huijbregts 1978). With this approach, a large number of independent realizations of a line process are generated; these lines emanate from a single origin and are randomly distributed over the interval $(0,2\pi)$. By projecting the values of the line processes onto the points in the domain where values are to be generated, the realization is obtained as a weighted sum of the line process contributions. The relationship between the covariance (or variogram) of the line process and that of the spatial process allows any prescribed covariance structure for the latter to be simulated. Moreover, since the spatial process is generated by adding up a large number of independent line processes, the spatial process will have an approximately Gaussian distribution.

The method of generating conditional realizations involves a combination of unconditional generation and Kriging. The observations are Kriged over the domain of interest; the object of conditional generation is to generate values of the parameter with the Kriging estimation variances at unmeasured points and the same spatial structure as the observations. A simple combination of unconditional generation and Kriging achieves this.

In generating conditional realizations of the transmissivity field of an aquifer in northern France, Delhomme (1979) made the conventional assumption that point values of transmissivity T have a lognormal distribution i.e. $\log T$ then has a Gaussian distribution. The effect of uncertainty in the transmissivity distribution on the estimation of the hydraulic head distribution for prescribed boundary conditions was demonstrated by running a numerical model of the aquifer with each conditional realization of the transmissivity distribution. Measurement errors in the observed transmissivities were also accounted for in the analysis.

Stochastic geometry, Markov random fields, and fractals

While the geostatistical approach to simulating media properties has been the focus of a considerable amount of research effort during the past decade, other approaches have also been explored, and with active UK involvement. Much of this research falls within the overall framework of stochastic geometry, which is defined as the study of objects placed in space randomly but with interactions (Ripley 1989). Aquifers or oil reservoirs may be characterized by a hierarchy of nested geological structures with different scales of heterogeneity; in oil reservoirs, for example, the spatial positioning of rock types (especially impermeable shales) can have a crucial influence in predicting extraction profiles for the reservoir. Consequently, efforts are being made to develop stochastic models which can reproduce the patterns of observed rock types. Markov random field (MRF) models depend on a discretization of the (three-dimensional) domain into blocks. Each block is assigned a rock type from a small number (4–10) of types, and the random field governs the interactions between the neighbouring rock types (Ripley 1989). To be realistic, the small blocks have to combine to form large regions of one type. Properties such as permeability or porosity can then be assigned to regions of one type using a stochastic model of the geostatistical type to characterize within-region heterogeneity. Ripley at the University of Strathclyde has been exploring the potential of MRF models, as well as some general spatial point processes, for characterizing oil reservoir rock types. Media realizations produced by such models have to be conditioned on the known rock types at the wells and should also reflect other, less certain, information from seismic data and the geologists' real world knowledge. Ongoing research is also concerned with developing efficient methods of simulation and parameter estimation for the models. With MRF models, some of the methods of parameter estimation have been found to be enormously sensitive to small departures from the model (Ripley 1988).

Research on characterizing media properties for oil reservoirs has also been conducted by Farmer at AEA Technology, Winfrith. This has included a review of methods for generating media realizations with a specified probability density function (pdf) and two-point correlation function: such methods are claimed to be successful only if the pdf is Gaussian (Farmer 1988). Markov methods are discussed as an alternative approach where a realization is constructed with a locally specified conditional probability distribution. In one dimension, the simplest Markov model is obtained by considering a sequence of states each of which is random but conditional on the preceding state. To describe heterogeneous oil reservoirs, two and three-dimensional generalizations of the Markov chain are required; such generalizations are possible (Besag 1974) and relevant background review material has been presented in Bartlett (1975), Whittle (1986), and Ripley (1987). To generalize a Markov chain to several dimensions, Farmer (1988) introduced pdf's known as Gibbs distributions which he uses to derive conditional pdf's in which the probability is conditioned on the values of neighbouring sites. Farmer then demonstrated that Markov random field (MRF) models, in which the conditional probabilities relating to a neighbouring set of sites (referred to as a clique), and models based on Gibbs distributions, are equivalent.

To generate MRF realizations, Farmer (1988) suggested the use of the Metropolis algorithm (Metropolis *et al.* 1953) which, in the MRF context, has been used to model the 2-D surface texture of natural materials. Although the Metropolis algorithm is one of a number of methods of generating random numbers from a multivariate pdf, its form is apparently suited to MRF methods where it is required to control the point histogram and the local conditional pdf.

In a more recent contribution, Farmer (1990) described a new approach to generating oil reservoir media realizations (termed 'Numerical Rocks') which is based on the frequency of rock types, the frequency of rock type pairs and the correlations, in several directions. A robust simulated annealing algorithm is described which avoids the solution of difficult inverse problems associated with the identification of the control parameters of MRF models. The algorithm is robust in the sense that a set of control parameters can be found readily which will reproduce accurately two-point histograms from a control pattern.

Any review of attempts to describe the complex geometry of nature would be incomplete without a reference to the world of fractals. Mandelbrot's fractal geometry provides both a description and a mathematical model for a host of seemingly complex forms found in nature (Mandelbrot 1983). Shapes such as coastlines, mountains, and clouds, although difficult to describe using classical Euclidean geometry, possess a remarkable simplifying invariance under changes of magnification. This statistical self-similarity is the essential attribute of fractals in nature; it can be quantified by a fractional dimension, a number which agrees with our intuitive notion of dimension, but need not be an integer (Voss 1988).

Although there is apparently some evidence to suggest that oil reservoir media are self similar (Hewett 1986), there does not seem to have been any major research effort to date in hydrogeology to explore the full potential of fractals as stochastic models of porous media. However, one of the main criticisms levelled at the conventional models of geostatistics (e.g. based on a spherical variogram) is that generated realizations are too smooth, particularly when properties are viewed on larger grid block sizes than those on which they were generated. This is a close analogue of a problem encountered some years ago when the increments of a self similar fractal model, fractional Brownian motion (fBm), were used to model long-term fluctuations in the flows of the River Nile (Mandelbrot and Wallis 1968). Conventional stochastic models were unable to explain this behaviour because their realizations were too smooth. FBm models have the peculiar property that, although stationary, the summation of the covariance function of fBm increments $\Sigma C(h)$ diverges. This reflects the presence of small but non-negligible correlations at high lags, which leads to pronounced low frequency behaviour in the process.

It is tempting to speculate that an fBm model might be used to explain the large scale variations in permeability encountered in oil-reservoirs and aquifers without the need to invoke models at different scales (e.g. an MRF model for variations between rock types and a conventional geostatistical model for within-rock variations). Some rainfall field models have already been formulated which are based on fBm concepts (Shah *et al.* 1986): this has been achieved by using an ARIMA (p,d,q) line process with fractional d (Hosking 1984) (which closely approximates the behaviour of fBm increments) within the turning bands framework. However, no attempt has yet been made to apply the model to heterogeneous porous media.

Fractured media

Interest in the modelling of flow and transport in fractured media has been stimulated largely by the problems of assessing the safety of radioactive waste disposal in fractured rock. Since a comprehensive chapter on transport in fractured rock is included in this volume (Chapter 13), only those aspects of flow and mass transport modelling which involve stochastic modelling and statistical estimation procedures are considered here. Nearly all of the research falling into this category is concerned with stochastic models for generation of discrete fracture networks, and with attempts to infer the parameters of these models from limited available data.

Some of the earliest work on the stochastic generation of fracture networks was conducted in the UK at AERE, Harwell, leading to the development of a computer code called NAMNET (Robinson 1982). Since then, several other notable contributions have resulted (Robinson 1983, 1984*a,b*, 1986) and these are reviewed here briefly together with selected contributions from the USA, Canada, and France.

Generation of two-dimensional fracture networks

Early research efforts were concentrated on the development and implementation of computationally feasible algorithms for the Monte Carlo simulation of two-dimensional fracture networks, and with the interpretation of flow and transport experiments conducted in these networks (e.g. Schwartz *et al.* 1983; Endo *et al.* 1984; Robinson 1984*a*; Smith and Schwartz 1984). The steps involved are as follows (Schwartz and Smith 1987):

(1) the definition of the flow domain and the boundary conditions;

(2) the probabilistic generation of one realization of the fracture network;

(3) the reduction of the generated network to what is termed the essential network;

(4) the solution of the flow and transport problems;

(5) the repetition of steps (2) to (4) and the collection of various output parameters to complete the Monte Carlo simulation.

The flow domain has invariably been chosen to be either square or rectangular, with the possibility of specifying known hydraulic heads along all four sides, or along the two ends with no flow boundaries along the sides. The first set of boundary conditions is more flexible, since flow can be defined with a specified orientation relative to the fracture sets.

The next step in the procedure is to generate a discrete fracture network. Fractures in two dimensions are normally represented as sets of linear features which are defined in terms of their midpoint, fracture length, attitude, and aperture. Some or all of these parameters are considered to be random variables. In the references cited above, the midpoints of fractures are located randomly within the domain. This distribution characterizes a homogeneous medium where fracturing is not controlled by structural or lithological variability (Priest and Hudson 1976). Fractures are assumed to occur in one or more distinct sets; for example, Schwartz *et al.* (1983) and Smith and Schwartz (1984) considered two sets oriented at 90° to each other while Robinson (1984*a*) considered an arbitrary number of fracture sets with variability in fracture attitudes.

Fracture lengths are typically treated as random variables with lengths sampled from a lognormal or negative exponential distribution. Lognormal distributions for fracture apertures are justified on the basis of a limited number of studies (e.g. Snow 1970).

Fracture density is determined by the number of fractures added to a given domain in relation to its overall size. While the most convenient assumption is that of a constant number of fractures in each set, it is not difficult to build in variability and spatial correlation into the density of fractures (Schwartz and Smith 1987).

The third step in the modelling procedure is to

reduce this network to what Schwartz and Smith (1987) termed the essential network; this involves simplifying the network by removing dead-end fracture segments and isolated clusters of fractures. This procedure, in addition to reducing the computational effort, can improve the stability of the solution of the flow equation.

Robinson (1983, 1984*b*) described a percolation theory approach to determining the connectivity of fracture networks. Lines with specified probability distributions for length and orientation are generated uniformly over a square. Each new line can form a new cluster (no intersections), extend an existing cluster, or unite two or more existing clusters. After each line is generated, modified clusters are checked to see if they satisfy the percolation criterion, which generally requires the cluster to connect the open boundaries of the region. Once this criterion is satisfied, line generation is stopped and the critical density recorded. The average number of intersections per line at percolation has also been found and varies only slightly over the cases considered. By calculating critical densities for a range of system sizes and using a finite-size scaling argument, a prediction of the infinite critical density is made. Extensions to three-dimensional systems of fixed size orthogonal planes are also considered.

The final step in the procedure is to simulate fluid flow and mass transport through the network. Typically, finite difference (Schwartz *et al.* 1983) or finite element (Robinson 1984*a*) techniques are used to calculate hydraulic heads at nodes which are defined as the points where fractures intersect. Implicit in this approach are assumptions of parallel-walled fractures whose apertures can be related to the quantity of flow by a cubic law expression, laminar flow conditions, a rigid fracture network, and no coupling between fluid density and flow (Schwartz and Smith 1987).

Mass transport is simulated using a particle tracking technique. Moving particles are added to one or more fractures at the start of the simulation, and advected through the network with a velocity that for individual fractures is determined from estimates of hydraulic head, aperture, and fixed constants such as fluid density and viscosity. Dispersion of the swarm of particles is generated by the geometry of the network, with either perfect or partial mixing assumed at intersections.

Generation of three-dimensional fracture networks

More recently, efforts have been made to generate three-dimensional fracture systems; however, the sizes of such systems and the number of realizations which can be generated are limited by computational considerations. The approach adopted for three-dimensional systems is similar to that described above for 2-D systems; however, each of the steps is more difficult than for the 2-D case. There is the problem of describing fracture shape: Smith *et al.* (1985) and Robinson (1986) assumed rectangular planar shapes, while Long *et al.* (1985) modelled fractures as disks. Such shapes are adopted primarily for computational convenience.

Simulation of flow and transport in a 3-D network is more difficult because of the problem of dealing with intersections of planes and parts of planes in space. The rough-walled character of individual fracture planes can be modelled by incorporating spatial variability in fracture apertures (Schwartz and Smith 1987). For certain correlation structures, it is then possible to simulate channelling of flow and mass within the network (Moreno *et al.* 1985).

Finite element techniques have been used by Robinson (1986) and Smith *et al.* (1985) to solve the steady state flow problem for 3-D networks. Because of common nodes at fracture intersections, the global matrix equation has a large bandwidth, and an iterative solution using conjugate gradient acceleration is advocated (Schwartz and Smith 1987). As in the 2-D case, particle tracking is used to simulate mass transport.

A comprehensive review of 3-D fracture network modelling is provided by Chiles (1989) which includes models based on random infinite fractures (planes), random finite fractures (polygons or discs), and locally self similar fractal networks. Generalizations of the basic models are considered by mixing, clustering, and regionalizing the parameters. Various techniques which might be employed for determining the parameters are also considered, together with a technique for conditional simulation.

Parameter estimation for fracture networks

Parameter estimation for 3-D fracture network models is recognized to be a very challenging problem which has only recently been addressed in the literature. Large scale, expensive field pro-

grammes are required to tackle the problem such as that which has been conducted in the Fanay-Augères mine in Limousin, France. The data from this experiment have formed the basis of two recent studies aimed at calibrating 3-D fracture network models (Hestir *et al.* 1989, Cacas *et al.* 1990*a,b*). The model described by Hestir *et al.* represented a development of the 3-D model structure outlined in the previous sub-section. Two sections of a drift wall in the Fanay-Augères mine were mapped, one was wet and one was dry. For each case, the fractures were divided into five different sets, and each set was modelled separately. The fractures in each set were represented as discs placed randomly in space, with a lognormal distribution used to describe disc diameters. For the location of discs, a point process called a parent-daughter process was used; the choice of point process was based on a comparison of variograms of observed data with projections of 3-D fracture networks onto 2-D planes. The orientation of the discs was characterized as a fluctuation about the mean orientation of the set, the spatial structure of which was described through a variogram. The model was shown to reproduce the spatial variability of fracture density and orientation, and also incorporated a measure of connectivity for fracture mesh simulations. The connectivities of the simulations are to be assessed to see if there is any correlation with the fact that one section of the drift is wet and one is dry.

UK case studies

The need for comprehensive radiological risk assessment procedures in radioactive waste disposal has been the main stimulus for applications of a stochastic modelling approach to contaminant transport problems in the UK. Two case studies are reported here, both relating to porous media and both of which are based on the geostatistical methodology described earlier. One case study, which has been carried out at the University of Newcastle, was based on a (then) potential site for low level radioactive waste disposal at Elstow, Bedfordshire (Mackay *et al.* 1988). Since the site was no longer under consideration for disposal by the time the study was underway, the case study essentially became a demonstration of how uncertainty in the pathways of radionuclides migrating to the biosphere from the site might be assessed by coupling geostatistical simulation with a numerical model of flow and transport through the geosphere. The primary pathway for migration of radionuclides to the biosphere was found to be the Kellaways Beds, a sandstone underlying the repository site in the overlying Oxford Clay. Variogram models for both the aquifer thickness and hydraulic conductivity were constructed. The variogram model for hydraulic conductivity had to be inferred from a literature review, while the model of thickness was based on a limited amount of field data. Unconditional realizations of hydraulic conductivity and thickness were generated using the turning bands method and, for each pair of realizations, the flow field was simulated. Particle tracking was then used to evaluate the migration pathways to the biosphere. Three primary escape zones to the biosphere were identified and a statistical analysis of the simulated migration data was undertaken to characterize the migration patterns of radionuclides in terms of their distributions in both space and time. Due to the uncertainty associated with the estimated variogram parameters, a sensitivity analysis of the effect of this uncertainty on the results was also included. Furthermore, a sampling methodology for risk analysis was defined by discretizing the domain into a number of one-dimensional migration pathways; the information generated by this stage of the analysis was then in a form suitable for incorporation into a probabilistic risk assessment (PRA) code (e.g. Laurens *et al.* 1990).

The case study, although restricted in the scope of the simulations which could be undertaken, was effective in demonstrating the benefits of a geostatistical approach to evaluating uncertainty and in identifying unresolved issues which directly affect the implementation of models in risk analysis.

Although based on the Wolfcamp aquifer of the Palo Duro basin, Texas, the second case study reported here had a significant UK involvement through Clark (Harper and Clark 1989). Before progressing to the application, the authors discussed two possible approaches to conditional simulation. The first of these, developed in the context of the inverse problem is based on the sampling of parameter interpolation errors with a covariance matrix V derived using Kriging. Although reason-

ably simple to implement, Harper and Clark (1989) referred to two practical problems associated with the procedure. It depended heavily on the assumption of multivariate normality of the errors, and on the accuracy of the underlying model of the random function i.e. the drift and the variogram. Second, a grid of 10×10 points required a (100×100) V matrix for the simulations.

The alternative approach based on the turning bands method was adopted by Harper and Clark for the case study. The Wolfcamp aquifer underlies a potential high level nuclear waste repository site in the Palo Duro Basin and plays an important part in any study of far field flow in this area. Groundwater travel times, travel paths and their associated uncertainties are considered to be key performance measures in an overall performance assessment uncertainty analysis. The main object of the case study was to quantify the variability in the potentiometric surface of the Wolfcamp aquifer as a component of this uncertainty analysis.

Potentiometric data were available at 85 locations within the aquifer domain. A geostatistical analysis of the data indicated that a linear trend and a spherical variogram provided an adequate description of the data. Using this model, 100 realizations of potentiometric head were generated over the aquifer and a particle tracking algorithm was used to follow the path of radionuclides released into the aquifer at one location. A five mile square grid was adopted for the particle tracking experiments as a compromise between computation time and reasonable travel path calculations. For each realization, the corresponding travel path was computed and stored, and the variability of the travel paths subsequently analysed. It was found that, while the majority of the travel paths followed the expected travel path associated with the Kriged or expected potentiometric surface, considerable variability around the expected travel path was still observed.

An extension of this approach to the multivariate case was discussed in which correlated variables can be employed to reduce the range of uncertainty in the simulated head values. This extended approach is to be tested using data from both the Wolfcamp and Pennsylvanian aquifers.

Effective parameter models

In the first section, the representative elementary volume (REV) was defined as the volume of media around a point, x, over which a parameter $p(x)$ can be averaged such that the variation of the resultant average parameter $\langle p(x) \rangle$ over the domain of interest is sufficiently smooth to permit its use in a simplified model of the system. The resultant parameter is an 'effective parameter' defined at the scale of the REV. There are two problems with this route for defining effective parameters which describe the macroscopic variation of the media properties. First, the functional form for averaging over the REV is not yet well understood for parameters such as hydraulic conductivity and dispersivity. Second, the application of any numerical technique for solving the partial differential equations of flow or transport introduces its own averaging volume, the grid block or grid element, which is usually considerably larger than that of the apparent scale of the REV.

Implicit in the term 'effective parameter' is the concept that its use in a model should permit the model to replicate the average behaviour of the system it is simulating. Unfortunately, this is not a precise definition since average behaviour over all possible dependent variables may not be required for many modelling problems. In a flow system, average behaviour may mean average head distribution, average flux distribution, or average discharge over the boundary of the system. Thus, an effective parameter should be defined in terms of the output variable of interest as well as in terms of the model in which it is embedded.

The concept of an effective parameter has a number of implications for both simulation studies and data collection. If variable media parameters are replaced in a model by effective parameters with smoothed distributions, bias may arise in the model results. The scale of the bias will reflect the model formulation, the conceptual assumptions used to derive the model and the method of solution used in the model. It may also depend on the form of the output from the simulation. An understanding of the nature of the bias, its influence on the results of the modelling, and the factors controlling its generation is necessary if model results are to be interpreted correctly. Equally, an understanding is needed of how to scale between local data values derived from field measurements and appropriate effective parameter values in the model. As part of this understanding, consideration must be given to the methods used to obtain field data. Hydraulic data are generally obtained from measurements of potentials taken during field testing by solving the inverse problem using an idealized model describing the test conditions. The choice of model used to analyse the results of the field test can significantly affect the data obtained. Since the parameter values derived from the test are effective parameter values for the model used to analyse the test, they may not be appropriate measures of the effective parameters needed as input to a simulation model. The determination of the best methods for field data collection, including appropriate analysis procedures, is an important research problem.

Two parameters have received the most attention in effective parameter studies; namely, hydraulic conductivity and dispersivity. Dispersion has been the subject of considerable interest as the scale of the dispersion coefficient appears to be related to either spatial scale and/or time. Methods for determining dispersion coefficients from small scale measurements are being sought.

Effective hydraulic conductivity

The early work of Freeze (1975) concluded that, in the analysis of idealized one-dimensional steady state and transient saturated flow systems, an equivalent homogeneous hydraulic conductivity could not be defined uniquely to replace the underlying heterogeneous hydraulic conductivity distribution. Although Freeze showed that the effective hydraulic conductivity was equal to the harmonic mean of the underlying hydraulic conductivity distribution for the steady state case, he found that the effective hydraulic conductivity in the transient case depended on time.

Dagan (1979) and Gutjahr *et al.* (1978) found from analytical studies that the geometric mean of the hydraulic conductivity distribution is a valid approximation to the effective hydraulic conductivity for two-dimensional steady state flow. For low variance in the underlying hydraulic conductivity (K) distribution the effective hydraulic conductivity ($\langle K \rangle$) can be expressed analytically in terms of the geometric mean (K_g):

1-D models $\langle K \rangle = K_g \, (1 - \sigma^2 \log(K)/2)$
2-D models $\langle K \rangle = K_g$
3-D models $\langle K \rangle = K_g \, (1 + \sigma^2 \log(K)/2)$

where σ^2 is the variance of the hydraulic conductivity distribution.

Dagan (1982a) showed for unsteady flow conditions that effective hydraulic conductivity depends on time. This finding is consistent with Freeze's earlier work. For problems with periodic flux variations and/or time dependent boundary conditions the effective hydraulic conductivity not only depends on time but also on the head gradient.

Gelhar and Axness (1983) studied dispersion in idealized three-dimensional heterogeneous media described by an anisotropic covariance. The point values of the hydraulic conductivity field were defined with a lognormal probability distribution function. Using spectral analysis techniques they developed analytical expressions for the six components of the effective hydraulic conductivity tensor.

Whilst analytical studies provide valuable insights into the form of the effective parameters that should be used for simulating flow and transport through heterogeneous media, they are usually restricted to solving problems with idealized media characteristics. Flow and transport through more realistic representations of heterogeneous media can be studied using numerical simulation methods. The Monte Carlo experiments described by Smith and Freeze (1979b) and El Kadi and Brutsaert (1985) showed the potential of numerical approaches for gaining an understanding of flow and transport in heterogeneous porous media. Cacas *et al.* (1990a,b) adopted a Monte Carlo approach to study the behaviour of flow in fractured media.

Following on from much of this early work, a number of practical studies using numerical methods and Monte Carlo procedures have been forthcoming in the United Kingdom. In the petroleum industry, specific attention has been focused on the impact of shale bodies in sand reservoirs on effective hydraulic conductivity. Begg *et al.* (1985)

developed a statistical method of calculating effective vertical permeability of a reservoir containing discontinuous shales. The approach adopted is based on the concept that the presence of impermeable shale units increases the tortuosity of the medium to flow. By evaluating the changes in stream tube geometry (in this case length and cross-sectional area) as a result of the presence of shales the effect on hydraulic conductivity can be established over the media. An alternative method is to carry out repeated simulations using a numerical model capable of representing individual shale units at the scale of the model mesh (Begg and King 1985) for different sand/shale geometries. The restriction imposed by this type of modelling procedure is that it introduces boundary impacts on the solution for the effective permeability. Moreover, it is computer intensive when used for solving large three-dimensional problems.

The heterogeneity of geological media can be viewed as comprising a number of distinct but possibly overlapping scales of heterogeneity. Each contributes to the characteristics of flow and transport through the media observed at the largest scale. The problem is then how to analyse the effective media properties for models using large grid blocks, given information on the variability in the media parameters arising from each of the underlying scales of heterogeneity. Numerical simulations using a single model to carry out the analysis would, in computational terms, be prohibitively expensive. A potential alternative is to adopt a procedure of hierarchical modelling which operates by nesting simulation models. Sub-grid models are used at a lower scale to define effective parameter values for the next scale of simulation. Monte Carlo experiments based on the effective parameter variability at the sub-grid scale provide the statistics for characterizing the spatial variability of the effective parameters at the next scale. The process can be repeated as many times as necessary. Mackay and Glendinning (1989) examined the problem of averaging the underlying hydraulic conductivity distribution at one mesh scale over a larger mesh scale. Using a simple two-dimensional model, the choice of averaging volume was addressed. Averaging volumes that extend beyond the dimensions of the grid cell were found to reproduce the flow distributions more accurately at the larger grid scale than by averaging over the volume of the grid cell alone. This work is still at a very early stage of development.

As part of a Monte Carlo simulation study of a two-dimensional transport problem in a thin heterogeneous aquifer, Mackay *et al.* (1988) showed that homogeneous permeabilities applied to a heterogeneous aquifer can potentially lead to substantial bias in the spatial distribution of contaminants compared to that which arises from the underlying heterogeneous distribution.

Binley *et al.* (1989*a,b*) used a fully three-dimensional model of variably saturated flow on a heterogeneous hillslope to explore the validity of simulating the flow using homogeneous hydraulic parameters. In this study they considered single realizations of random patterns of saturated hydraulic conductivity and attempted to fit homogeneous hydraulic conductivities such that surface and subsurface flow hydrographs were reproduced. No single values of effective hydraulic conductivity could be found that were capable of reproducing both surface and subsurface flow responses for low permeability soils. Better reproduction of flow was achieved for high permeability soils. The effective parameter values needed to reproduce the desired output characteristics were found to be dependent on the simulated flow event.

Effective dispersion

The convection dispersion equation (2) assumes that the nature of dispersion in geological media is Fickian (i.e. the spreading of the contaminant about the mean advective movement of the fluid is linearly proportional to the concentration gradient):

$$J = D \frac{\partial c}{\partial x} \qquad (18)$$

where J is the dispersive flux and D is the dispersion coefficient. However, this model of dispersion was adopted only as an initial working hypothesis for simulating transport over large distances in heterogeneous porous media (Bear 1972). Since the late 1970s, considerable work has been carried out to verify the range of validity of this model for dispersive transport in porous media.

The dispersion coefficient (D) represents the sum of mechanical dispersion and molecular diffusion. Mechanical dispersion is dependent on the Darcian velocity of the fluid flowing through the medium as well as the medium and fluid properties. One dominant area in the field of dispersion research during the last ten years has concerned the apparent scale dependence of dispersion. The greater the distance travelled by a tracer plume, the larger the dispersivity value needed to simulate the dispersion process (Anderson 1979). This apparent scale dependence is a function of the heterogeneity of the medium. Scale dependence presents two problems for simulation. First, it renders almost all efforts to obtain useable field data from small scale tests useless, and second, it raises serious difficulties in evaluating the validity of a simulation model describing the growth of a contamination plume from a point source. The reason for the latter difficulty arises from the fact that, in the early stages of plume development, the contamination has only sampled (i.e. travelled through) a small section of the media. The heterogeneity of the media encountered in the early stages is potentially very different from the mean heterogeneity of the media as a whole. Consequently, the dispersive component of mixing at early times may be expected to be different from that at later times.

The recognition that mechanical dispersion is a function of the heterogeneity of the medium has suggested to many researchers (Gelhar *et al.* 1979; Gelhar 1984; Dagan 1982*b*) that a stochastic approach to the analysis of dispersion should be employed. Both Gelhar and Dagan used analytical methods to evaluate the dispersion coefficient. The analyses presented by all these authors showed the following important features (Gelhar and Axness 1983):

(1) the mean transport becomes Fickian for large times;

(2) the dispersion coefficient for large times is in the form of a product of a mean velocity and a dispersivity;

(3) the asymptotic value of dispersivity is related to statistical properties of the medium;

(4) the approach to the asymptotic dispersive pro-

cess is slow and significant non-Fickian transport can occur early in the transport process.

These results suggest that if the transport path is long enough compared to the underlying length scale of variability of the media then a Fickian model is valid. Unfortunately, if the variability of the media is not stationary (i.e. it does not reach a limiting value as the domain of interest enlarges) then such conditions are unlikely to exist. Indeed, even if stationarity of the variability is apparent, the dispersive transport may not be Fickian. Matheron and de Marsily (1980) showed that dispersive transport is not attained, in general, for the case of a stratified aquifer system with flow parallel to the bedding.

Recent developments by Dagan (1990) and Neuman *et al.* (1987) have extended this earlier work in two ways: first, by increasing the generality of the asymptotic results for dispersivity in statistically anisotropic heterogeneous media and, second, by unifying the various mathematical procedures that have been used to analyse the transport of solutes in heterogeneous media.

Monte Carlo methods based on numerical models of contaminant transport have also been used to study the problem of defining effective dispersivities in heterogeneous media. The combination of numerical methods for solving the transport of solutes through heterogeneous media and geostatistical techniques for generating media realizations has the advantage over analytical methods of permitting the examination of a much wider class of models of media heterogeneity. However, as previously noted the computational cost of numerical simulations with sufficiently large numbers of grid blocks to allow scale effects to be observed is very high. Consequently, far fewer papers describing the results of this approach have been forthcoming. Perhaps the best known work is the paper by Smith and Schwartz (1980) which described an experimental Monte Carlo study of dispersion in two-dimensional heterogeneous media. It is sufficient to note that the results of this approach confirm the analytical results of Gelhar and Axness (1983).

In the UK, Mackay and Glendinning (1989) described a Monte Carlo study of the early stages of plume development in a porous medium. This study differs from the work of Smith and Schwartz in that consideration is given to plume development in three-dimensional porous media and more emphasis is placed on the distributional properties of the generated concentration fields.

Computer simulations were used to examine the development of a contaminant plume in a 3-metre cube of porous media with spatially varying hydraulic conductivity. A block centred finite difference model comprising 30^3 elements was used for the flow simulations. The hydraulic conductivity fields were characterized by a lognormal distribution and spherical spatial correlation. The results of the experiments showed, as might be expected, that there is little evidence to support the hypothesis that the concentration field converges to a Fickian distribution for the distance and time-scale considered for the experiment. Highly non-normal distributions were found to occur in individual realizations. The results of this exercise raise the question of whether the averaging implicit in the Fickian model ever produces worthwhile estimates of concentration distribution, given the extreme variance of point concentrations between different media realizations.

Farmer (1988) reviewed a wide body of literature relating to the problem of scale dependence in general and to dispersion in particular. Whilst Farmer's treatise is highly mathematical in its presentation it does lead to some interesting conclusions for initiating constructive research on this problem. Farmer suggested that the hierarchical modelling approach to simulation described earlier in this section is potentially the most viable route to calculating the behaviour of contaminant transport over a large spatial domain. He also proposed the adoption of a two stage process only; higher numbers of stages in the nesting procedure should be permissible if the effective parameter models derived at each stage are valid.

One significant development in the United Kingdom that is now underway in this area is an experimental research programme being undertaken by the Fluid Processes Research Group of the British Geological Survey. The aim of the programme is to develop a new tracer test procedure for measuring dispersion. The new tracer test permits fine scale resolution of the heterogeneity of the aquifer's hydraulic properties as well as providing detailed information on the three-dimensional geometry of the tracer plume generated over time during an injection test. The data from this test are to be used to examine procedures, both deterministic and stochastic, for quantifying the dispersive nature of the aquifer medium. The experiment can be repeated at a number of points over the domain at low cost. Using the multiple test data it should be possible to evaluate the large scale dispersive characteristics of the medium from the small scale tests. This study (Williams *et al.* 1988) is at a preliminary stage of development.

Outstanding problems and future prospects

The need to employ simulation models to evaluate the impacts of contamination in aquifer systems and to permit engineering and management decisions to be made about problems of aquifer restoration and maintenance is becoming increasingly important. Whilst deterministic models of flow and transport are being used to support groundwater quality management, they suffer from two major disadvantages. Field data are usually sparse for describing the hydrogeological characteristics of the contaminated region and the processes of transport in heterogeneous media at the aquifer scale are still not well understood. To overcome these problems research must continue in a number of areas:

(1) assessment of existing field techniques and the development of new field techniques for quantifying the spatial characteristics of hydraulic properties of both fractured and porous media;

(2) development of realistic models describing the heterogeneity of both porous and fractured media and their parameters inferred from limited hard and soft data;

(3) development of robust estimators for effective parameters at the scales required for simulation of contamination events. This research should provide information about the range of applicability of effective parameter models used for different flow and transport problems;

(4) development of standard methods of assigning

risk estimates to flow and contaminant transport calculations that allow engineering decisions to be made which take account of the risk.

International research on the problem of uncertainty in the results of simulations of both flow and transport in heterogeneous media have led, over the last fifteen years, to substantial advances in the understanding of the uses and abuses of current modelling techniques. There are still many problems to be solved and further substantial research effort is required. Such research is likely to be highly cost effective in reducing the risk of choosing the wrong options in the management and restoration of contaminated groundwater systems. In this context, a major aim must be to bring stochastic modelling techniques developed for studying hydrogeological problems to a state where they can be applied confidently by practising hydrogeologists.

Acknowledgements

The authors wish to thank all those mentioned in this review who responded so willingly to requests for information on their research.

References

Anderson, M. P. (1979). Using models to simulate the movement of contaminants through groundwater flow systems. In *CRC Critical Reviews in Environmental Control*, **9**, 97–156.

Bartlett, M. S. (1975). *The statistical analysis of spatial pattern*. Chapman and Hall, London.

Bear, J. (1972). *Dynamics of fluids in porous media*. Elsevier, New York.

Begg, S. H. and King, P. R. (1985). *Modelling the effects of shales on reservoir performance: calculation of effective vertical permeability*. Soc. Pet. Eng. Reservoir Simulation Symposium, SPE 13529, Richardson, Texas.

Begg, S. H., Chang, D. M., and Haldorsen, H. H. (1985). *A simple statistical method of calculating the effective vertical permeability of a reservoir containing discontinuous shales*. Soc. Pet. Eng. 6th Ann. Tech. Conf., SPE 14271, Richardson, Texas.

Besag, J. (1974). Spatial interaction and the statistical analysis of lattice systems. *J. Roy. Stat. Soc.*, **B36**, 192–236.

Binley, A., Elgy, J., and Beven, K. (1989*a*). A physically based model of heterogeneous hillslopes, 1. Runoff production. *Water Resour. Res.*, **25**, 1219–26.

Binley, A., Beven, K., and Elgy, J. (1989*b*). A physically based model of heterogeneous hillslopes, 2. Effective hydraulic conductivities. *Water Resour. Res.*, **25**, 1227–34.

Cacas, M. C., Ledoux, E., de Marsily, G., Tillie, S., Barbreau, A., Durand, E., Feuga, B., and Peaudecerf, P. (1990*a*). Modelling fracture flow with a stochastic discrete fracture network: calibration and validation, 1. The flow model. *Water Resour. Res.*, **26**, 479–90.

Cacas, M. C., Ledoux, E., de Marsily, G., Barbreau, A., Calmels, P., Gaillard, B., and Margritta, M. (1990*b*). Modelling fracture flow with a stochastic discrete fracture network: calibration and validation, 2. The transport model. *Water Resour. Res.*, **26**, 491–500.

Carrera, J. and Neuman, S. P. (1986*a*). Estimation of aquifer parameters under transient and steady-state conditions, 1. Maximum likelihood method incorporating prior information. *Water Resour. Res.*, **22**, 199–210.

Carrera, J. and Neuman, S. P. (1986*b*). Estimation of aquifer parameters under transient and steady-state conditions, 2. Uniqueness, stability and solution algorithms. *Water Resour. Res.*, **22**, 211–27.

Carrera, J. and Neuman, S. P. (1986*c*). Estimation of aquifer parameters under transient and steady-state conditions. 3. Application to synthetic field data. *Water Resour. Res.*, **22**, 228–42.

Chiles, J. P. (1989). Three-dimensional geometric modelling of a fracture network. In *Geostatistical, sensitivity and uncertainty methods for groundwater flow and radionuclide transport modelling* (ed. B. E. Buxton), pp. 361–85. Battelle Press, Ohio.

Clark, I., Basinger, K. L., and Harper, W. V. (1989). MUCK—a novel approach to Co-Kriging. In *Geostatistical, sensitivity and uncertainty methods for groundwater flow and radionuclide transport modelling* (ed. B. Buxton), pp. 473–93. Battelle Press, Ohio.

Clifton, P. M. and Neuman, S. P. (1982). Effects of Kriging and inverse modeling on conditional simulation of the Avra Valley aquifer in southern Arizona. *Water Resour. Res.*, **18**, 1215–34.

Cooley, R. L. (1982). Incorporation of prior information of parameters into nonlinear regression groundwater flow models, 1. Theory. *Water Resour. Res.*, **18**, 965–76.

Cooley, R. L. (1983). Incorporation of prior information of parameters into nonlinear regression

groundwater flow models, 2. Applications. *Water Resour. Res.*, **19**, 662–76.

Cooley, R. L., Konikow, L. F., and Naff, R. L. (1986). Nonlinear regression groundwater flow modeling of a deep regional system. *Water Resour. Res.*, **22**, 1759–78.

Dagan, G. (1979). Models of groundwater flow in statistically homogeneous porous formations. *Water Resour. Res.*, **15**, 47–63.

Dagan, G. (1981). Analysis of flow through heterogeneous random aquifers by the method of embedding matrix, 1. Steady flow. *Water Resour. Res.*, **17**, 107–21.

Dagan, G. (1982*a*). Analysis of flow through heterogeneous random aquifers, 2. Unsteady flow in confined formations. *Water Resour. Res.*, **18**, 1571–85.

Dagan, G. (1982*b*). Stochastic modeling of groundwater flow by unconditional and conditional probabilities, 2. The solute transport. *Water Resour. Res.*, **18**, 835–48.

Dagan, G. (1990). Transport in heterogeneous porous formations: spatial moments, ergodicity and effective dispersion. *Water Resour. Res.*, **26**, 1281–90.

Delhomme, J. P. (1979). Spatial variability and uncertainty in groundwater flow parameters: a geostatistical approach. *Water Resour. Res.*, **15**, 269–80.

El-Kadi, A. I. and Brutsaert, W. (1985). Applicability of effective parameters for unsteady flow in non-uniform aquifers. *Water Resour. Res.*, **21**, 183–98.

Endo, H. K., Long, J. C. S., Wilson, C. R., and Witherspoon, P. A. (1984). A model for investigating mechanical transport in fracture networks. *Water Resour. Res.*, **20**, 1390–400.

Farmer, C. L. (1988). The generation of stochastic fields of reservoir parameters with specified geostatistical distributions. In *Mathematics in oil production* (ed. Sir Sam Edwards and P. R. King), pp. 235–52. Clarendon Press, Oxford.

Farmer, C. L. (1990). Numerical rocks: the mathematical generation of reservoir geology. In *The mathematics of oil recovery* (ed. F. J. Fayers and P. R. King). Oxford University Press, Oxford.

Freeze, R. A. (1975). A stochastic-conceptual analysis of one-dimensional groundwater flow in nonuniform homogeneous media. *Water Resour. Res.*, **11**, 725–41.

Gelhar, L. W. (1984). *Stochastic analysis of flow in heterogeneous media*. Fundamentals of transport phenomena in porous media: NATO ASI Series E, No. 82, 673–717.

Gelhar, L. W. (1986). Stochastic subsurface hydrology from theory to applications. *Water Resour. Res.*, **22**, 1355–455.

Gelhar, L. W. (1987). Stochastic analysis of solute transport in saturated and unsaturated porous media. In *Advances in transport phenomena in porous media* (ed. J. Bear and M. Y. Corapcioglu). NATO ASI Series E: Applied Sciences, 128, pp. 657–700. Martinus Nijhoff, Dordrecht.

Gelhar, L. W. and Axness, C. L. (1983). Three-dimensional stochastic analysis of macrodispersion in aquifers. *Water Resour. Res.*, **19**, 161–80.

Gelhar, W. L., Allan, L. G. and Naff, R. L. (1979). Stochastic analysis of macrodispersion in a stratified aquifer. *Water Resour. Res.*, **15**, 1387–97.

Gutjahr, A. L., Gelhar, L. W., Bakr, A. A., and Macmillan, J. R. (1978). Stochastic analysis of spatial variability in subsurface flows. Part II: Evaluation and application. *Water Resour. Res.*, **14**, 953–60.

Guerillot, D., Rudkiewicz, J. L., Ravenne, C., Renard, G., and Galli, A. (1990). An integrated model for computer aided reservoir description: from outcrop study to fluid flow simulations. *Rev. Inst. Fran. Pet.*, **45**, 71–7.

Harper, W. V. and Clark, I. (1989). Travel path uncertainty—a case study in conditional simulation. *Geostatistics*, **2**, 685–97.

Hestir, K., Long, J., Chiles, J. P., and Billaux, D. (1989). Some techniques for stochastic modelling of three-dimensional networks. In *Geostatistical, sensitivity and uncertainty methods for groundwater flow and radionuclide transport modelling* (ed. B. E. Buxton), pp. 495–519. Battelle Press, Ohio.

Hewett, T. A. (1986). Fractal distributions of reservoir heterogeneity and their influence on fluid transport. American Geophysical Union Fall Meeting, San Francisco.

Hosking, J. R. M. (1984). Modeling persistence in hydrological time series using fractional differencing. *Water Resour. Res.*, **20**, 1898–908.

Journel, A. G. (1989). Imaging of spatial uncertainty: a non-Gaussian approach. In *Geostatistical, sensitivity and uncertainty methods for groundwater flow and radionuclide transport* (ed. B. E. Buxton), pp. 585–99. Battelle Press, Ohio.

Journel, A. G. and Huijbregts, C. J. (1978). *Mining geostatistics*. Academic Press Inc., London.

Kitanidis, P. K. and Vomvoris, E. G. (1983). A geostatistical approach to the inverse problem in groundwater modelling (steady state) and one-dimensional simulations. *Water Resour. Res.*, **19**, 677–90.

Laurens, J. M., Thompson, B. G. J., and Summer-ling, T. J. (1990). *The development and application of an integrated radiological risk assessment procedure using time-dependent probabilistic risk analysis*. Proc. OECD/NEA/IEA/CEC symp. Safety assessment of radioactive waste repositories, pp. 627–38. OECD/NEA, Paris.

Long, J. C. S., Gilmour, P., and Witherspoon, P. A. (1985). A model for steady fluid flow in random three-dimensional networks of disc-shaped fractures. *Water Resour. Res.*, **21**, 1105–15.

Mackay, R. and Glendinning, R. H. (1989). *Asessment of representations of dispersion for inclusion in transport models*. Water Resources Systems Research Unit Report No. 18, prepared for UKAEA, Harwell. Department of Civil Engineering, University of Newcastle upon Tyne.

Mackay, R., Porter, J. D., Williams, G. M., Ross, C. A. M., and Noy, D. (1986). Modelling mass transport in the saturated zone—a case study. In *Water quality modelling in the inland natural environment*, pp. 259–75. BHRA, Cranfield, UK.

Mackay, R., Cooper, T. A., Porter, J. D., O'Connell, P. E., and Metcalfe, A. V. (1988). *Geostatistical analysis of prevailing groundwater conditions and potential solute migration at Elstow, Bedfordshire*. DoE/RW/88.082, London.

Mandelbrot, B. B. (1983). *The fractal geometry of nature* (2nd edn). Freeman, New York.

Mandelbrot, B. B. and Wallis, J. R. (1968). Noah, Joseph and operational hydrology. *Water Resour. Res.*, **4**, 909–18.

de Marsily, G. (1986). *Quantitative hydrogeology*. Academic Press, London.

Matheron, G. (1971). *The theory of regionalised variables and its applications*. Paris School of Mines, Cah. Cent. Morphologie Math., 5. Fontainebleau.

Matheron, G. and de Marsily, G. (1980). Is transport in porous media always diffusive? A counter example. *Water Resour. Res.*, **16**, 901–17.

Metropolis, N., Rosenbluth, A. W., Rosenbluth, M. N., Teller, H. A., and Teller, E. (1953). Equation of state calculations by fast computing machines. *J. Chem. Phys.*, **21**, 1087–92.

Moreno, L., Neretnieks, I., and Eriksen, T. (1985). Analysis of some laboratory tracer runs in natural fissures. *Water Resour. Res.*, **21**, 951–8.

Neuman, S. P. (1980). A statistical approach to the inverse problem of aquifer hydrology, 3. Improved solution method and added perspective. *Water Resour. Res.*, **16**, 331–46.

Neuman, S. P. (1982). Statistical characterization of aquifer heterogeneities. An overview. In *Recent trends in hydrogeology* (ed. T. N. Narasimhan).

Geol. Soc. Amer. Spec. Pap. **189**, pp. 81–102. Boulder, Colorado.

Neuman, S. P. and Yakowitz, S. (1979). A statistical approach to the inverse problem of aquifer hydrology, 1. Theory. *Water Resour. Res.*, **15**, 845–60.

Neuman, S. P., Winter, C. L., and Newman, C. N. (1987). Stochastic theory of field-scale Fickian dispersion in anisotropic porous media. *Water Resour. Res.*, **23**, 453–66.

Peck, A., Gorelick, S., de Marsily, G., Foster, S., and Kovalevsky, V. (1988). *Consequences of spatial variability in aquifer properties and data limitations for groundwater modelling practice.* IAHS Pub. No 175, Wallingford, UK.

Priest, S. D. and Hudson, J. A. (1976). Discontinuity spacing in rock. *Int. Jour. Rock Mech. Min. Sci.*, **13**, 135–48.

Ripley, B. D. (1987). *Stochastic simulation.* Wiley. New York.

Ripley, B. D. (1988). *Statistical inference for spatial processes.* Cambridge University Press, Cambridge.

Ripley, B. D. (1989). Stochastic geometry in reservoir modelling (Abstract). In *Statistics earth and space sciences*, 50–51. Katholieke Universität, Leuven.

Robinson, P. C. (1982). *NAMNET—network flow program.* AERE R10510, AERE, Harwell, UK.

Robinson, P. C (1983). Connectivity of fracture systems—a percolation theory approach. *J. Phys. A: Math. Gen.*, **16**, 605–14.

Robinson, P. C. (1984a). *Connectivity flow and transport in network models of fractured media*, TP1072. AERE, Harwell.

Robinson, P. C. (1984b). Numerical calculations of critical densities for lines and planes. *J. Phys. A: Math. Gen.*, **17**, 2823–30.

Robinson, P. C. (1986). *Flow modelling in three-dimensional fracture networks.* AERE R11965. AERE, Harwell.

Schwartz, F. W. and Smith, L. (1987). An overview of the stochastic modelling of dispersion in fractured media. In *Advances in transport phenomena in porous media* (ed. J. Bear and M. Y. Corapcioglu). NATO ASI Series E: Applied Sciences, 128, pp. 727–50. Martinus Nijhoff, Dordrecht.

Schwartz, F. W., Smith, L., and Crowe, A. S. (1983). A stochastic analysis of macroscopic dispersion in fractured media. *Water Resour. Res.*, **19**, 1253–65.

Shah, S. M., O'Connell, P. E., and Hosking, J. R.

M. (1986). *Generation of random rainfall fields using a fractional Brownian process. American Geophysical Union Fall Meeting*, San Francisco.

Smith, L. and Freeze, R. A. (1979a). Stochastic analysis of steady state groundwater flow in a bounded domain, 1. One-dimensional simulations. *Water Resour. Res.*, **15**, 521–8.

Smith, L. and Freeze, R. A. (1979b). Stochastic analysis of steady state groundwater flow in a bounded domain, 2. Two-dimensional simulations. *Water Resour. Res.*, **15**, 1543–59.

Smith, L. and Schwartz, F. W. (1980). Mass transport, 1. A stochastic analysis of macrodispersion. *Water Resour. Res.*, **16**, 303–13.

Smith, L. and Schwartz, F. W. (1984). An analysis of the influence of fracture geometry on mass transport in fractured media. *Water Resour. Res.*, **20**, 1241–52.

Smith, L., Mase, C. W., and Schwartz, F. W. (1985). A stochastic model for transport in networks of planar fractures. *Greco 35 Hydrogéologie*, Ministère de la Récherche et la Technologie, C.N.R.S. pp. 523–36.

Snow, D. T. (1970). The frequency and apertures of fractures in rock. *Jour. Rock Mech. Min. Sc.*, **7**, 23–40.

Voss, R. F. (1988). Fractals in nature; from characterization to simulation. In *The science of fractal images* (ed. H. O. Peitgen and D. Saupe), pp. 21–69. Springer Verlag, Berlin.

Wagner, B. J. and Gorelick, S. M. (1986). A statistical methodology for estimating transport parameters: theory and applications to one-dimensional advective-dispersive systems. *Water Resour. Res.*, **22**, 303–15.

Wagner, B. J. and Gorelick, S. M. (1987). Optimal groundwater quality management under parameter uncertainty. *Water Resour. Res.*, **23**, 1162–74.

Warren, J. E. and Price H. S. (1961). Flow in heterogeneous porous media. *Soc. Pet. Eng.*, **1**, 153–69.

Whittle, P. (1986). *Systems in stochastic equilibrium.* Wiley, Chichester.

Williams, G. M., Noy, D. J., Jackson, P. D., and Mackay, R. (1988). *Theoretical potential of radial injection tracer tests with 3-D pressure and solute monitoring.* Report WE/88/26. British Geological Survey, Keyworth.

15. Noble gases and radioelements in groundwaters

J. N. Andrews

The past two decades have witnessed a world-wide growth in interest in the application of the noble gases and the natural radioelements to hydrogeological problems. Much attention has been focused on their use for assessing groundwater age and radionuclide migration, particularly for groundwaters in deep sedimentary formations and for flow in fractured crystalline rocks. Some objectives of these studies have been to understand the relationship between groundwater flow and oil/gas maturation in hydrocarbon reservoirs; to determine groundwater residence times and resources in aquifers in semi-arid zones and to evaluate crystalline rocks as potential repositories for nuclear wastes. Noble gas and radioelement investigations are complementary because of the ubiquitous occurrence of the natural radioelements in crustal rocks and because reactions induced by natural radioactivity cause changes in noble gas composition. It is also beneficial to link these studies with investigations of the major and minor element chemistry of groundwater systems. This chapter outlines the principles of noble gas and radioelement applications in hydrogeology and includes some typical examples.

Noble gases in groundwaters

The noble gases have been applied to the study of processes which affect groundwater geochemistry in various geological settings. The major advantages of noble gases for such purposes are the absence of interactions which could affect the chemical state of the dissolved gas and the well defined inputs for the atmospheric noble gases.

The objectives of these studies are to assess or evaluate:

(1) groundwater migration and groundwater age;

(2) the palaeoclimate which prevailed during infiltration from estimates of the equilibration temperatures between the noble gases and the atmosphere (that is the estimation of recharge temperatures);

(3) geochemical processes in the shallow crust, for example, the release of ^{40}Ar by mineral alteration or the progress of denitrification by measurement of the N_2/Ar ratio in solution;

(4) deep crustal and mantle processes, for example, the use of $^3He/^4He$ ratios to evaluate the crustal migration of mantle-derived ^3He and to study mantle convection.

For these applications it is essential that the groundwater is sampled in such a manner that its gas content remains representative of that existing in the aquifer. The solution of biogenic gases (carbon dioxide and methane) and temperature increase due to the geothermal gradient may cause groundwater to become supersaturated with gas. Such gas generally remains in solution because of large hydrostatic pressures in the aquifer but may readily be lost by depressurization during the sampling procedure. For the estimation of recharge temperatures, the dissolved atmospheric gases must be conserved in solution and any additional gas dissolution during groundwater migration must also be assessed. Radiogenic ^{40}Ar dissolution, for example, may be identified by a change in the ^{40}Ar/^{36}Ar ratio.

Noble gas sources

The isotopes of the noble gases which are produced by cosmogenic or by *in-situ* production processes can generally be applied to groundwater migration problems. Some nuclides have also been produced in anthropogenic sources, such as during nuclear weapon testing (^{3}H–^{3}He), in nuclear power reactors (^{37}Ar) and during nuclear fuel reprocessing (^{85}Kr). Anthropogenic production is much more significant than cosmogenic production for these last two nuclides. *In-situ* production of some radionuclides occurs within geological formations due to cosmic proton, muon, and neutron induced reactions at shallow depths and due to the *in-situ* neutron flux at depths greater than 50 m (Andrews *et al.* 1989*a*). The *in-situ* neutron flux arises from alpha particle induced neutron emission (α,n) reactions caused by alpha particle irradiation of light elements in the rock matrix during radioactive decay of the U and Th series nuclides present. The nuclides ^{37}Ar and ^{39}Ar, especially, may occur in groundwater because of such underground production whereas present estimates of the fission yield of ^{81}Kr suggest that its production is probably insignificant compared with the amount of dissolved cosmogenic ^{81}Kr (Andrews *et al.*, 1991*a*).

Both ^{3}He and ^{4}He are produced by *in-situ* processes, the former by neutron irradiation of lithium and the latter by decay of the natural radioelements. The ^{3}He/^{4}He ratio may in some cases reflect the production ratio of these isotopes within a geological formation. However, atmospheric He, contains significant amounts of both cosmogenic and mantle-derived ^{3}He and deep crustal sources may be dominated by mantle stored He which has a relatively high content of ^{3}He. The isotopic ratio of dissolved He in groundwater may be the result of admixture of these different sources.

Solution of gases by groundwater

The major mechanisms for noble and permanent gas dissolution by groundwaters are (i) equilibration with the atmosphere during recharge; (ii) solution, as the hydrostatic pressure increases, of additional or 'excess' air bubbles entrained during infiltration; and (iii) the solution of gases produced by radiogenic and biogenic mechanisms within the aquifer. The dissolution of atmospheric gases by air equilibration is dependent upon the temperature and partial pressures of atmospheric gases in the soil atmosphere and is controlled by Henry's law in the case of dilute solutions. The 'excess' air component of dissolved gas may vary from 10 per cent to almost 100 per cent of the air-equilibration component and must be determined experimentally (Mazor 1972; Andrews and Lee 1979; Heaton and Vogel 1981). Its magnitude is probably dependent upon the morphology of flow channels in the unsaturated zone and upon the frequency of recharge which combine to permit air and air-saturated water to be alternately admitted to the flow paths. Any entrained air dissolves under increased hydrostatic pressures as it migrates to greater depth with the groundwater. Relatively large amounts of excess air may occur in flow systems within fractured rocks.

During groundwater migration, the radiogenic gases, principally ^{4}He and ^{40}Ar, may be dissolved from the rock matrix, thus supplementing the amounts of these gases dissolved at recharge. Radiogenic ^{4}He content increases in the direction of groundwater flow (Andrews and Lee 1979; Bath *et al.* 1979; Heaton and Vogel 1979; Andrews *et al.* 1985; Torgersen and Clarke 1985) in predominantly porous media aquifers. Biogenic CO_2 produced within the soil horizon is dissolved by groundwaters in the unsaturated zone and biogenic or thermogenic CH_4 and N_2 may also be generated in the aquifer. The generation of these gases during groundwater migration and the effect of temperature increase due to the geothermal gradient determine the supersaturation of water relative to gas solubilities at their atmospheric partial pressures. Many groundwaters exsolve gases when depressurized to atmospheric pressure because of such supersaturation.

The gas contents of groundwaters are also influenced by structural relationships. Fault-aided diffusion may enhance the amounts of gas in solution, especially for the relatively diffusive noble gases ^{4}He and ^{222}Rn (Scholz *et al.* 1973; Bulashevich and Bashorin 1973; Banwell and Parizek 1985). Crustal stresses prior to earthquakes are known to cause

Table 15.1 Isotopic composition of atmospheric noble gases

	volume % of atmosphere	element composition (moles %)
^{3}He	6.8×10^{-10}	10^{-4}
^{4}He	5.239×10^{-4}	100
Total He	5.239×10^{-4}	(pp = 5.239×10^{-6} atm.)
^{20}Ne	1.65×10^{-3}	91.05
^{21}Ne	4.7×10^{-7}	0.01
^{22}Ne	1.62×10^{-2}	8.94
Total Ne	1.818×10^{-3}	(pp = 1.818×10^{-5} atm.)
^{36}Ar	3.15×10^{-3}	0.33
^{38}Ar	5.9×10^{-4}	0.06
^{40}Ar	0.934	99.6
Total Ar	9.34	(pp = 9.34×10^{-3} atm.)
^{78}Kr	4.0×10^{-7}	0.354
^{80}Kr	2.59×10^{-6}	2.27
^{82}Kr	1.317×10^{-5}	11.56
^{83}Kr	1.316×10^{-5}	11.55
^{84}Kr	6.48×10^{-5}	56.90
^{86}Kr	1.98×10^{-5}	17.37
Total Kr	1.139×10^{-4}	(pp = 1.139×10^{-6} atm.)
^{124}Xe	8.3×10^{-9}	0.096
^{126}Xe	7.7×10^{-9}	0.090
^{128}Xe	1.7×10^{-7}	1.919
^{129}Xe	2.3×10^{-6}	26.44
^{130}Xe	3.5×10^{-7}	4.08
^{131}Xe	1.8×10^{-6}	21.18
^{132}Xe	2.3×10^{-6}	26.89
^{134}Xe	9.0×10^{-7}	10.44
^{136}Xe	7.6×10^{-7}	8.89
Total Xe	8.6×10^{-6}	(pp = 8.60×10^{-8} atm.)

Abundance data from (Glueckauf 1946; Glueckauf and Kitt 1956)

enhancement of He/Ar and N_2/Ar ratios in groundwaters associated with fault zones (Sugisaki 1978, 1981), suggesting that N_2 as well as He is migrating from the deep crust in such situations.

Noble gas solubilities

The isotopic composition of atmospheric noble gases and their partial pressures in the atmosphere (Table 15.1) have been accurately determined by Glueckauf (1946) and by Glueckauf and Kitt (1956). The amount of each gas dissolved on equilibration of water with the atmosphere may be calculated from Henry's law:

$$n_i = p_i/k_i \tag{1}$$

where n_i is the number of moles of gas i dissolved in 1 mole of water at partial pressure, p_i, and k_i is the Henry's law constant (atm^{-1}) for the temperature of equilibration. This can be expressed more conveniently in the form:

$$S_i = \beta_i p_i/1000 \tag{2}$$

where S_i is the amount of gas i dissolved (cm^3 STP

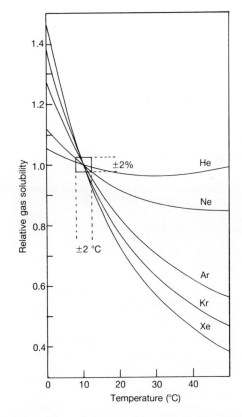

Fig. 15.1. The variation of noble gas solubilities with temperature relative to their solubility at 10°C. A change of +/− 2 per cent in gas solubility corresponds to an equilibration temperature change of up to +/− 2°C.

cm^{-3}H$_2$O) at its atmospheric partial pressure p_i (atm.) and β_i is the Bunsen coefficient for the gas (cm^3 STP l^{-1} atm.$^{-1}$). The Bunsen coefficient and the Henry's law constant are related by the equation:

$$\beta_i = 1.244 \times 10^6 \times \varrho / k_i \tag{3}$$

where ϱ is the solution density (g cm^{-3}) at the equilibration temperature. The numerical constant in this equation expresses the dissolved gas volume at 0°C and 1 atmosphere (STP). Gas solubilities vary with temperature (Fig. 15.1) and both Henry's law and Bunsen coefficients therefore depend upon temperature. The temperature dependence of these constants for atmospheric gases in distilled water has been determined experimentally (Morrison and Johnstone 1954; Benson and Parker 1961; Benson and Krause 1976) and measurements for saline waters have also been reported (Weiss 1970, 1971). These data are all in reasonable agreement over the temperature range at which groundwater recharge occurs; the data of Benson and Krause (1976) are a particularly consistent set for the determination of groundwater recharge temperatures as described below.

The nitrogen, oxygen, and noble gas solubilities may be calculated from the equation (Benson and Krause 1976):

$$\ln (1/k_i) = A_i \left(\frac{T_i}{T-1} \right) + 36.855 \left(\frac{T_i}{T-1} \right)^2 \tag{4}$$

where k_i is the Henry's law constant (atm^{-1}) for gas i, A_i and T_i are constants (Table 15.2(a)) and T is the absolute temperature of the solution. The corresponding Bunsen coefficients may be calculated from eqn (3).

Saline solutions

The Bunsen coefficient, β_s, for a saline solution is related to that for pure water, β_0, by the equation:

$$\ln (\beta_0 / \beta_s) = M\,k_s \tag{5}$$

where M is the Cl$^-$ molarity of the solution and k_s is the salting coefficient. The salting coefficient is independent of solution molarity but is dependent upon temperature according to the equation (Wilson 1986):

$$k_{si} = a_i(T/100)^3 + b_i(T/100)^2 + c_i(T/100) + d_i \tag{6}$$

Table 15.2 Solubility data for atmospheric gases

(a) Constants for eqn 5 (Benson and Krause 1976)

	Gas(i)	A$_i$	T$_i$
sp + 1	He	41.824	131.42
	Ne	41.667	142.50
	Ar	40.404	168.87
	Kr	39.781	179.21
	Xe	39.273	188.78
	N$_2$	41.712	162.02
	O$_2$	40.622	168.85

(b) Constants for eqn 6 (Wilson 1986)

Gas(i)	a$_i$	b$_i$	c$_i$	d$_i$
He	−0.2304	2.2951	−7.7154	9.0060
Ne	−0.3022	3.6278	−13.6641	16.8309
Ar	−0.4050	3.8471	−12.3389	13.6921
Kr	−0.1124	1.3282	−5.1432	6.8403
Xe	−0.1611	1.9007	−7.3019	9.5072
N$_2$	−0.2427	2.1504	−6.6100	7.4294
O$_2$	−0.3616	3.3454	−10.5501	11.6687

where $T(K)$ is the solution temperature and the constants a_i, b_i, c_i, and d_i are given in Table 15.2(b). For groundwaters with salinities less than 1 g l^{-1}, the salting out effect reduces gas solubilities by less than 1 per cent.

Solution in hydrocarbons

Noble gas solubilities in crude oils have recently been determined (Kharaka and Specht 1988). For He and Ne the solubility in oil is increased relative to aqueous solubilities by factors of approximately 2 and 1.6 respectively; for Ar and Kr by a factor of 4–5 and for Xe by a factor of about 13. These are important new data which are required for the interpretation of noble gas data in oilfield brines.

Noble gas diffusion

The noble gases do not generally participate in chemical interactions. The only exceptions to this are the heavy gases, Kr and Xe, which can form 'clathrate' compounds which may cause geochemical trapping of these gases in some environments. In the case of groundwater solutions, noble gas migration may only occur by diffusion and/or transport processes. In many cases the rate of diffusive advection is very much less than that due to

transport of the dissolved species by the groundwater. However, diffusion from within a rock matrix into an adjacent aqueous phase or diffusion from an adjacent geological formation into an aquifer can be important controls on noble gas sources, especially for radiogenic He and Ar which are stored within crystalline minerals. The noble gas diffusion constants are fundamental parameters for assessing diffusive migration. At present, there are few data available, particularly for He diffusion in rock-forming minerals and within whole rocks. In the case of whole rocks, diffusion occurs not only through the mineral lattices or defects within them, but also along grain boundaries and microfractures within the rock. Table 15.3 summarizes diffusion constant data for He, Ar, and ^{222}Rn.

Crustal rocks can lose He by diffusion from shallow depths. For a layer of rock with uniform radioelement content, the He concentration, $C_{z,t}$, at depth z and time t after the rock emplacement may be calculated from the equation (Andrews 1985):

$$C_{z,t} = Gt \{1 + \exp(-2z/\sqrt{\pi Dt})\} \qquad (7)$$

where G is the He generation rate and D is the He diffusion coefficient in the rock. The concentration

Table 15.3 Some diffusion coefficients (cm^{2}s^{-1}) for the noble gases

Medium	^{4}He $D(20°C)$	^{4}He $D(200°C)$	^{40}Ar $D(100°C)$	^{40}Ar $D(800°C)$	^{222}Rn $D(20°C)$
Air[1]					0.1
Water[1]	10^{-5}				10^{-5}
Crystal lattice[1]					10^{-20}
Concrete[2]					10^{-5}
Pyrex glass[3]	10^{-8}	4.5×10^{-7}			
Silica[4]	2.4×10^{-8}	9.4×10^{-7}	10^{-12}		
Basaltic glass[5,6]	9.2×10^{-17}	4.2×10^{-11}			
Granite[7]	10^{-16}	10^{-10}			
Feldspar[8]			10^{-28}	10^{-11}	
NaCl[9]	3.6×10^{-16}	6.9×10^{-11}			

[1] Tanner (1964)
[2] Culot *et al.* (1976)
[3] Thomson and Wardle (1955)
[4] Shelby (1972)
[5] Jambon and Shelby (1980)
[6] Kurz and Jenkins (1981)
[7] Andrews and Hussain (Unpublished)
[8] Dalrymple and Lanphere (1969)
[9] Rogers *et al.* (1954)

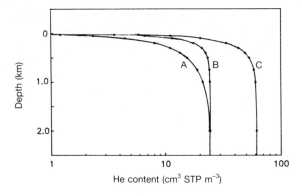

Fig. 15.2. Depth profiles for ⁴He diffusion from a uniform He generating rock with the U and Th contents (0.45 and 1.45 p.p.m. respectively) of average sandstone, calculated for the following conditions:

Profile	Formation age (Ma)	Diffusion coefficient (m²a⁻¹)
A	100	3.15×10^{-3}
B	100	3.15×10^{-4}
C	250	3.15×10^{-4}.

of He stored within the rock varies with depth and with the age of the formation (Fig. 15.2). It becomes effectively constant at depths greater than $4\sqrt{Dt}$ and diffusive He loss can occur only from shallower depths. The surface flux of He can be shown to be $2G\sqrt{Dt/\pi}$ and the He loss is effectively generated within the surface layer of thickness $2\sqrt{Dt/\pi}$. Helium loss from great depth is prevented by the supply of He generated in the uppermost part of the formation. The radiogenic He content of a formation at depth, [He], may be calculated from the equation:

$$[\text{He}] = GT \qquad (8)$$

where T is the age of the formation, if radiogenic He is stored without loss.

Sampling techniques for noble gas analysis

The conditions of sampling must be carefully controlled to ensure that the sample and subsequent gas analysis are truly representative of the *in-situ* gas contents of the groundwater. As previously mentioned, many groundwaters become supersaturated with gases which are generally held in solution by the greater hydrostatic pressure at the depth of the aquifer. The entrainment of excess air at recharge can also cause supersaturation as it is dissolved at depth. The amount of such excess air which is present in most groundwaters is small. It may be represented by an index which is defined as the ratio of the groundwater Ne-content to the Ne-solubility at the recharge temperature for the groundwater. Values of this index are generally less than two and all the dissolved gases can be held in solution by a hydrostatic pressure of less than 1 bar (10 m head). Groundwaters may also dissolve biogenic and/or thermogenic methane and nitrogen during circulation and the amount of these gases may be such that very high hydrostatic pressures are needed to retain all of the gases in solution. For example, if a groundwater had been equilibrated with atmospheric nitrogen and subsequently dissolved methane so that the CH_4/N_2 ratio became 10:1, a hydrostatic pressure of 9 bars (90 m head) would be necessary to retain all the gases in solution at 10°C (Andrews and Wilson 1987). Brines in the Canadian Shield contain large amounts of dissolved methane (Fritz *et al.* 1987) and gases are continuously exsolved from wells depressurized to atmospheric pressure.

The estimation of recharge temperatures (and hence palaeoclimate) from noble gas measurements is unaffected by the presence of these additional gases provided that all gases are held in solution by the hydrostatic pressure in the aquifer and by the sampling procedure. This generally requires the samples to be taken under pressure and also that the well has previously been pressurized and flushed under pressure so that any zone of degassed water is removed. This requirement is often the most difficult sampling condition to satisfy. Subsequently, samples may be collected at pressures up to 30 bars in soft annealed copper tubing which is closed by swage-clamps. It is possible to reconstruct the *in-situ* gas contents of a groundwater which has undergone degassing if the flow rates and compositions of the water and gas phases have been determined (Andrews and Wilson 1987). This reconstruction must assume that the collected gas and water phases are in equilibrium with respect to gas dissolution. In practice, it is often difficult to ensure that such equilibrium has been attained and to measure the flow rates of both phases. The outgassing tendency of gas-rich brines, for example, makes it very difficult to estimate the gas contents of the undisturbed formation fluids. The amounts of N_2 and CH_4 present may be expressed relative to the groundwater Ar content which is generally a conservative species derived mainly from the atmosphere. Suppression of outgassing for $N_2/Ar = 60$ and $CH_4/Ar = 10$, for example, requires pressures of 7 and 16 bars at 10°C and 60°C, respectively. The *in-situ* CH_4 content of groundwater in the Milk River Aquifer (Alberta, Canada) has been estimated from its degassing effect on N_2/Ar and Ne/Ar ratios (Andrews *et al.* 1991*b*).

Recharge temperatures

The temperature at which air-equilibration of a groundwater occurs in the unsaturated zone is close to the mean annual air temperature for the geographical region of recharge. It is therefore possible to derive the 'recharge temperature' of a

Table 15.4 Noble gas compositions in air and in solution

(a) Volumes relative to Ar in air and in air-saturated water (ASW)

Gas	air	ASW, 10°C
He	560	122
Ne	1950	531
Ar	1000000	1000000
Kr	122	238
Xe	9	34

(b) Gas concentrations in ASW and ASW containing excess air (ASW/EA) (10^{-8} cm³ STP cm⁻³H_2O)

	ASW, 10°C	ASW/EA, 10°C EAI = 1.2	ASW/EA, 10°C EAI = 2.0
He	4.75	5.94	10.70
Ne	20.65	24.78	41.30
Ar	38910.0	41030.0	49520.0
Kr	9.24	9.50	10.54
Xe	1.32	1.34	1.43

Excess air index (EAI) for ASW/EA is the ratio of the Ne content of the water sample to that of ASW at 10°C.

groundwater from quantitative analysis of the noble gases in solution. A difficulty arises because of the almost universal presence of excess air in solution, which occurs because of entrainment of air with the groundwater as it moves towards the saturated zone. The volumetric ratios in which noble gases are dissolved by air-equilibration and those in air itself (Table 15.4) are sufficiently different to permit the correction of analytical data for the presence of excess air. Small aliquots of noble gases in their air-abundance ratios may be subtracted sequentially from the measured gas contents until the estimated equilibration temperatures for the remaining Ne, Ar, Kr, and Xe are in best agreement. The average of these temperatures is the best estimate of the groundwater recharge temperature. An analytical precision of better than $+/- 2$ per cent is necessary to estimate recharge temperatures within $+/- 2$°C (see Fig. 15.1). Helium is omitted from this procedure because of the dissolution of radiogenic ^4He from the aquifer matrix. The $^{40}Ar/^{36}Ar$ ratio must also be determined so that Ar contents may be corrected for any radiogenic ^{40}Ar dissolved.

The groundwater recharge may be placed in its palaeoclimatic context using the recharge temperature derived from noble gas measurements (Andrews and Lee 1979; Bath *et al.* 1979; Andrews *et al.* 1985). The stable isotope composition ($^2H/^1H$ and $^{18}O/^{16}O$) of a groundwater also reflects the climatic conditions at recharge (Bath 1984) and the relationships between the estimated recharge temperature from noble gases and the isotopic composition of groundwaters have been determined for several aquifers (Andrews *et al.* 1984, 1985; Rudolph *et al.* 1984). Figure 15.3 shows plots of groundwater recharge temperature against ^{18}O composition for some European aquifers. All of these aquifers have zones which were recharged under cooler climatic conditions than presently prevail which indicates that some of the groundwaters were recharged in the interstadial prior to the last glaciation. A strong continental rain-out effect on the isotopic composition is also evident in Fig. 15.3; groundwater becomes isotopically much lighter over continental Europe than over maritime Europe. Because of such effects, the isotopic composition of groundwater alone cannot be used to determine its recharge temperature, and noble gas measurements are needed for an

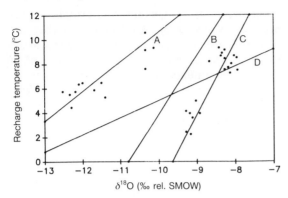

Fig. 15.3. The relationship between the noble gas derived recharge temperature and oxygen isotope composition for recent and palaeo-groundwaters from A: the Innviertel (Miocene) of Upper Austria; B: Triassic sandstone near Nuremberg, Germany, (Rudolph *et al.* 1984) and C: Triassic sandstone from the East Midlands, UK. Because of the maritime climate, the recent waters from the East Midlands are close to the intersection of line C with the Dansgaard line (D) for modern oceanic precipitation. A strong continental effect on the isotopic composition of modern and palaeowaters is evident for the other aquifers.

absolute estimate of this temperature. In each of the aquifers considered, isotopic lightening is accompanied by a decrease in recharge temperature. The results indicate that the continental effect during the last interstadial in Europe was qualitatively similar to the present continental effect for recharge temperature and isotopic composition. The palaeoclimatic record has been established in detail by isotopic studies on foraminifera (Hays *et al.* 1976) and groundwater recharge temperatures can therefore be used to place groundwater residence times within possible climatic eras.

Terrestrial helium

The isotopes ^{3}He and ^{4}He are both produced by radiogenic processes in terrestrial rocks. ^{3}He is generated in the Earth's crust by the action of neutrons on Li, the crustal neutron flux being the result of (α,n) reactions on light nuclei. ^{4}He is generated by decay of U and Th and their daughter nuclides. These reactions may be summarized as:

$$^6\text{Li}(n,\alpha)^3\text{H} \xrightarrow{\beta} {}^3\text{He} \qquad (9)$$

$$^{238}\text{U} \rightarrow {}^{206}\text{Pb} + 8\ {}^4\text{He} \qquad 10)$$

$$^{235}\text{U} \rightarrow {}^{207}\text{Pb} + 7\ {}^4\text{He} \qquad (11)$$

$$^{232}\text{Th} \rightarrow {}^{208}\text{Pb} + 6\ {}^4\text{He} \qquad (12)$$

The ^{3}He/^{4}He ratio of atmospheric He is 1.3×10^{-6} (Coon 1949) and this relatively high abundance of ^{3}He was explained (Libby 1946) as a consequence of the reaction of cosmic-ray produced neutrons with nitrogen in the atmosphere:

$$^{14}\text{N}(n,{}^3\text{H})^{12}\text{C} \text{ followed by } {}^3\text{H} \rightarrow {}^3\text{He} + e^- \quad (13)$$

The solar wind, however, is a more significant source of atmospheric ^{3}He and losses of ^{3}He from the atmosphere occur by thermal and polar wind escape (Johnson and Axford 1969). Evidence for an increased ^{3}He/^{4}He ratio in mantle-stored He (Craig and Lupton 1976) and for excess ^{3}He in deep ocean waters adjacent to oceanic ridges (Clarke *et al.* 1969; Craig *et al.* 1975; Jenkins *et al.* 1980) led to the recognition of a significant flux of ^{3}He from the oceans into the atmosphere (Teegarden 1978). This flux of mantle-derived ^{3}He from the oceans, with the solar-wind ^{3}He, approximately balances ^{3}He losses from the atmosphere. There are several reviews of ^{3}He/^{4}He isotopic ratio variations in terrestrial samples (Rankama 1963, 1954; Tolstikhin 1978; Mamyrin and Tolstikhin 1984; Ozima and Podosek 1983).

All groundwaters contain dissolved He which entered solution because of equilibrium with the atmosphere at recharge. Additional radiogenic He accumulates in an amount which increases with groundwater age, as a consequence of radioelement decay in the aquifer, and diffusion from underlying formations (Andrews and Lee 1979; Heaton and Vogel 1979). The ^{3}He/^{4}He ratio of He in groundwaters decreases from that characteristic of atmospheric He for sources close to the recharge zone to much lower values as radiogenic He is progressively dissolved during groundwater migration. If the amount of radiogenic He dissolved greatly exceeds the amount of atmospheric He acquired at recharge, the ^{3}He/^{4}He ratio approaches a limiting value which is characteristic of the source rock for the radiogenic He.

Characteristic $^3He/^4He$ ratios for various rock types

The $^3He/^4He$ ratio of the radiogenic He produced within a rock matrix has been shown to be independent of the U content of the rock, and almost independent of its Th content, but to be strongly influenced by the Li-content of the rock (Andrews 1985). Radiogenic ^4He in ultramafic and basaltic rocks, which contain less than a few p.p.m. of Li, has a $^3He/^4He$ isotopic ratio of about 10^{-9}. Granitic rocks, which commonly have Li contents in the range 40–80 p.p.m., generate radiogenic He with a $^3He/^4He$ ratio in the range 10^{-8} to 5×10^{-8}. For similar Li-contents, the $^3He/^4He$ ratio in sedimentary rocks is generally somewhat less than for granites. Provided that sufficient data on the Li-contents of the various stratigraphic horizons are available, measurements of the limiting $^3He/^4He$ ratio of He in migrating groundwaters may be used to identify the genetic source of radiogenic He.

Variations in the $^3He/^4He$ ratio in groundwaters have been applied to earthquake prediction (Mamyrin *et al.* 1979; Nagao *et al.* 1980). Such variations reflect a change in the He source and may be caused by a contribution from mantle He in deeply faulted or thin crustal regions. Mantle He has been identified in geothermal fluids associated with volcanic activity in a region of thin crust (Nagao *et al.* 1981; Hooker *et al.* 1985). The $^3He/^4He$ isotope ratio for He in groundwaters from the Stripa granite shows that the radiogenic He has the isotopic composition expected for the average Li-content of the rock (Nordstrom *et al.* 1985). The isotopic composition of dissolved He in groundwaters from the Austrian Molasse shows that the He can only have been generated within the granitic basement and hence that it has migrated by diffusion processes (Andrews *et al.* 1985). The $^3He/^4He$ ratio for these groundwaters decreased with increasing dissolved He content and was attributed to mixing of atmospheric He with radiogenic He. The limiting $^3He/^4He$ ratio was equal to that calculated for the Li content of the granitic basement.

Tritiogenic ^3He

The isotopic composition of helium in recent groundwaters may also be influenced by the decay of tritium which was introduced following ther-

monuclear weapon testing in the atmosphere. The decay of such ^3H to ^3He has been used as the basis of a sensitive ^3H analysis method (Clarke *et al.* 1976) and the $^3H/^3He$ ratio has been used to study the ages of recent ocean and groundwater masses (Tolstikhin and Kamenskiy 1969; Top *et al.* 1980).

Radiogenic He accumulation in groundwaters

Considerable amounts of He in excess of that due to equilibration with the atmosphere at recharge have been observed in groundwaters from fractured rocks. The source of this excess He is the generation of ^4He by radioactive decay of U, Th, and their alpha-emitting daughters in the rock matrix. A linear relationship between excess ^4He and groundwater conductivity (or salinity) was found in the crystalline metamorphic basement in South Carolina (Marine 1979) and groundwater age was calculated according to the equation:

$$t = [He] \; \phi \varrho^{-1} \{1.19 \times 10^{-13}[U] \\ + 2.88 \times 10^{-14}[Th]\}^{-1} \text{ years} \quad (14)$$

where [He] is the groundwater He content, cm^3 STP $cm^{-3}H_2O$, ϕ is the aquifer porosity and ϱ is its bulk density (g cm^{-3}), [U] and [Th] are the U and Th contents (p.p.m) of the rock matrix. An age of 840 ka for the groundwater (calculated from eqn (14)) represents a minimum groundwater age because it is an inherent assumption of eqn (14) that all of the He generated leaves the mineral phases of the rock and is dissolved in the fracture or interstitial fluids. The age estimate is also conditional upon the absence of any diffusive movement of He into or out of the aquifer. Similarly calculated brine ages for the Texas Panhandle (Zaikowski *et al.* 1984) are more than an order of magnitude greater than hydrological estimates.

Increasing He contents with ^{14}C-age of groundwaters in intergranular media aquifers have been demonstrated (Andrews and Lee 1979; Bath *et al.* 1979; Heaton and Vogel 1979) and it was shown that groundwater ages calculated from eqn (14) were much greater than ^{14}C-derived ages. It was suggested that this could arise because of diffusive movement of ^4He into the aquifer from other geological horizons (Andrews and Lee 1979). Such diffusion into intergranular aquifers has since been identified (Andrews *et al.* 1985; Torgersen and Clarke 1985) and groundwater ages were estimated on the basis of dissolution of the diffusive flux of He by the groundwater. If the rate of groundwater movement is greater than the rate of He diffusion in the retrograde direction, the groundwater must dissolve the diffusive flux of He from underlying formations during its migration. Groundwater age may then be estimated from the He content of the groundwater, [He], by the equation:

$$[He] = t_r F / (\phi h) \qquad (15)$$

where F is the diffusive flux of He into the aquifer, t_r is the groundwater residence time, ϕ is the aquifer porosity and h is the formation thickness. Use of this equation requires estimates of the He flux into the aquifer and of the porosity and thickness of the formation. Clearly, the necessity to adopt average values which describe the geochemical and hydrological conditions over extensive regions, must limit the reliability of such groundwater ages. For the Miocene aquifer of the Austrian Molasse the He flux estimate was based on the calculated diffusive loss from the granitic basement (Andrews *et al.* 1985) whereas for the Great Artesian Basin the flux was equated with the He production rate throughout the crustal column (Torgersen and Clarke 1985). In both cases the derived groundwater ages were in broad agreement with the perceived hydrological situation. Mazor and Bosch (1987) reviewed applications of noble gases in sedimentary basins and concluded that He could be used as a groundwater age indicator.

In the case of fractured rocks, diffusion processes are generally more difficult to quantify because of the absence of any well-defined groundwater flow path. Accumulation of ^{4}He in fluids in fractures may occur because of (i) diffusion of stored radiogenic ^{4}He from the mineral phases of the rock into the fluid in the fractures and (ii) diffusive movement of ^{4}He along fluid-filled fractures from greater depths in the formation. The first of these processes depends upon lattice diffusion coefficients, grain boundary diffusion, the frequency of fractures, and the crystallization age of the rock. The diffusion coefficient for He in water is several orders of magnitude greater than that in natural glasses or crystalline minerals so that diffusion in fracture fluids can permit rapid He migration from depth. At all depths He solu-

tion in fracture fluids is controlled by short-range diffusion across the mineral–water interface. The He concentration gradient with depth may also be modified by He transport in mobile fracture fluids. It is extremely difficult to model accurately the ^{4}He concentration/depth profile because of the number of parameters involved and because the frequency of conductive fractures may vary considerably within particular rock masses.

Very large amounts of dissolved ^{4}He, up to 3×10^{-3}cm^3 STP cm^{-3}H$_2$O, have been reported in groundwaters from fractured granites at Stripa, Sweden (Andrews *et al.* 1989c). Groundwater He contents were found to be similar to those expected for diffusive loss from the intrusion with the diffusion coefficient of water. The granite has therefore lost He by diffusion processes which involve an aqueous phase. The rock matrix has many continuous and discontinuous microfractures (Nordstrom *et al.* 1985) which are likely to be water-filled and would facilitate He diffusion. At depths greater than 800 m, groundwater He contents were approximately equal to those calculated for diffusional loss according to eqn (7) with water as the diffusive medium. At shallower depths (less than 350 m) groundwaters contained much less He than suggested by the calculated diffusion profile and these low He contents were attributed to additional He-loss by aqueous transport in circulating groundwaters. The conductive fracture system is likely to be much more extensive in the weathered granite margins and this would facilitate such He transport.

Radiogenic argon

The biotites and feldspars of crystalline rocks generally retain most of the ^{40}Ar which is generated by ^{40}K decay within the mineral. Diffusive loss of radiogenic ^{40}Ar at temperatures up to 200°C is negligibly small (Dalrymple and Lanphere 1969). The ^{40}Ar content of a 650 Ma old mineral containing 5 per cent K is 4.0×10^{-4} cm^3 STP cm^{-3} which almost equals the Ar content of groundwater equilibrated with the atmosphere at 10°C. No concentration gradient across the fluid/rock boundary can exist for such conditions. Diffusion across the boundary is only possible for Ar concentrations in the mineral phase corresponding to ages greater than 650 Ma and/or K contents greater than 5 per cent.

The presence of excess ^{40}Ar in groundwaters may be detected by measurement of the ^{40}Ar/^{36}Ar isotopic ratio. Ratios which are greater than that for atmospheric Ar (295.5) indicate that radiogenic ^{40}Ar has been added. The presence of radiogenic Ar in groundwaters from granites (Edmunds *et al.* 1984; Andrews *et al.* 1989*c*) are most readily explained as a consequence of mineral alteration processes which would release ^{40}Ar from the lattice without involving diffusion processes. Argon isotopic ratios of up to 800, observed in oil and formation waters in the Molasse Basin of Austria (Andrews *et al.* 1986*b*), were attributed to migration from a high temperature environment where diffusive release of ^{40}Ar occurred.

Nitrogen/argon ratio

Any radiogenic ^{40}Ar dissolved during groundwater migration may be readily identified by measurement of the ^{40}Ar/^{36}Ar ratio and subtracted from the total dissolved Ar content of the groundwater. The remaining atmospheric argon is a conservative species in the groundwater and any changes in the N_2/Ar(atmospheric) ratio may be used to identify nitrogen production by nitrate reduction (Vogel *et al.* 1981) or by thermogenic processes. Nitrate reduction has been investigated for groundwaters in the Kalahari (Heaton *et al.* 1983) and in the Triassic sandstones in the UK (Wilson *et al.* 1990). Thermogenic nitrogen has been found in oil-field brines (Andrews *et al.* 1986*b*).

Natural radionuclides in groundwaters

The natural radioelements, U, Th, and K are widely distributed in crustal rocks and consequently may be dissolved by water–rock interaction in aquifer systems. The radionuclide, ^{40}K, is a rare (0.0118 per cent) isotope of K and its solution, which occurs without any isotopic fractionation, is controlled by the geochemical behaviour of K. The combination of low abundance with long half-life causes the activity of ^{40}K in solution to be negligibly small, even for groundwaters with high K contents (up to several hundreds of p.p.m.). Nat-

ural U and Th form stable Pb isotopes through sequential alpha- and beta-decay processes which involve several chemical species. The activity of these species in solution is controlled by both their radiochemical and geochemical properties. Uranium is more soluble as U^{VI} than in the reduced state U^{IV} and is complexed in solution by the presence of HCO_3^- ions. Thorium, in contrast, is very insoluble and the concentration of ^{232}Th in natural waters is extremely low (< 0.01 p.p.b.). Dissolved Ca commonly acts as a carrier for Ra isotopes in solution but radioisotopes of Pb and Bi are not geochemically stabilized and consequently have very short residence times in solution. The long-lived nuclide ^{210}Pb, for example, never attains equilibrium in solution with its precursor, ^{222}Rn, because its residence time in solution is less than 8 days (Andrews *et al.*, 1989*b*) which is much too short to allow significant equilibrium ingrowth. In the lacustrine environment, Barnes *et al.* (1979) found that ^{210}Pb residence time was determined by that of the particulate matter on which it was rapidly adsorbed. Radon dissolution is independent of chemical controls because it is a noble gas. The varied chemical behaviour and half-lives of the radionuclides that are dissolved in groundwaters allow geochemical processes with timescales from hours to many thousands of years to be studied. Consequently, there has been much recent interest in the application of radionuclide measurements to the estimation of radionuclide migration rates, groundwater residence times, and hydrogeological parameters such as porosity or fracture apertures. Some of these applications are discussed in more detail below.

Uranium solution processes in groundwaters

Natural radionuclide dissolution is controlled by the chemical character of the groundwater, that is by its pH, redox potential, salinity, and the dissolved species present. Uranium in solution, for example, is stabilized as carbonate complexes of both U^{IV} and U^{VI}, and the importance of these complexes is dependent upon the pH, Eh, and bicarbonate content of the water (Langmuir 1978). Radioactive equilibrium is established throughout the ^{238}U decay series within 1.25 Ma in closed systems. Most rock matrices have remained closed to U loss for much longer times and the ^{234}U/^{238}U activity ratio (AR) is at equilibrium (AR = 1.0) within the bulk of the rock matrix. This equilibrium can only be disturbed close to the water–rock interface by U dissolution processes. However, the AR of dissolved U in groundwaters is often enhanced; values up to 5 are common and extreme values in excess of 10 occur less frequently. Such disequilibrium may be a consequence of either preferential solution of ^{234}U atoms at the water–rock interface or of alpha-recoil induced solution of ^{234}Th. Preferential solution of ^{234}U rather than ^{238}U atoms may occur because the former are decay products of ^{238}U and are consequently present in lattice regions which have suffered recoil damage during the decay process. Such damage involves lattice displacement and is not readily annealed. It is also probable that the ^{234}U atoms are oxidized to the more soluble U^{VI} oxidation state during the recoil process (Rosholt *et al.* 1963). ^{238}U atoms which decay in the rock surface close to the rock–water interface may result in the ejection of alpha-recoil ^{234}Th atoms into the solution (Kigoshi 1971). These short-lived ^{234}Th atoms either decay in solution or are deposited on the rock surface where they decay to form ^{234}U atoms. The latter are situated at the rock–water interface where they may readily be dissolved.

For a uniform distribution of U on the rock surface, the rate of ^{234}Th recoil solution, ^{234}Th$_{rec}$, due to ^{238}U decay at the rock–water interface, may be calculated from the equation:

$$^{234}\text{Th}_{rec} = 0.0122 \, \varrho \, [\text{U}]_r \times 0.25 \, R \; (\text{s}^{-1}\text{cm}^{-2}) \quad (16)$$

where $[\text{U}]_r$ is the natural U content (p.p.m.) of the rock, ϱ is the rock density (g cm^{-3}), R is the recoil range (cm) of ^{234}Th in the rock matrix, 0.25 is the fraction of recoil atoms from within the recoil range of the surface that enter solution and 0.0122 is the ^{238}U specific activity (s$^{-1}\mu$g^{-1}) in natural uranium, which equals the equilibrium specific activity of ^{234}U.

For estimation of the total ^{234}Th solution rate (recoil flux), the extent of rock surface, S, in contact with unit volume of groundwater must be known. The solution of ^{222}Rn is also a surface dependent process and it is possible to use groundwater ^{222}Rn concentrations to estimate the specific

surface at the rock–water interface (see below). The change in ^{234}U/^{238}U activity ratio, AR$_t$, after recoil has proceeded for time t is dependent upon the U contents, $[\text{U}]_r$ and $[\text{U}]_s$, of the rock surface and the solution, respectively. For a groundwater which has become reducing so that preferential leaching of ^{234}U has ceased, the AR changes with time according to the equation (Andrews *et al.* 1982):

$$\text{Ar}_t = 1 + (\text{AR}_i - 1)\text{e}^{-^{234}\lambda t} + 0.25 \, \varrho \, SR(1 - \text{e}^{-^{234}\lambda t}) \, [\text{U}]_r/[\text{U}]_s \quad (17)$$

where AR$_i$ is the initial activity ratio of the dissolved uranium as the groundwater enters the reducing zone (at $t = 0$) and $^{234}\lambda$ is the decay constant of ^{234}U. An initially enhanced activity ratio may be a consequence of either preferential solution of ^{234}U relative to ^{238}U or of alpha-recoil induced solution of ^{234}Th. Because the recoil process requires time for subsequent ^{234}U-ingrowth to become significant, enhanced ARs for groundwaters close to recharge must generally be caused by preferential ^{234}U-solution. The recoil process becomes more significant as the residence time of the groundwater increases after chemical solution has ceased. The occurrence of U deposition as the groundwater migrates towards more reducing conditions cannot change the AR of dissolved U because any isotope effect for these high mass nuclides is negligibly small. Equation (17) shows that groundwaters with very low U contents are more likely to undergo AR increase due to the alpha-recoil process than those with high U contents. Such an AR increase acts to counterbalance the decay of any excess ^{234}U which was dissolved by preferential solution under the more oxidizing conditions closer to recharge. For conditions where U solution by leaching processes has ceased, eqn (17) may be used to calculate groundwater age from the observed ^{234}U/^{238}U activity ratio, the initial activity ratio characteristic of the leaching process and the specific surface at the rock–water interface.

Activity ratio of U leachates

Rock matrices which have been closed systems for 1.25 Ma or more can only contain U with the equilibrium ^{234}U/^{238}U AR. Rosholt (1982) showed that the majority of 64 granitic rocks examined had such equilibrium ARs. Any enhancement of the ARs for U dissolved by rock–water interaction over a short time (very much less than the half-life of ^{234}U) can only be caused by preferential solution of ^{234}U because there is insufficient time for the recoil-process to become significant. The results of leaching under laboratory conditions of crushed samples of granite (Rosholt 1982) and limestone (Bonotto and Andrews, unpublished) have shown that enhanced ^{234}U/^{238}U ARs in the range 1.3 to 2.0 are caused by direct chemical solution. The direct solution of ^{234}U from the rock matrix is therefore significantly easier than that of ^{238}U.

Activity ratio and U content changes during groundwater evolution

Groundwater which has dissolved U with some characteristic AR may undergo further changes in U content and/or AR because of subsequent geochemical or hydrogeological processes. These radiochemical changes may sometimes be used to assess possible hydrogeological processes in the aquifer. The effects of such changes on the initial AR and U content of a groundwater are illustrated in Fig. 15.4. Dilution of the groundwater with U-free water or deposition of U because of a redox change towards reducing conditions have the same effect and cause the groundwater to move towards the left in the diagram. Mixing with a lower U content groundwater (which would be most likely to have a higher AR) causes a displacement into the upper quadrant of the figure. Groundwater ageing without any associated redox or other geochemical changes permits excess ^{234}U decay and/or ^{234}U enhancement by alpha-recoil, corresponding respectively to downward or upward movements on the diagram. The relative importance of the two processes may be assessed using eqn (17). A change towards more reducing conditions causes a decrease in U content and permits an alpha-recoil increase in AR, so that the groundwater follows a trajectory into the upper left quadrant of the diagram.

A parent groundwater which has acquired its U by leaching processes (and is therefore likely to be relatively young and oxidizing) would move to the

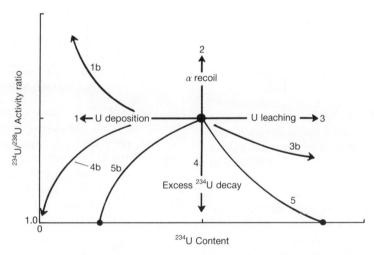

Fig. 15.4. U content and ^{234}U/^{238}U activity ratio changes for the following geochemical processes:
1 Deposition of U induced by redox change
1b Deposition of U with alpha-recoil solution of ^{234}U
2 Alpha-recoil solution of ^{234}U
3 U dissolution with characteristic AR
3b U dissolution with decreasing AR as solution progresses
4 Excess ^{234}U decay
4b U deposition and excess ^{234}U decay
5 Mixing with groundwater with higher U content and lower AR
5b Mixing with groundwater with lower U content and lower AR.

right on the diagram if the leaching process continued with the same characteristic AR. If the preferential leach rate for ^{234}U relative to ^{238}U changes, the groundwater could follow trajectories above or below this line; a decrease in the leach rate ratio being the more probable process because of progressive depletion of ^{234}U in mineral surfaces. Mixing with a groundwater which has a high U content (and most likely to have a lower AR) would follow a mixing line into the lower right quadrant of the diagram.

Activity ratio and groundwater mixing

Figure 15.4 shows qualitatively that mixing between two end-member groundwater types is identifiable by consequent changes in the activity ratio and content of the dissolved uranium. Provided that U is conserved in the groundwater mixture, it can be shown from mass balance considerations that the U content, $[U]_{mix}$, and its activity ratio, AR_{mix}, in the groundwater mixture may be calculated from the equations:

$$[U]_{mix} = f[U]_1 + (1 - f)[U]_2 \qquad (18)$$

and

$$AR_{mix} = \frac{f[U]_1 R_1 + (1 - f)[U]_2 R_2}{f[U]_1 + (1 - f)[U]_2} \qquad (19)$$

where $[U]_1$ and $[U]_2$ are the U concentrations in groundwater species 1 and 2 respectively; R_1 and R_2 are the corresponding activity ratios and f is the volumetric fraction of species 1. Substitution of $r = [U]_2/[U]_1$ in eqn (19) yields:

$$AR_{mix} = \frac{fR_1 + (1 - f)rR_2}{f + (1 - f)r} \qquad (20)$$

Figure 15.5 shows the AR and U content changes for mixtures of groundwater species with ARs of 2 (species 1) and 1 (species 2) and for various values of the U concentration ratio, r. These mixing lines are not linear except for the special case where the U concentrations in both groundwater species are the same ($r = 1$). Exact evaluation of groundwater mixing is rarely possible because the requirement for conservation of U in solution is so readily violated. For example, redox changes may cause either further dissolution or deposition of U and

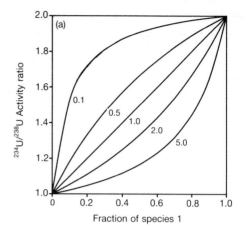

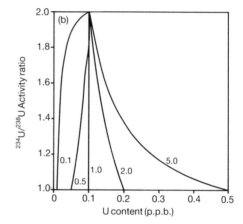

Fig. 15.5. ^{234}U/^{238}U Activity ratio changes for mixing between groundwater species with AR (species 1)=2 and AR (species 2)=1 and for values of the U concentration ratio, $r=[U]_2/[U]_1$ as indicated on the curves. (a) AR of groundwater mixture plotted against fraction of species 1 (b) AR plotted against U content of groundwater mixture for $[U]_1 = 0.1$ p.p.b.

the AR is subject to change by the recoil and decay processes.

Some application of U content and AR measurements in groundwaters

Applications of U isotopic variations to hydrological problems have been reviewed by Osmond and Cowart (1976) and Osmond *et al.* (1983). The influence of redox changes on U hydrogeochemistry were well demonstrated by the study of the

Carrizo Sandstone in Texas (Cowart and Osmond 1974). Groundwater U contents decreased abruptly at a redox boundary where the oxidizing character of the infiltrated groundwater changed to reducing conditions. This was accompanied by an increase in AR as the groundwater entered the reducing zone, explained as a consequence of the high [234]Th recoil rate produced by U deposition at the redox boundary. The AR subsequently declined as the groundwater migrated beyond the redox boundary. When compared with hydrological ages, this decline was found to be more rapid than that corresponding to radioactive decay of the excess [234]U. Comparison of ^{14}C ages with AR decline has shown that U does not behave conservatively in this zone but is 'retarded' by adsorption processes on the aquifer matrix (Pearson *et al.* 1983). A rare situation in which such retardation does not occur and AR decline by excess [234]U decay directly yielded groundwater ages in agreement with hydrological estimates, is reported for the Milk River Aquifer in Alberta, Canada (Ivanovich *et al.* 1991).

The use of U concentrations and ARs to assess groundwater mixing was first reported by Osmond *et al.* (1968) for a karstic aquifer in Florida. It was possible to identify source areas for major springs but mixing between different source areas could only be semi-quantitatively estimated. Estuarine mixing between Mississippi River water and sea water was well demonstrated by AR and U concentration changes (Ku *et al.* 1977).

Several studies in British aquifers have also demonstrated the application of U hydrogeochemistry. In the Lincolnshire Limestone, a marked redox boundary coincides with a large decrease in U concentration and an AR increase (Andrews and Kay 1982). The subsequent decline in AR as the groundwater migrates beyond the redox boundary was attributed to mixing between the rapidly moving fracture water and stored intergranular fluids. It was also shown that these intergranular fluids had an AR close to equilibrium although the alpha-recoil equilibrium AR would have been much greater. A possible explanation for this apparent contradiction is that authigenic deposits of U-free calcite have sealed the oolith surfaces against alpha-recoil ejection of [234]Th into solution.

Uranium concentration in solution was found to

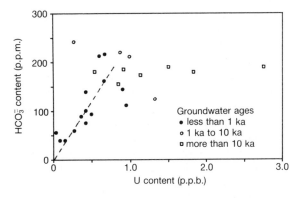

Fig. 15.6. The U content *v.* bicarbonate content of groundwaters from the Triassic sandstones in the East Midlands. U contents of groundwaters younger than 1 ka (filled circles) correlate with bicarbonate content. No correlation exists for older groundwaters.

correlate with bicarbonate content (Fig. 15.6) of groundwater recharged within the last 1 ka in the Triassic sandstones of the East Midlands (Andrews and Kay 1983). Well defined correlations with bicarbonate content have also been found in waters from major rivers of the world (Mangini *et al.* 1979; Scott 1982). The mean AR in river water is in the range 1.2–1.3 which is close to the AR for dissolved U in ocean water. This suggests that the AR in sea water reflects the average preferential leach ratio for U dissolution in the terrestrial environment. Langmuir (1978) has shown that carbonate complexes of the uranyl ion are important for the stabilization of U in solution. Diagrams of Eh against pH, which show the stability fields for the various carbonate complexes, are useful for determining the redox conditions at which U is deposited from solution as uraninite. Cuttell *et al.* (1988) used U contents and ARs to distinguish five groundwater groups in Permo-Triassic sandstones of the Lower Mersey Basin. These groundwater types ranged from recent freshwaters to sea water but U content, AR, and salinity were not the result of conservative mixing of such end-members. It was shown that parts of the basin have been flushed by freshwaters several times since the end of the last glaciation and that rock–water interaction was an important cause of the non-conservative nature of mixing processes.

There have been several studies of U geochemistry in the fracture flow systems in granitic rocks (Andrews *et al.* 1982, 1989*b*; Edmunds *et al.* 1984; Ivanovich and Kay 1983; Gascoyne 1989) and modelling of radionuclide transport in such environments has been reviewed by Latham and Schwarcz (1989). It has generally been found that U is easily mobilized under oxidizing conditions and consequently shallow groundwaters contain high U contents with ARs close to or only slightly greater than equilibrium. At depths up to 300 m, U contents are generally less than in the shallow environment and ARs are significantly greater than the equilibrium value. At great depth, where reducing conditions have existed for considerable periods of time, U contents are low and ARs are lower than at intermediate depths. These generalizations are well illustrated by the study at Stripa, Sweden (Andrews *et al.* 1982, 1989*b*). In the shallow zone, groundwater U contents are controlled by the micro-distribution and mineralogy of U within the rock matrix, rather than solely by its total U content. This is well demonstrated by comparison of shallow groundwaters in the Carnmenellis granite (Edmunds *et al.* 1984), which contain up to 5 p.p.b. U, with those from the Stripa granite which contain up to 90 p.p.b. U, although the Stripa granite contains only 2.4 times more U than the Carnmenellis granite.

222Rn dissolution by groundwaters

The nuclide ^{222}Rn (half-life 3.8 days), is a member of the ^{238}U decay series and is the immediate decay product of ^{226}Ra (half-life 1620 a). In the matrix of most rocks, the latter is in equilibrium with its precursors ^{230}Th and ^{238}U. The solution of ^{222}Rn by interstitial or fracture fluids in a rock matrix is controlled by alpha-recoil and diffusion processes at the water–rock interface (Andrews and Wood 1972).

The activity, A_t, of ^{222}Rn released into a surrounding air or water phase for a constant flux of Rn at the rock–water interface, increases with time, t, according to the equation:

$$A_t = A_e \left(1 - e^{-^{222}\lambda t}\right) \qquad (21)$$

where A_t and A_e are, respectively, the ^{222}Rn activities in the air or water phase at time t and at equilibrium (after more than 5 half-lives of ^{222}Rn)

and $^{222}\lambda$ is the decay constant of ^{222}Rn. The mechanisms which can cause ^{222}Rn release from a planar rock surface are (i) ^{222}Rn recoil on alpha-decay of ^{226}Ra atoms situated within the ^{222}Rn recoil range (0.036 μm) of the surface; (ii) ^{222}Rn diffusion through the crystalline lattice from production sites below the rock surface; and (iii) ^{222}Rn diffusion along crystal defects, grain boundaries, or microfractures from greater depths below the rock surface.

The concentration gradient which results from ^{222}Rn diffusion in an isotropic medium may be evaluated from the equation:

$$C_x = C_0 e^{-x/L} \qquad (22)$$

where C_x, C_0 are the ^{222}Rn concentrations at distances x and 0 from some reference in the diffusion direction and L is the diffusion length for ^{222}Rn (cm). This diffusion length is defined as:

$$L = \sqrt{D/^{222}\lambda} \qquad (23)$$

where D is the ^{222}Rn diffusion coefficient (cm^2 s^{-1}). The ^{222}Rn concentration is reduced by a factor of 0.37 over a distance equal to one diffusion length and only 5 per cent of ^{222}Rn atoms can migrate a distance of five diffusion lengths from their production sites.

The recoil mechanism alone cannot account for observed ^{222}Rn contents of groundwaters and ^{222}Rn release into water is generally greatest for rocks with more than one mineral phase present (Andrews and Wood 1972). Crystalline lattice diffusion constants are extremely small and the corresponding diffusion length is less than the ^{222}Rn recoil range in silicates, so that lattice diffusion cannot contribute significantly to ^{222}Rn release. The ^{222}Rn diffusion coefficient in Carboniferous Limestone has been shown (Zereshki 1983) to be much greater than for lattice diffusion and was attributed to intergranular diffusion in this rock. For crystalline rocks, thermal and isostatic changes associated with the processes of crystallization, cooling, and erosion of overburden can cause extensive microfracturing as a result of stress relief. Such microfractures have apertures of about 10 μm and although they are generally discontinuous within the matrix, can provide tortuous ^{222}Rn diffusion paths from below the rock surface.

The ^{222}Rn content of intergranular or fracture fluids reaches an equilibrium value after a ground-water residence time of 25 days according to eqn (21). The equilibrium ^{222}Rn content ([Rn], in becquerels, Bq) in intergranular fluids may be calculated from the equation:

$$[Rn] = 12.2\, E\varrho\, [U]_r / \phi \quad (Bq\, l^{-1}) \qquad (24)$$

where ϱ is the bulk density (g cm^{-3}), ϕ is the porosity, $[U]_r$ is the U content of the rock (p.p.m.), and the factor E is the fractional efficiency for ^{222}Rn release from the rock matrix. The value of E may be as large as 0.1 for a poorly cemented sandstone and can approach 0.2 for an unconsolidated sand aquifer.

The ^{222}Rn flux from a plane rock surface is mainly due to diffusion along grain boundaries or microfractures. As the diffusion length for ^{222}Rn in air is 218 cm whilst in water it is only 2.18 cm, the distance which ^{222}Rn may migrate from below the rock surface depends upon whether the microfractures or interstitial space is air or water filled. The ^{222}Rn concentration in groundwater from a fracture flow system may be calculated from the equation (Andrews *et al.* 1986a):

$$[Rn] = 2F / w \quad (Bq\, m^{-3}) \qquad (25)$$

where F is the ^{222}Rn flux (atoms m^{-2} s^{-1}) from the fracture surfaces and w is the fracture aperture (m). Hence it may be shown that the specific surface, S, of the fracture system may be obtained from the equation:

$$S = [Rn] / F \quad (m^{-1}) \qquad (26)$$

Some ^{222}Rn concentrations in UK groundwaters

The ^{222}Rn content of groundwater is controlled both by the U content of the aquifer and by the effect of its physical properties on the ^{222}Rn release process. The importance of the latter is well illustrated by comparison of the ^{222}Rn contents (Fig. 15.7(a)) in groundwaters from the Carboniferous Limestone and Jurassic Midford Sands (Andrews and Wood 1972). Although these formations have similar U contents, the ^{222}Rn contents of the groundwaters are much greater for the Midford Sands, which are an intergranular flow aquifer of uncemented sands, than for the Carboniferous

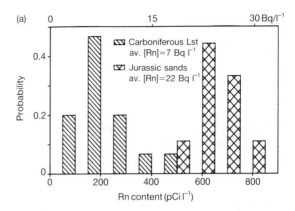

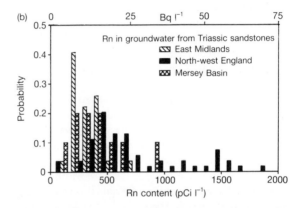

Fig. 15.7. The distribution of ^{222}Rn contents of groundwaters from Carboniferous Limestone (fracture flow), Jurassic sands (intergranular flow) and Triassic sandstones (mixed flow systems).

Limestone, which is a karst system with ^{222}Rn solution dominated by fracture flow.

The ^{222}Rn contents (Andrews and Lee 1979; Cuttell *et al.* 1988) of groundwater from some Triassic sandstones are compared in Fig. 15.7(b). Although rock U contents are similar, the distribution of ^{222}Rn contents in these aquifers have different ranges and modal values. Such differences for similar lithology and U content, may be explained as a consequence of variation in rock fracturing within the aquifers. Relatively high ^{222}Rn contents occur in intergranular flow systems but movement of groundwater from an intergranular to a fracture system can permit subsequent decay of ^{222}Rn within the fracture.

In granitic terrains, groundwater readily penetrates the granite to shallow depths and an active fracture flow system is largely controlled by local topography. At depths greater than a few hundred metres, natural circulation is much slower and may be controlled by sea level changes during glacial cycles. More rapid induced circulation occurs at depth around deep mines which have created large hydraulic gradients between the surface and mine excavations. The ^{222}Rn content of groundwater moving within a fracture may be estimated from eqn (25) if the fracture width and flux are known or alternatively, the mean fracture width in the flow system may be estimated from the groundwater ^{222}Rn content if the ^{222}Rn flux from the granitic surfaces is known. This flux has been measured directly for some highly radioactive granites by Andrews *et al.* (1986*a*). The mean ^{222}Rn content in groundwater from the Carnmenellis granite (Edmunds *et al.* 1984) is about 340 Bq l^{-1}. This is much greater than for sedimentary lithologies, in part because of the greater U content of the granite but principally because of the large ^{222}Rn flux across fracture surfaces in granitic rocks. The corresponding estimated fracture width for the Carnmenellis granite is 90–250 μm. It is interesting to note that this is comparable to the fracture aperture range of 30–260 μm found for the Stripa granite (Andrews *et al.* 1989*b*).

HDR reservoir modelling by ^{222}Rn

The dissolution of ^{222}Rn during the forced circulation of water through the fracture system in the granitic reservoir of a hot dry rock (HDR) geothermal doublet in Cornwall (Chapter 17) has been used to estimate the average fracture width and surface area of the reservoir (Andrews *et al.* 1986*a*). Flow in the various fractures of the reservoir was assumed to obey Darcy's law for fracture flow. The fluid flow rate in a single fracture, q, could then be calculated from the equation (Fox and McDonald 1978):

$$q = \frac{\Delta P \, w^3 d}{12 \eta l} \qquad (27)$$

where ΔP is the pressure differential along the fracture of width w, depth d and length l, and η is the fluid viscosity. The mean fluid transit time, t,

for plug flow in the fracture is given by the equation:

$$t = \frac{wld}{q} \qquad (28)$$

from which on substituting for q from equation (27):

$$t = \frac{12\eta}{\Delta P}\left(\frac{l}{w}\right)^2. \qquad (29)$$

The reservoir was envisaged as consisting of n fracture types, each defined by a characteristic length, width, and fluid transit time. Equation (29) shows that the transit time is determined by the length/width ratio of the flow path for specified fluid viscosity and pressure differential. Partial flow rates, q_i, and their corresponding fluid transit times, t_i, for the various flow paths comprising the reservoir were determined from an inert tracer (Na-fluorescein) recovery curve. The ^{222}Rn content, $[Rn]_i$, for fluids arriving from each flow-path type was then calculated from the ^{222}Rn flux, F, and the set of equations:

$$[Rn]_i = \frac{2F}{w_i}(1 - e^{-^{222}\lambda_i}) \quad (Bq \ m^{-3}) \qquad (30)$$

and the mean Rn content, $[Rn]_m$, of the produced fluid was obtained from the summation (for constant w):

$$[Rn]_m = \sum_{i=1}^{n} \frac{q_i \, [Rn]_i}{Q}$$

$$= \frac{2F}{wQ} \sum_{i=1}^{n} q_i(1 - e^{-^{222}\lambda_i}) \quad (Bq \ m^{-3}) \qquad (31)$$

where $Q = \sum_{i=1}^{n} q_i = $ total flow.

The fracture width, w, may be evaluated from equation (31) on substituting the experimentally determined ^{222}Rn content of the return fluids and values of q_i and t_i obtained by analysis of a tracer recovery curve. Corresponding values of fracture lengths, l_i were calculated from equation (29) and the vertical extents of the fracture types, d_i, from equation (28). The total fracture surface area and fluid volume in the reservoir was then obtained by summation.

The Rn modelling of the reservoir was subsequently combined with a model for diffusive loss of He from the reservoir (Andrews and Hussain 1989). It was shown that both the Rn and He contents of the produced fluids were related to the fracture surface area of the reservoir.

Radium solution processes

The possible processes for solution of ^{226}Ra by groundwater are

(1) chemical solution by rock-etch processes;

(2) decay of ^{230}Th in solution; and

(3) alpha-recoil of ^{226}Ra on decay of ^{230}Th which has been formed by ^{234}U decay in the rock surface.

Chemical dissolution of ^{226}Ra should increase in importance with increasing groundwater salinity, provided that sufficient barium or calcium ions are present in solution. The residence time of ^{226}Ra in solution depends upon congruent and incongruent solution of barium and calcium as well as upon its half-life (1620 years). Radium is present in solution predominantly as the Ra^{2+} ion and complex species are unimportant even in saline solutions (Trivedi 1990). The formation of ^{226}Ra due to decay of ^{230}Th in solution is negligible because ^{230}Th is rapidly removed from solution by adsorption processes. The activity of ^{226}Ra dissolved by the recoil mechanism increases with time according to the equation:

$$^{226}A_t = {}^{226}A_e(1 - e^{-^{226}\lambda t})$$ (32)

where $^{226}A_t$ and $^{226}A_e$ are the ^{226}Ra activities at time, t, and at equilibrium, respectively, and $^{226}\lambda$ is the decay constant for ^{226}Ra. The ^{226}Ra activity in solution should, if alpha-recoil is the dominant mechanism, become constant after about 8000 years. However, the incongruent solution of calcium carbonate results in the exchange of calcium

between solution and rock-carbonate and co-precipitation of ^{226}Ra with calcium may prevent its activity ever attaining the equilibrium value in solution.

The isotope, ^{228}Ra (half-life 7.6 a), is produced by decay of ^{232}Th. The ^{226}Ra/^{228}Ra activity ratio in solution is therefore controlled by the relative accessibility of U and Th to aqueous leaching. Whereas U may occur in secondary mineralization, Th is associated with accessory minerals. Comparison of the ^{226}Ra/^{228}Ra activity ratio with the U/Th activity ratios in the whole rock and in fracture mineralizations may be used to identify the predominant sites for water–rock interaction (Andrews *et al.* 1989*b*).

Groundwater ^{226}Ra contents

The ^{226}Ra contents of groundwater from many geological environments are generally less than their ^{222}Rn contents by several orders of magnitude (Andrews *et al.* 1989*b*; Andrews and Wood 1972; Andrews and Lee 1979; Cuttell *et al.* 1988). Relatively high ^{226}Ra contents are generally found in saline groundwaters such as geothermal waters and oilfield brines (Bloch and Key 1981; Andrews *et al.* 1986*b*).

Conclusions

The amounts of the noble gases, other than He, which are dissolved in groundwaters are principally determined by air-equilibration during recharge. At sufficient depth in flow systems, the concentrations of these noble gases generally behave conservatively even when large amounts of biogenic and thermogenic gases such as N_2 and CH_4 are present. The contents of radiogenic ^{4}He in groundwaters can be used qualitatively to establish the pattern of groundwater flow and to identify those groundwaters with the longest residence times. Departures from the expected ^{4}He concentration/depth profile for diffusive loss from a fractured crystalline rock can indicate the most conductive part of the flow system. Residence times may be estimated for intergranular flow systems from groundwater ^{4}He contents and the diffusive flux of ^{4}He into the aquifer.

Noble gas contents may be used to estimate groundwater recharge temperatures which place groundwater age in its palaeoclimatic context. For groundwaters infiltrated under variable climatic conditions, noble gas recharge temperatures are correlated with the stable isotope composition of the groundwater.

Excess radiogenic ^{40}Ar in groundwaters from crystalline rocks is more likely to arise because of mineral alteration processes than because of diffusive loss from K minerals. It may also indicate proximity to thermal events or the migration of thermogenic gases. The nitrogen/argon ratio for dissolved gases in groundwaters can indicate N_2 production by nitrate reduction or thermogenic degradation of organic matter. Outgassing of noble gases with CH_4 and N_2 occurs in aquifers containing large amounts of biogenic or thermogenic gases. The residual noble gases in solution may be used to estimate the gas concentrations originally present in the aquifer.

The U content of groundwater is a strong indicator of the redox character of the groundwater. The $^{234}U/^{238}U$ activity ratios, in conjunction with U contents, may be used to identify different groundwater sources and mixing processes. Groundwater ages may be estimated from excess ^{234}U decay in situations where the U system is geochemically closed.

The ^{222}Rn contents of groundwaters may be used to estimate fracture apertures for flow in fractured crystalline rocks and to assess the relative importance of intergranular and fracture flow in sedimentary aquifers. Modelling of an HDR reservoir is an example of this application.

Radium solution is primarily controlled by groundwater geochemistry and ^{226}Ra contents are significantly high only in very saline waters. The $^{226}Ra/^{228}Ra$ activity ratio is an indicator of the relative accessibility of U and Th to leaching processes.

Acknowledgement

This chapter is published by permission of the Director of the Postgraduate Research Institute for Sedimentology, in the University of Reading; it is PRIS Contribution Number 125.

References

Andrews, J. N. (1985). The isotopic composition of radiogenic helium and its use to study groundwater movement in confined aquifers. *Chem. Geol.*, **49**, 339–51.

Andrews, J. N. and Hussain, N. (1989). Radon and helium modelling of an HDR geothermal reservoir. In *Water–rock interaction, WRI–6*, (ed. D. L. Miles), pp. 19–22. Balkema, Rotterdam.

Andrews, J. N. and Kay, R. L. F. (1982). $^{234}U/^{238}U$ activity ratios of dissolved uranium in groundwaters from a Jurassic limestone aquifer in England. *Earth Planet, Sci. Lett.*, **57**, 139–51.

Andrews, J. N. and Kay, R. L. F. (1983). The U contents and $^{234}U/^{238}U$ activity ratios of dissolved uranium in groundwaters from some Triassic sandstones in England. *Isotope Geoscience*, **1**, 101–17.

Andrews, J. N. and Lee, D. J. (1979). Inert gases in groundwater from the Bunter Sandstone of England as indicators of age and palaeoclimatic trends. *J. Hydrol.*, **41**, 233–52.

Andrews, J. N. and Wilson, G. B. (1987). The composition of dissolved gases in deep groundwaters and groundwater degassing. In *Saline water and gases in crystalline rocks*, (ed. P. Fritz and S. K. Frape). Geological Association of Canada Special Paper 33, 245–52.

Andrews, J. N. and Wood, D. F. (1972). Mechanism of radon release in rock matrices and entry into groundwaters. *Trans. Inst. Min. Metall.*, (B)**81**, 198–209.

Andrews, J. N., Giles, I. S., Kay, R. L. F., Lee, D. J., Osmond, J. K., Cowart, J. B. et al. (1982). Radioelements, radiogenic helium and age relationships for groundwaters from the granites at Stripa, Sweden. *Geochim. Cosmochim. Acta* **46**, 1533–43.

Andrews, J. N., Balderer, W., Bath, A. H., Clausen, H. B., Evans, G. V., Florkowski, T. et al. (1984). Environmental isotope studies in two aquifer systems: a comparison of groundwater dating methods. In *Isotope hydrology 1983*, pp. 535–76. IAEA Vienna.

Andrews, J. N., Goldbrunner, J. E., Darling, G., Hooker, P., Wilson, G. B., Youngman, M. J. et al. (1985). A radiochemical, hydrochemical and dissolved gas study of groundwaters in the Molasse basin of Upper Austria. *Earth Planet. Sci. Lett.*, **73**, 317–32.

Andrews, J. N., Hussain, N., Batchelor, A. S. and Kwakwa, K. (1986a). ^{222}Rn solution by circulating groundwaters in a 'hot dry rock' geothermal reservoir. *Appl. Geochem.*, **1**, 647–58.

Andrews, J. N., Youngman, M. J., Goldbrunner, J. E., and Darling, W. G. (1986*b*). The geochemistry of formation waters in the Molasse basin of Upper Austria. *Environmental Geology and Water Sciences*, **10**, 43–7.

Andrews, J. N., Davis, S., Fabryka-Martin, J., Fontes, J-Ch., Lehmann, B. E., Loosli, H. H. *et al.* (1989*a*). The in-situ production of radioisotopes in rock matrices with particular reference to the Stripa granite. *Geochim. Cosmochim. Acta.*, **53**, 1803–15.

Andrews, J. N., Ford, D. J., Hussain, N., Trivedi, D., and Youngman, M. J. (1989*b*). Natural radioelement solution by circulating groundwaters in the Stripa granite. *Geochim. Cosmochim. Acta*, **53**, 1791–802.

Andrews, J. N., Hussain, N., and Youngman, M. J. (1989*c*). Atmospheric and radiogenic gases in groundwaters from the Stripa Granite. *Geochim. Cosmochim. Acta*, **53**, 1831–41.

Andrews, J. N., Florkowski, T., Lehmann, B. E. and Loosli, H. H. (1991). Underground production of radionuclides in the Milk River Aquifer. *Appl. Geochem.*, **6**.

Andrews, J. N., Loosli, H. H., Drimmie, R. J., and Hendry, M. J. (1991) Dissolved gases in the Milk River Aquifer (Alberta). *Appl. Geochem.*, **6**.

Banwell, G. M. and Parizek, R. R. (1985). Relationship between ^{4}He, ^{222}Rn and lineaments. *Eos, Trans. AGU*, **66**, 1115.

Barnes, R. S., Birch, P. B., Spyridakis, D. E., and Schell, W. R. (1979). Changes in the sedimentation histories of lakes using lead-210 as a tracer of sinking particulate matter. In *Isotope hydrology 1978*, Vol. II, pp. 875–98. IAEA, Vienna.

Bath, A. H. (1984). Stable isotopic evidence for palaeo-recharge conditions of groundwater: palaeoclimates and palaeowaters; A collection of environmental isotope studies. In *Isotope hydrology 1983*, pp. 169–86. IAEA, Vienna.

Bath, A. H., Edmunds, W. M. and Andrews, J. N. (1979). Palaeoclimatic trends deduced from the hydrochemistry of a Triassic sandstone aquifer, United Kingdom. In *Isotope hydrology 1978*, Vol. II, pp. 545–68. IAEA Vienna.

Benson, B. B. and Krause, D. (1976). Empirical laws for dilute aqueous solutions of non-polar gases. *J. Chem. Phys.*, **64**, 639–709.

Benson, B. B. and Parker, P. D. M. (1961). Relations among the solubilities of nitrogen, argon and oxygen in distilled water and sea water. *J.Phys. Chem.*, **65**, 1489–96.

Bloch, S. and Key, R. M. (1981). Modes of formation of anomalously high radioactivity in oil-field brines.

Bull. Amer. Assoc. Pet. Geol., **65**, 154–9.

Bulashevich, Yu. P., and Bashorin, V. N. (1973). On the detection of faults along the Sverdlovsk DSS profile from high concentrations of helium in underground water. *Phys. Solid Earth.*, **9**, 185–9.

Clarke, W. B., Begg, M. A., and Craig, H. (1969). Excess ^{3}He in the sea: evidence for terrestrial primordial helium. *Earth Planet. Sci. Lett.*, **6**, 213–20.

Clarke, W. B., Jenkins, W. J., and Top, Z. (1976). Determination of tritium by mass spectrometric measurement of ^{3}He. *Int. J. Appl. Radiat. Isot.*, **27**, 515–22.

Coon, J. H. (1949). ^{3}He isotopic abundance. *Phys. Rev.*, **75**, 1355–7.

Cowart, J. B., and Osmond, J. K. (1974). ^{234}U and ^{238}U in the Carrizo sandstone aquifer of south Texas. In *Isotope techniques in hydrology*, Vol. II, 131–49. IAEA, Vienna.

Culot, M. V., Olson, H. G., and Schiager, K. J. (1976). Effective diffusion coefficient of radon in concrete, theory and method for field measurements. *Health Phys.*, **30**, 263–70.

Cuttell, J. C., Ivanovich, M., Tellam, J. H., and Lloyd, J. W. (1988). Uranium-series isotopes in the groundwater of the Permo-Triassic sandstone aquifer, Lower-Mersey Basin, UK. *Appl. Geochem.*, **3**, 255–71.

Craig, H., and Lupton, J. E. (1976). Primordial neon, helium and hydrogen in oceanic basalts. *Earth Planet. Sci. Lett.*, **31**, 369–85.

Craig, H., Clarke, W. B., and Begg, M. A. (1975). Excess ^{3}He in deep water on the East Pacific Rise. *Earth Planet. Sci. Lett.*, **26**, 125–32.

Dalrymple, G. B. and Lanphere, M. A. (1969). *Potassium-argon dating*. W. H. Freeman, San Francisco.

Edmunds, W. M., Andrews, J. N., Burgess, W. G., Kay, R. L. F., and Lee, D. J. (1984). The evolution of saline and thermal groundwaters in the Carnmenellis granite. *Min. Mag.*, **48**, 407–24.

Fox, R. W., and McDonald, A. T. (1978). *Introduction to fluid mechanics*. Wiley, New York.

Fritz, P., Frape, S. K., and Miles, M. (1987). Methane in the crystalline rocks of the Canadian Shield. In *Saline water and gases in crystalline rocks*, (ed. P. Fritz and S. K. Frape), pp. 211–24. Geological Association of Canada Special Paper 33.

Gascoyne, M. (1989). High levels of uranium and radium in groundwaters at Canada's underground Research laboratory, Lac du Bonnet, Manitoba, Canada. *Appl. Geochem.*, **4**, 577–91.

Glueckauf, E. (1946). A micro-analysis of the He and Ne contents of air. *Proc. Roy. Soc. A*, **185**, 98–119.

Glueckauf, E. and Kitt, G. P. (1956). The Kr and Xe contents of atmospheric air. *Proc. Roy. Soc. A*, **234**, 557–65.

Hays, J. D., Imbrie, J., and Shackleton, N. J. (1976). Variations in the Earth's orbit: pacemaker of the ice ages. *Science*, **194**, 1121–32.

Heaton, T. H. E., and Vogel, J. C. (1979). Gas concentration and ages of groundwaters in Beaufort Group sediments, South Africa. *Water SA*, **5**, 160–70.

Heaton, T. H. E. and Vogel, J. C. (1981). 'Excess Air' in groundwater. *J. Hydrol.*, **50**, 201–16.

Heaton, T. H. E., Talma, A. S., and Vogel, J. C. (1983). Origin and history of nitrate in groundwater in the western Kalahari. *J. Hydrol.*, **62**, 243–62.

Hooker, P. J., Bertrami, R., Lombardi, S., O'Nions, R. K., and Oxburgh, E. R. (1985). Helium-3 anomalies and crust-mantle interaction in Italy. *Geochim. Cosmochim. Acta*, **49**, 2505–14.

Ivanovich, M., and Kay, R. L. F. (1983). Uranium series disequilibrium: application to studies of the groundwater system at Altnabreac, Caithness, UK. *Harwell Report* AERE-R10847.

Ivanovich, M., Frohlich, K., and Hendry, M. J. (1991). Uranium series radionuclides in fluids and solids from Milk River Aquifer, Alberta, Canada. *Appl. Geochem.*, **6**.

Jambon, A., and Shelby, J. E. (1980). Helium diffusion and solubility in obsidians and basaltic glass in the range 200–300°C. *Earth Planet. Sci. Lett.*, **51**, 206–14.

Jenkins, W. J., Rona, P. A., and Edmond, J. M. (1980). Excess ^{3}He in deep water over the Mid-Atlantic Ridge at 26°N: evidence for hydrothermal activity. *Earth Planet. Sci. Lett.*, **49**, 39–44.

Johnson, H. E. and Axford, W. I. (1969). Production and loss of ^{3}He in the earth's atmosphere. *J. Geophys. Res.*, **74**, 2433–8.

Kharaka, Y. K. and Specht, D. J. (1988). The solubility of noble gases in crude oil at 25–100°C. *Appl. Geochem.*, **3**, 137–44.

Kigoshi, K. (1971). Alpha-recoil ^{234}Th: dissolution into water and the ^{234}U/^{238}U disequilibrium in nature. *Science.*, **173**, 47–8.

Ku, T.-L., Knauss, K. G., and Mathieu, G. G. (1977). Uranium in open ocean: concentration and isotopic composition. *Deep Sea Res.*, **24**, 1005–17.

Kurz, M. D., and Jenkins, W.J. (1981). The distribution of helium in oceanic basalt glasses. *Earth Planet. Sci. Lett.*, **53**, 41–54.

Langmuir, D. (1978). Uranium solution mineral equilibria at low temperatures with applications to sedimentary ore deposits. *Geochim. Cosmochim. Acta*, **42**, 547–69.

Latham, A. G., and Schwarcz, H. P. (1989). Review of the modelling of radionuclide transport from U-series disequilibria and of its use in assessing the safe disposal of nuclear waste in crystalline rock. *Appl. Geochem.*, **4**, 527–37.

Libby, W. F. (1946). Atmospheric helium three and radiocarbon from cosmic radiation. *Phys. Rev.*, **69**, 671–3.

Mamyrin, B. A., and Tolstikhin, I. N. (1984). *Helium isotopes in nature*. Elsevier, Amsterdam.

Mamyrin, B. A., Tolstikhin, I. N., and Khabavin, L. V. (1979). ^{3}He/^{4}He ratios in earthquake forecasting. *Geochem. Int.*, **16**, 42–4.

Mangini, A., Sonntag, C., Bertsch, G., and Muller, E. (1979). Evidence for a higher natural uranium content in world rivers. *Nature.*, **278**, 337–9.

Marine, I. W. (1979). The use of naturally occurring helium to estimate groundwater velocities for studies of geologic storage of radioactive waste. *Water Resour. Res.*, **15**, 1130–6.

Mazor, E. (1972). Palaeotemperatures and other hydrological parameters deduced from noble gases dissolved in groundwaters, Jordan Rift Valley, Israel. *Geochim. Cosmochim. Acta*, **36**, 1321–36.

Mazor, E. and Bosch, A. (1987). Noble gases in formation fluids from deep sedimentary basins: a review. *Appl. Geochem.*, **2**, 621–7.

Morrison, T. J., and Johnstone, N. B. (1954). Solubilities of inert gases in water. *J. Chem. Soc.*, 3441–6.

Nagao, K., Takaoka, N., Wahita, H., Matsuo, S., and Fujii, N. (1980). Isotopic composition of rare gases in the Matsushiro earthquake fault region. *Geochem. J.*, **14**, 63–9.

Nagao, K., Takaoka, N., and Matsubayashi, O. (1981). Rare gas isotopic compositions in natural gases of Japan. *Earth Planet. Sci. Lett.*, **53**, 175–88.

Nordstrom, D. K., Andrews, J. N., Carlsson, L., Fontes, J.-Ch., Fritz, P., Moser, H., and Olsson, T. (1985). *Hydrogeochemical investigations in boreholes—geochemical and isotope characterisation of the Stripa groundwaters*. Technical report 85–06. SKB, Stockholm.

Osmond, J. K., and Cowart, J. B. (1976). Uranium disequilibrium in groundwater. *Atomic Energy Rev.*, **14**, 621–79.

Osmond, J. K., Rydell, H. S., and Kaufman, M. I. (1968). Uranium disequilibrium in groundwater: an isotope dilution approach in hydrologic investigations. *Science*, **162**, 997–9.

Osmond, J. K., Cowart, J. B., and Ivanovich, M. (1983). Uranium isotopic disequilibrium in groundwater as an indicator of anomalies. *Int. J. Appl. Radiat. Isot.*, **34**, 283–308.

Ozima, M., and Podesek, F. A. (1983). *Noble gas geochemistry*. Cambridge University Press, Cambridge.

Pearson, F. J., Noronha, C. J., and Andrews, R. W. (1983). Mathematical modelling of natural ^{14}C, ^{234}U and ^{238}U in a regional groundwater system. *Radiocarbon*, **25**, 291–300.

Rankama, K. (1954). *Isotope geology*. Pergamon Press, London.

Rankama, K. (1963). *Progress in isotope geology*. Wiley, New York.

Rogers, W. A., Buritz, R. S., and Alpert, D. (1954). Diffusion coefficient, solubililty and permeability for helium in glass. *J. Appl. Phys.*, **25**, 868–75.

Rosholt, J. N. (1982). Mobilisation and weathering. In *Uranium series disequilibrium; applications to environmental problems*, (ed. M. Ivanovich and R. S. Harmon.) pp. 167–80. Clarendon Press, Oxford.

Rosholt, J. N., Shields, W. R., and Garner, E. L. (1963). Isotopic fractionation of uranium in sandstone. *Science*, **139**, 224–6.

Rudolph, J., Rath, H. K., and Sonntag, C. (1984). Noble gases and stable isotopes in ^{14}C-dated palaeowaters from central Europe and the Sahara. In *Isotope hydrology 1983*, pp. 467–77. IAEA, Vienna.

Scholz, C. H., Sykes, L. R., and Aggarwal, Y. P. (1973). Earthquake prediction: a physical basis. *Nature*, **181**, 803–10.

Scott, M. R. (1982). The chemistry of U and Th series nuclides in river waters. In *Uranium series disequilibrium: applications to environmental problems*, (ed. M. Ivanovich and R. S. Harmon), pp. 181–201. Clarendon Press, Oxford.

Shelby, J. E. (1972). Helium migration in glass forming oxides. *J. Appl. Phys.*, **43**, 3068–72.

Sugisaki, R. (1978). Changing He/Ar and N_2/Ar ratios of fault air may be earthquake predictors. *Nature.*, **275**, 209–11.

Sugisaki, R. (1981). Deep seated gas emissions by the earth tide: a basic observation for earthquake prediction. *Science.*, **212**, 1264–6.

Tanner, A. B. (1964). Radon migration in the ground: a review. In *The natural radiation environment*, (ed. A. J. S. Adams and W. M. Lowder), pp. 161–90. University Press, Chicago.

Teegarden, B. J. (1978). Cosmic ray production of deuterium and tritium in the earth's atmosphere. *J. Geophys. Res.*, **72**, 4863–8.

Thomson, S. J., and Wardle, G. (1955). The diffusion of helium in sodium chloride. *Trans. Faraday Soc.*, **50**, 1051–6.

Tolstikhin, I. N. (1978). Some recent advances in isotope geochemistry of light rare gases. In *Advances in earth and planetary sciences* (ed. E. C. Alexander, Jr., and M. Ozima), Vol. 3, pp. 33–62. Scientific Societies Press, Tokyo, Japan.

Tolstikhin, I. N., and Kamenskiy, I. L. (1969). Determination of groundwater ages by the T-^{3}He method. *Geochem. Int.*, **6**, 810–11.

Top, Z., Clarke, W. B., Eismont, W. C., and Jones, E. P. (1980). Radiogenic helium in Baffin Bay bottom water. *J. Marine Res.*, **38**, 435–52.

Torgersen, T., and Clarke, W. B. (1985). Helium accumulation in groundwater I: An evaluation of sources and the continental flux of crustal ^{4}He in the Great Artesian Basin, Australia. *Geochim. Cosmochim. Acta*, **49**, 1211–18.

Trivedi, D. (1990). The mobility of U and Th series radionuclides in groundwaters. Ph.D. Thesis, University of Bath.

Vogel, J. C., Talma, A. S., and Heaton, T. H. E. (1981). Gaseous nitrogen as evidence for denitrification in groundwater. *J. Hydrol.*, **50**, 191–200.

Weiss, R. F. (1970). The solubility of nitrogen, oxygen and argon in water and seawater. *Deep Sea Research*, **17**, 721–35.

Weiss, R. F. (1971). Solubility of helium and neon in water and seawater. *J. Chem. Eng. Data*, **16**, 235–41.

Wilson, G. B. (1986). Isotope geochemistry and denitrification processes in groundwaters. Ph.D. Thesis, University of Bath.

Wilson, G. B., Andrews, J. N., and Bath, A. H. (1990). Dissolved gas evidence for denitrification in Lincolnshire Limestone groundwaters, Eastern England. *J. Hydrol.*, **113**, 51–60.

Zaikowski, A., Kosanke, B. J., and Hubbard, N. (1984). Progress on radiometric dating of Wolfcamp brines using ^{4}He and ^{40}Ar. *Proc. Mat. Res. Soc. Symp.*, **26**, 943–9.

Zereshki, A. (1983). The solution of ^{222}Rn by groundwaters. Ph.D. Thesis, University of Bath.

16. Geochemical characteristics of groundwater in granites and related crystalline rocks

W. M. Edmunds and D. Savage

Introduction

The hydrogeological map of the United Kingdom indicates that large areas of the land mass are classified as regions generally without significant groundwater except at shallow depths. These areas contain nearly all the rocks of Lower Palaeozoic age or older, as well as most igneous and metamorphic rocks, collectively referred to here as basement rocks. Why then should a chapter of this volume be devoted to groundwater in basement and, specifically, granitic and related crystalline rocks? It is true that only a few public water supplies in Britain are derived from basement rocks but, on the other hand, many rural communities in areas where they do crop out, rely on private wells, springs, and shallow boreholes for perennial supplies. The baseflow of rivers and streams in upland hard-rock areas also depends significantly on groundwater derived from weathered bedrock, and baseflow may attain up to 50 per cent of the total flow (Institute of Hydrology 1980; Gustard et al. 1987).

Because basement rocks, especially granites, have been regarded as virtually impermeable at depth, they have been targets of interest during the past two decades as potential sites for the disposal or storage of radioactive wastes. A large amount of research has been carried out on the physical properties of granites and especially the evolution of fluid chemistry as a part of such programmes in the United Kingdom (Kay and Bath 1982), Sweden (Nordstrom et al. 1989a; Fontes et al. 1989), and also in Canada, Switzerland, USA, and other industrial nations. The results of these studies have generally indicated that granites have important fracture, or microfracture, permeability, to depths of up to 1000 m, and contain pore or matrix fluids.

Regions of high heat flow in low-permeability rocks have also proved attractive for testing the creation of reservoirs for hot dry rock geothermal development. A large amount of geochemical research has been carried out on the nature of fluids occurring naturally within the Carnmenellis granite in Cornwall and also during circulation and testing of an intermediate depth (2.5 km) reservoir at Rosemanowes Quarry in Cornwall (Chapter 17 and Edmunds et al. 1989). Similar work is also being carried out at Fenton Hill, New Mexico, in the USA, by the Los Alamos National Laboratory (Grigsby et al. 1989; Grigsby and Tester 1989), and also in continental Europe (Kappelmayer and Gerard 1989).

In this chapter it is not possible to cover all aspects of groundwater chemistry in basement rocks and therefore the discussion will be restricted mainly to granitic and crystalline rocks, such as gneisses, of a similar geochemical character. The hydrogeochemical characteristics are largely the result of the physical and chemical characteristics of the rocks themselves. It should also not be forgotten that the chemistry of groundwaters will also be dependent upon the geological history of the rock concerned. For example, periods of uplift and subaerial weathering, marine transgressions, faulting, and thermal convection may serve to introduce 'new' fluids into a formation. It is the combination of these processes which is illustrated by the chemical composition of the fluids analysed today. The chapter is concerned with a review of granite water–rock interaction at low (0–80°C) temperature, emphasizing results from recent research programmes in the UK. The review covers both the processes in shallow groundwaters as well as the geochemical characteristics of waters derived from granites in the UK at various depths and several different environments (Fig. 16.1).

Physical characteristics of granitic rocks

Granitic and related crystalline rocks usually have low porosities ($< 10^{-2}$), and permeabilities ($< 10^{-8}$ m s^{-1}), which govern their hydraulic behaviour. Although matrix porosity (as opposed to fracture porosity—see Norton and Knapp 1977) is responsible for the bulk of the water stored within basement rocks, the major proportion of groundwater flow occurs within the network of fractures within the rock (Black 1987). Hydraulic potentials in such rocks are restricted to gravity-driven or thermally-driven flow systems. The latter is possible if heat production, heat flow, and permeability are high enough (Fehn 1985). Groundwaters in basement rocks in areas of low topographic relief and low geothermal gradient may be expected to contain stagnant, or very slowly-moving groundwater. In the absence of thermally-driven flow, groundwaters in basement rocks may be 'decoupled' from flow systems in overlying or adjacent sedimentary strata, which may lead to a different evolution of the chemical composition of the groundwaters they contain (Couture and Seitz 1986). Nevertheless, on a geological time-scale, groundwaters in such areas may be subject to tectonic or related disturbances that produce circulation patterns and reactions.

At shallow depths, the alteration of primary granitic minerals along fractures can give rise to enhanced storage of groundwater. The distribution and frequency of fractures is the overriding factor in controlling the development of zones of weathering. The presence of fault zones and lineaments is important for the creation of zones of higher than normal permeability and storage in shallow granitic terrains.

In summary, the physical characteristics of many basement rocks at depth may lead to the isolation of groundwater from external influences and the consequent preservation of the chemical characteristics of the groundwaters for extremely long periods of time in comparison with pore fluids in sedimentary basins. In near-surface environments, groundwater flow in granitic/basement rocks will be governed by the extent of weathering of primary minerals, controlled by tectonic features, the hydraulic head, and chemical reactivity of the rock-forming minerals.

Chemical characteristics of granitic rocks

In terms of bulk composition, there may be little to distinguish basement from younger cover rocks. Indeed 'granitic' basement is very similar in major element composition to many sandstones. The radioelement content of basement rocks may be important in governing heat flow, and hence the possibility of thermally-driven convection. The essential identifying feature of basement rocks is their mineralogical composition. Although the bulk composition may be variable, basement material consists of igneous or metamorphic rocks which are built up of minerals stable at high temperatures and/or pressures and which are intrinsically unstable at low temperatures in the presence of a fluid phase. It is the irreversible dissolution of these phases which contributes to the salinity of entrained groundwaters and may govern the major chemical characteristics of the fluids. The precise nature of these phases may be particularly important in governing the trace element chemistry of the groundwaters. The large concentrations of Li in groundwaters in the Carnmenellis granite for example are almost certainly due to the dissolution of biotite, and perhaps muscovite, in the granite over extended periods.

The physical evidence for water–rock reaction over a wide time-span of geological time is evident both on the micro- and macro-scale. Reactions between fluid and rock within the pores of the matrix may be evident as grain-boundary solids (Plate 1), whereas water–rock reaction along major fractures may result in alteration zones ranging from millimetres to metres in thickness. The great cross-course (zone of altered granite) in the north of the Carnmenellis granite, is up to 100 m wide and is the product of hydrothermal ($\sim 100°C$) alteration of the granite, consisting almost entirely of illitic alteration of feldspars (Bromley 1989). Typically, alteration minerals are

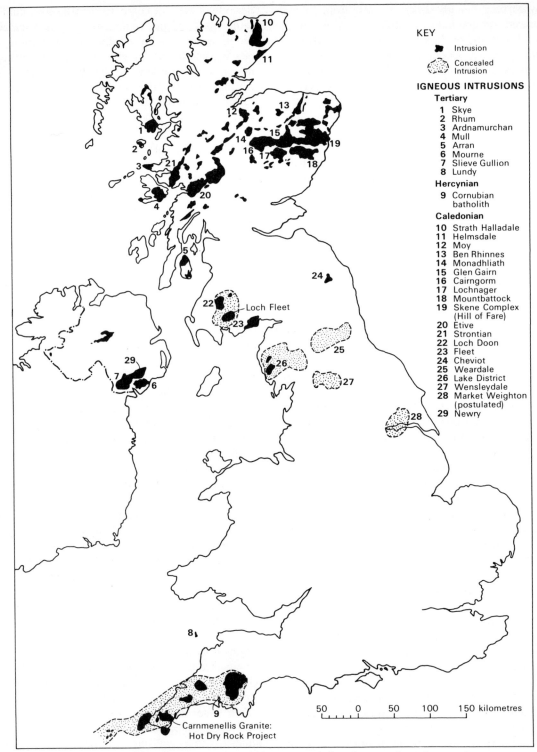

Fig. 16.1. Location of granite areas in the UK with localities referred to in the text.

polymorphs of silica, clays, zeolites, oxides, hydroxides, sulphides, sulphates, carbonates, fluorides and phosphates. The chemical composition of groundwater may be considered to be the net mass balance between 'initial' fluid and solid compositions and the product solid composition, i.e.

initial fluid + initial solid
$$= \text{final solid} + \text{final fluid} \quad (1)$$

As a result, some chemical components (e.g. Si, Al) may be 'mopped up' by the final (product) solids, whereas others may be concentrated in the fluid phase (e.g. Na, Li). These latter components may be sensitive indicators of the nature and extent of water–rock reaction. The style of alteration will also be governed by the effective water–rock ratio during water–rock reaction. A high water–rock ratio will be characterized by the production of hydrous phases and the stabilization of minerals relatively depleted in alkali and alkaline–earth cations, (e.g. clays, micas), whereas low water–rock ratios will be characterized by 'alkaline' or 'hyperalkaline' minerals such as feldspars or zeolites.

Although the water–rock reactions implied from chemical analyses of groundwaters in granitic rocks are irreversible and hence non-equilibrium processes, it is also true that thermodynamics can be used as a predictive and interpretive tool to explain some of the phenomena observed. For example Giggenbach (1984, 1988) was able to explain a number of chemical features of groundwaters from a purely thermodynamic standpoint namely, the low concentration of Mg in relation to other components and the decrease of this element with temperature; the increase of K with increasing temperature; and the decrease of Ca with increasing temperature. Giggenbach extended his analysis to predict the style of alteration associated with the direction of movement of fluids through the crust. Consequently, descending (heating) fluids should be characterized by the alteration of 'average' crustal rock to produce Na–Ca–Mg clays, micas, and carbonates, whereas ascending (cooling) fluids are characterized by the deposition of K and Si in K-feldspar and the silica polymorphs (Plate 2). In the UK, Richards *et al.* (1989*b*) used Giggenbach's approach to predict the composition

of groundwater likely to be found at a depth of 6 km in the Carnmenellis granite, as part of a design study for the development of a 'hot dry rock' geothermal system.

Kinetics of water–rock reactions in granitic basement

Although the nature of secondary solid products may be governed by thermodynamic stability, many chemical characteristics of groundwaters are controlled by kinetic processes. For example, the predominance of Ca over Na in many groundwaters in basement rocks is probably a result of the greater rate of dissolution of calcic as opposed to sodic plagioclases, which are the principal aluminosilicates in most rocks. Consequently, it is useful to understand the relative rates of dissolution of the principal rock-forming minerals at low temperatures in order to be able to interpret the chemical features of groundwaters.

Many laboratory studies of the dissolution of the major rock-forming minerals have now been carried out. Although some of the hypotheses concerning the absolute rate and mechanism of dissolution may differ from author to author, there is a general consensus concerning *relative* rates and mechanisms of reaction. In general, at low temperature, at a pH near neutrality and with conditions far from chemical equilibrium, the rate of reaction of the rock-forming silicates is governed by reaction at the mineral–fluid interface. Thus, the reaction is a surface-dependent process and mass transport in the fluid phase is not rate-limiting. In the last decade a number of reviews of the relative dissolution rates of rock-forming minerals have been carried out (Lasaga 1984; Helgeson *et al.* 1984; Murphy and Helgeson 1989; Brady and Walther 1989). A synthesis of this work (essentially that of Brady and Walther 1989) is presented in Table 16.1.

It may be seen from Table 16.1 that the least stable minerals (olivine, pyroxene, nepheline) are anhydrous and are undersaturated with respect to silica. These are the phases which are stable at the highest temperatures and have crystallized in volcanic lavas or intrusive igneous rocks. The most stable minerals in Table 16.1 are those which may actually precipitate at low temperatures and pres-

Table 16.1 Relative dissolution rates of principal rock-forming silicates at 25°C and near-neutral pH (after Brady and Walther 1989)

Mineral	$-\log$ rate (mol cm^{-2} s^{-1})
Nepheline (NaAlSiO$_4$)	13.3
Forsterite (Mg$_2$SiO$_4$)	14.2
Diopside (CaMgSi$_2$O$_6$)	15.0
Andalusite (Al$_2$SiO$_5$)	15.1
Enstatite (MgSiO$_3$)	15.2
Anorthite (CaAl$_2$Si$_2$O$_8$)	15.5
Analcime (NaAlSi$_2$O$_6$.H$_2$O)	15.7
Albite (NaAlSi$_3$O$_8$)	15.9
Quartz (SiO$_2$)	15.9
Chrysotile (Mg$_3$Si$_2$O$_5$(OH)$_4$)	16.3
Pyrophyllite (Al$_2$Si$_4$O$_{10}$(OH)$_2$)	16.3
Kaolinite (Al$_2$Si$_2$O$_5$(OH)$_4$)	17.4

sures in a hydrous environment. Therefore, from a mineral stability point of view, the greatest amount of water–rock reaction may be expected in mafic igneous rocks such as gabbros, syenites, and basalts and the smallest degree of reaction in slates, schists, gneisses, and granites. In practice, this degree of reaction is also related to the amount of fluid flow in the system, which is dependent upon the physical properties of the rocks. The trace element content of the groundwaters may be influenced strongly by the trace element content of primary minerals. Thus Rb and Ba are often derived from K-feldspar; Sr from plagioclase; Li, Rb, and Cs from biotite and muscovite; Ti and V from pyroxene; and Ni and Co from olivine.

By comparison with research conducted abroad, there have been very few experimental studies of water–rock reaction at low temperatures conducted in the UK. A few studies have been concerned with the disposal of radioactive wastes in basement rocks (Savage 1986; Savage *et al.* 1986*a*; Ragnarsdóttir 1989), but most have been concerned with the development of hot dry rock geothermal systems, principally relating to the test facility at Rosemanowes Quarry in the Carnmenellis granite. These experimental studies have been conducted in the field (McCartney 1986, 1987; Richards *et al.* 1989*a*), in the laboratory (Savage *et*

al. 1985, 1987, 1989), or both (Edmunds *et al.* 1985; Andrews *et al.* 1987; Edmunds *et al.* 1988). Field studies have concerned transient single-pass or re-circulating flow systems in the various reservoir systems at Rosemanowes in the temperature range 25°C–100°C. Laboratory studies have been performed at elevated temperatures to investigate biotite dissolution at 80°C (Edmunds *et al.* 1985), or the interaction of granite with stream-water, NaCl fluids, seawater, and pH buffers in both closed and flow-through systems in the temperature range 80°C–250°C (Savage *et al.* 1985, 1986*b*, 1987, 1989; Edmunds *et al.* 1988).

A unique feature of these studies has been the intercomparison of field and laboratory data concerning the rate of water–rock reaction (Richards *et al.* 1989*a*, Richards and Savage 1989). This work has revealed comparable (within an order of magnitude) rates of reaction for plagioclase feldspar between the field and laboratory studies, which contrasts with previous intercomparisons (using groundwater catchment data), which have recognized rates in the field as being much less than those determined in the laboratory (e.g. Velbel 1986).

Characteristics of groundwaters in UK granites

During the past decade, a number of hydrogeological studies have been carried out which permit an overview of basement (granite) groundwater chemistries to be made. They may be considered in two groups. *Shallow groundwater environments* have been investigated for environmental reasons including the effect of acid rain (Edmunds and Kinniburgh 1986) where baseflow chemistry is an important buffer. *Deep groundwaters* have been investigated in relation to geothermal and radioactive waste objectives. Representative data from each group of studies are given in Tables 16.2 and 16.3. The data in Table 16.2 are, with the exception of Loch Fleet, the average values of *n* samples of stream baseflow or shallow groundwaters from the places shown. Data for the Ben Rhinnes, Cairnsmore of Fleet, Cairngorm, Carnmenellis, and Mourne Mountains are from regional hydrogeochemical surveys carried out in these areas by BGS (unpublished data). The Strath Halladale

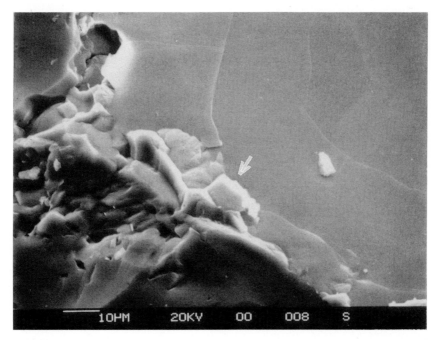

Plate 1. Evidence for water-rock reaction in the 'matrix' porosity of basement rocks. SEM photomicrograph of fluorite (arrowed) precipitated at the grain-boundary between biotite and quartz. The calcium and fluorine necessary for fluorite precipitation were derived from the dissolution of adjacent biotite and plagioclase grains. The sample of granite was taken from drill-core from 1780 m depth in the Carnmenellis Granite, Cornwall, UK. Photograph from Savage *et al.* (1987)

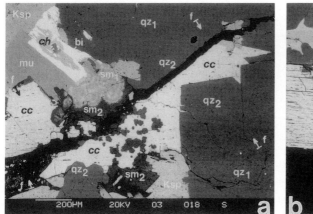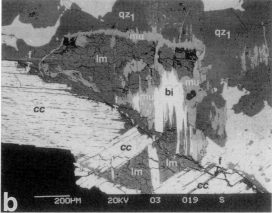

Plate 2. Different styles of alteration associated with 'ascending' (cooling) and 'descending' (heating) fluids in basement rocks. The SEM photomicrographs show two veins taken from drill-core from over 2000 m depth in the Carnmenellis Granite, Cornwall, UK. The veins are delineated by 'f' symbols with arrows pointing towards the vein centre. Both veins show alteration characteristic of a low water/rock ratio. (a) is a vein with alteration characteristic of an ascending fluid, showing overgrowths of quartz (qz_2 on qz_1), overgrowths of K-feldspar (Ksp), the replacement of biotite (bi) by smectite clay (sm_1, sm_2), and the precipitation of calcite (cc). (b) is a vein with alteration characteristic of a descending fluid showing replacement of muscovite (mu) by laumontite (lm) and calcite (cc). Both photographs are from Savage *et al.* (1987).

Table 16.2 Representative data for shallow groundwaters (or river baseflow) draining UK granites (data as mg l^{-1})

	Ben Rhinnes	Fleet	Strath Halladale shallow boreholes	Strath Halladale springs	Loch Fleet	Carnmenellis	Mourne
n		93	17	16	1	13	11
pH	6.79	6.34	6.35	6.19	7.2	5.44	5.97
Na	7.0	5.1	17.1	14.8	6.4	14.7	6.07
K	0.68	0.29	2.2	1.38	0.35	3.18	0.35
Ca	3.54	1.8	21.4	7.7	36.1	12.2	1.3
Mg	1.30	0.80	12.6	4.6	7.4	2.69	0.85
HCO$_3$	14.2	2.6	164	60	142	32.4	2.3
SO$_4$	3.7	4.5	3.6	8.8	6.8	14.9	6.5
Cl	8.8	7.6	23.6	20.7	8.1	25	7.2
NO$_3$-N	0.05	0.02	0.09	0.09	< 0.04	1.28	0.3
Sr	0.027	0.009	0.280	0.092	0.086	0.056	0.008
Ba	0.031	0.003	0.120	0.043	0.002	0.006	0.001
Li	—	—	0.016	< 0.2	0.017	0.0011	0.0014
B	0.008	0.004	< 0.5	0.03	0.006	0.005	0.012
Si	4.3	1.35	12.6	9.9	6.4	2.49	3.13
Mn	0.014	0.033	1.18	0.30	0.802	0.554	0.009
Al	0.145	0.114	< 0.1	< 0.10	< 0.003	0.122	0.199
Zn	0.010	0.007	0.086	—	< 0.002	0.047	0.007
Fe$_{TOT}$	—	0.053	5.20	4.74	0.116	2.00	0.043
F	—	—	< 0.2	0.16	0.54	—	—
Total mineralization	43.8	24.3	264	133	215	112	34.7
% balance	3.5	4.3	−7.2	−9.6	0.2	−4.8	1.8
SI$_{barite}$	−1.17	−2.00	−0.77	−0.74	−2.22	−1.37	−2.49
SI$_{calcite}$	−3.12	−4.69	−1.79	−2.79	−0.82	−3.61	−5.05
SI$_{quartz}$	0.43	0.02	0.80	0.79	0.65	0.19	0.21
SI$_{chalcedony}$	−0.11	−0.55	0.25	0.25	0.09	0.35	−0.31
mK/Na	0.057	0.033	0.076	0.055	0.032	0.127	0.034
mMg/Ca	0.61	0.51	0.97	0.98	0.34	0.36	1.08
mNa/Ca	1.72	2.47	0.69	1.67	0.15	1.05	4.06

Table 16.3   Representative data for deep groundwaters from UK granites (mg l^{-1})

	South Crofty mine	Wheal Jane	Hot dry rock circulation return water	South Crofty mine	Wheal Jane	Cambokeels Weardale	Strath Halladale
Date	17.10.86	26.10.84	15.12.87	15.3.84	13.3.84	4.89	19.07.81
Depth (OD)	635	400	2400	580	150	spring	223
Temp °C	41.1	47.8	35	40	21.6	16	7.3
pH	6.68	8.3	8.15	6.65	3.5	6.89	7.4
Na	7370	2090	23.4	3520	93	3511	32
K	289	132	4.9	153	12	181	1.5
Ca	2950	1300	25.5	1840	191	2482	37
Mg	71	13	1.26	55	43	53.7	1.4
HCO$_3$	58	54	46	60	12.7	150	201
SO$_4$	122	114	16.3	129	1390	106	10
Cl	19500	6090	35	9280	179	12200	24
NO$_3$-N	—	—	4.8	< 1.0	< 1.0	—	< 0.04
Sr	71	23	0.174	30	1.87	—	1.4
Ba	1.6	0.76	0.0028	0.94	0.052	—	0.068
Li	227	64	0.066	107	2.7	42.0	0.015
B	27.8	7.9	0.034	13.9	< 0.5	—	< 0.5
Si	16	16	5.42	16.0	11.9	4.85	7.9
Mn	4.0	0.62	0.012	3.5	19.7	3.75	0.17
Al	< 0.1	< 0.1	0.061	< 0.1	25	< 0.1	< 0.1
Zn	—	< 0.02	0.005	< 0.02	125	—	0.034
Fe$_{TOT}$	1.6	0.028	0.066	4.2	346	0.2	0.45
F	—	3.0	0.90	3.0	44	< 0.1	0.4
As	—	0.065	0.40	0.033	2.1	—	—
Cs	—	4.58	0.056	4.28	3.9	—	—
Rb	—	2.2	0.026	2.5	0.21	—	—
Total mineralization	31660	9930	180.3	15250	2470	18740*	332
K/Na	0.024	0.037	0.123	0.025	0.075	0.030	0.027
SI$_{calcite}$	0.19	1.35	−0.05	−0.22	−4.99	0.28	−0.48
SI$_{barite}$	0.31	0.11	−1.88	0.00	0.77	—	−0.54
SI$_{quartz}$	0.62	0.42	0.13	0.56	0.67	−0.40	−0.74
SI$_{chalcedony}$	0.20	0.03	−0.31	0.19	−0.19	−0.12	0.19
SI$_{fluorite}$	—	0.73	−1.18	0.80	0.50	−2.31	−1.39
m K/Na	0.023	0.037	0.586	0.026	0.076	0.030	0.027
m Mg/Ca	0.039	0.06	0.081	0.049	0.47	0.036	0.062
m Na/Ca	2.17	1.40	0.80	1.66	0.42	1.23	0.75

* An analysis in 1988 was reported as 38 111 mg l^{-1}.

granite data are taken from Kay and Bath (1982) and the Loch Fleet from Cook *et al.* (1987; 1991.) These localities are indicated in Fig. 16.1.

The total mineralization in shallow groundwaters in granites is generally low but the concentrations of most elements are significantly enhanced above rainfall chemistry. This effect has been demonstrated (Cook *et al.* 1987; 1991) for the Loch Fleet catchment (Galloway) after allowing for evaporation using the bromide concentration in rain as a conservative parameter; summary results are given in Table 16.4. For most elements the granite acts as a net source of solutes resulting from weathering reactions, but at the ambient conditions there is little or no addition of NO_3, Ba, Al, Zn, and B.

An important feature of groundwaters in several granitic regions is demonstrated by the data in Table 16.4. Quite high alkalinities are found at Loch Fleet and Strath Halladale for groundwaters from shallow boreholes, although this may not be so apparent in river baseflow samples. The high alkalinities are related to the presence of secondary vein carbonates which, although a very minor phase, are deposited along fractures and react with circulating groundwaters, such that their chemistry is dominated by calcium and bicarbonate ($\pm$ magnesium). Dissolved carbonate derived from vein calcite is likely to be a common characteristic of many groundwaters from granitic terrains. (It has been found in those areas where boreholes have been drilled). In shallower environments, the hydrogeological pathways are in weathered zones; decalcification has occurred and, as a result, more acid waters prevail.

Granitic regions more commonly give rise to acidic surface and shallow groundwaters (Edmunds and Kinniburgh 1986) where the pH may be as low as 4.0 during high flows, but only moderately acid waters (pH 6.0–6.8) may be more typical of baseflow where groundwater has an important buffering effect. Examples of such waters are given in Table 16.2 from the Ben Rhinnes Granite (Invernessshire), Fleet (Galloway), Strath Halladale (Sutherland), Carnmenellis (Cornwall), and the Mourne Mountains (Co. Down). The pH is typically maintained by very small bicarbonate (alkalinity) concentrations, and the acid neutralizing capacities (ANC) of these streams are typically very low, making them vulnerable to fluctuations in input

Table 16.4 An analysis of groundwater from a shallow borehole near Loch Fleet (Galloway) which has been adjusted on the basis of rainfall chemistry to show the net addition from geochemical reactions (as mg l^{-1})

Constituent	Loch Fleet Borehole 2	Net contribution from water-rock interaction
Na	6.4	3.4
K	0.35	0.16
Ca	36.1	35.8
Mg	7.4	6.6
HCO$_3$	142	142 –
SO$_4$	6.8	3.6
Cl	8.1	2.7
NO$_3$-N	< 0.04	< 0.3
Si	6.4	6.4
Sr	0.086	0.083
Ba	0.002	−0.005
Fe	0.116	0.111
Mn	0.802	0.798
B	0.006	−0.004
Li	0.0169	0.017
F	0.54	0.535

acidity. As pH falls, the concentrations of metals such as aluminium and copper may rise sharply.

In areas free of carbonate minerals the chemistry of groundwater in granitic terrains will be dominated by acid hydrolysis reactions of the primary silicate minerals expressed in a general way as:

$$\text{Na, Ca, Mg, K-silicate} + H_2O + H^+ = \text{kaolinite} + H_4SiO_4 + Na^+ + Ca^{2+}, K^+, Mg^{2+} \quad (2)$$

The solid or aqueous reaction products give some evidence as to the nature and extent of the reactants and therefore groundwaters in different granites should give rise to different chemistries according to their component minerals and mineral compositions (Garrels and Mackenzie 1971). The amount of CO_2 in the soil solution is important for controlling both the resultant alkalinity and the extent of silicate mineral hydrolysis by the production of H+ as input to equation (2):

$$CO_2(g) + H_2O = H^+ + HCO_3^- \qquad (3)$$

Weathering rates on granitic catchments in the UK have been studied by Creasey *et al.* (1986). Comparisons have been made between two catchments in north-east Scotland at Glendye (situated on a two feldspar-biotite granite of late Caledonian age) and Peatfold (situated on a quartz-biotite norite, also of late Caledonian age). The weathering output from the latter catchment is about ten times that of the granitic catchment. This is an expression of the distinctly different mineralogical compositions of the sites. However, 75 per cent and 50 per cent, respectively, of the element output can be attributed to the breakdown of plagioclase with lesser amounts derived from pyroxene, amphibole, and biotite.

Analyses of deep groundwaters from granites are only available from three areas of the UK— the Carnmenellis granite (Cornwall), Strath Halladale (Sutherland), and from the concealed Weardale granite (Durham). In and near the Strath Halladale granite, groundwaters have been sampled from boreholes to depths of 300 m in crystalline rocks (mainly granites). They are essentially of low salinity (< 500 mg l^{-1} total mineralization), and dominated by calcium and bicarbonate. The chemistry is otherwise influenced by hydrolysis reactions which occur within the upper tens of metres of the granite. In the Carnmenellis granite in Cornwall, groundwaters have been sampled in tin mines to depths of 700 m and through artificial circulation in the hot dry rock borehole network to depths of 2400 m, thus providing one of the most comprehensively sampled areas of granite groundwaters anywhere in the world. These groundwaters are often saline and thermal, and salinities range up to 31 600 mg l^{-1} total mineralization. Their occurrence has been fully documented by Edmunds *et al.* (1984, 1985, 1987, 1988). Recently, groundwater very similar to the saline water from the Carnmenellis granite has been reported (Manning and Strutt 1991) from a spring discharging at a site just above the concealed Weardale granite which is of Caledonian age. Although this spring is only slightly thermal (16°C), the maximum reported salinity (38 000 mg l^{-1}) exceeds that of values reported from Cornwall. There is evidence

that fluids in the sedimentary cover contribute to the salinity.

Water–rock reactions in the Carnmenellis granite

The main characteristic of the deep groundwaters within the Carnmenellis granite is their high salinity which is considered to have arisen almost entirely from water–rock interaction within the granite or its nearby envelope. The principal cation compositions are summarized in Fig. 16.2 on which are also plotted some of the other data from Table 16.2. The principal feature of the saline groundwater is the depletion in Na relative to chloride (Fig. 16.3), unlike most saline formation waters in which the ratio is near unity as a result of entrapment of ancient sea water or solution of formation evaporites. In granites, the majority of the salinity is derived from mineral sources and the net production of Cl is considerably above that of Na. A possible source of Cl is considered to be the release of lattice chloride, especially from biotite, which in Cornwall (Fuge and Power 1969) may contain more than 1 per cent chloride. An alternative or additional 'source' of chloride, if the water–rock ratio remained low, could be the net removal of water molecules due to the hydrolysis of primary silicates, thus 'concentrating' the remaining fluid with respect to chloride. The incorporation of Na into secondary hydrous minerals would also increase the Cl/Na ratio of the resultant fluid. Chloride would remain in solution and increase in concentration with time, although other cations may be removed by reactions to produce secondary minerals thus increasing chloride/metal ion ratios.

Fluid inclusions have been proposed as a likely source of salinity in groundwaters from some shield areas (Nordstrom *et al.* 1989*b*). In the case of the Carnmenellis granite this explanation is considered unlikely, even though fluid inclusions may contain a significant component of the chloride stored in the granite. The occurrence of fluid inclusions in the granite has been summarized by Smedley *et al.* (1989). The inclusions range in salinity from 4–27 wt per cent NaCl and can be shown to have originated over the temperature range 110–300°C with some originating at temperatures as low as

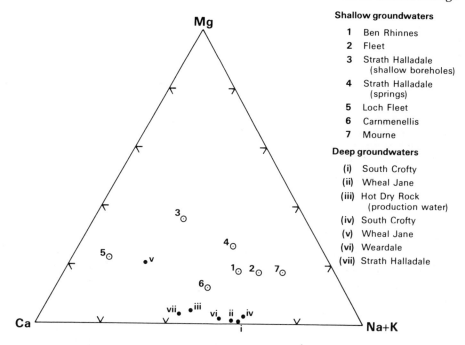

Shallow groundwaters

1	Ben Rhinnes
2	Fleet
3	Strath Halladale (shallow boreholes)
4	Strath Halladale (springs)
5	Loch Fleet
6	Carnmenellis
7	Mourne

Deep groundwaters

(i)	South Crofty
(ii)	Wheal Jane
(iii)	Hot Dry Rock (production water)
(iv)	South Crofty
(v)	Wheal Jane
(vi)	Weardale
(vii)	Strath Halladale

Fig. 16.2. Trilinear cation plot of shallow and deep groundwaters from UK granitic rocks.

70°C. The stable isotope compositions indicate a meteoric origin but they exhibit a distinct oxygen shift ($\delta^{18}O$ +2.2–4.4) in contrast to the groundwaters. To date, the chemical information is limited but the fluid inclusions tend to be slightly more Ca-enriched than present day groundwaters. The preferred model for evolution of fluid inclusions within the Carnmenellis granite is that they have encapsulated the history of the water–rock interactions over the whole history of the pluton. Silicate hydrolysis has produced abundant silica whenever the fluid circulation has initiated new reactions. This silica has then been deposited as unreactive quartz which has locked up the fluids over the wide span of temperatures and reaction conditions.

The sodium and calcium in the saline Carnmenellis groundwater (Table 16.3) are considered to be derived by reaction with plagioclase feldspar and the resultant Na/Ca ratio of the fluid, slightly enriched in Ca, can be explained by preferential reaction of the less stable anorthite-rich component in the zoned plagioclase. The Na/K ratio is particularly high, 20–40 in comparison to the bulk granite ratio (0.6), and reflects preferential incorporation of potassium into secondary clays, such as illite and mixed layer smectite-illite.

Magnesium (together with iron) is also produced during the hydrolysis of biotite and any other mafic minerals but will tend to react to form secondary Na-Mg-Ca clays or, at higher temperatures, chlorite (Fig. 16.3).

The concentration of lithium in not only the saline but also some shallow groundwaters provides important evidence for the evolution of the groundwater. Lithium is present in granitic rocks predominantly in biotite, where it resides in the octahedral layer along with aluminium. During hydrolysis reactions, the biotite may react congruently as in equation (2), or may break down incongruently, adjusting its composition and either remaining structurally a biotite or forming lower temperature minerals such as chlorite or mixed layer clays. In this process various elements which were incorporated in the lattice formed at high temperatures, are released in solution. Lithium is released in this way along with fluoride (and chloride) and is not incorporated into the predominantly

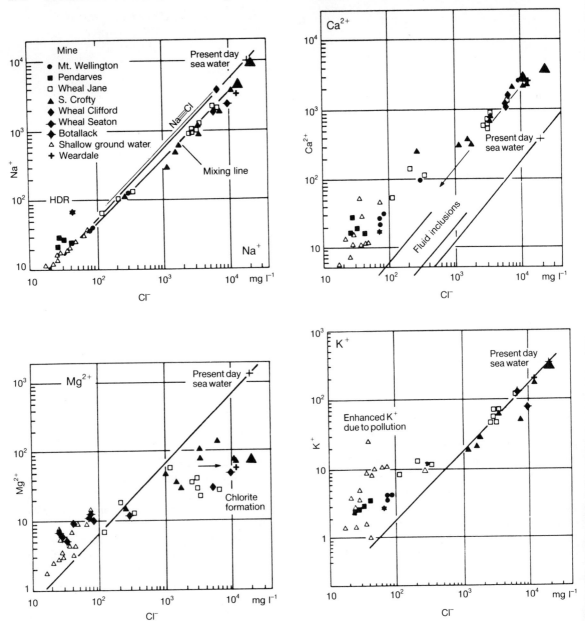

Fig. 16.3. Element/chloride plots showing cation variation relative to salinity for groundwaters from Cornwall; saline water from Weardale also included.

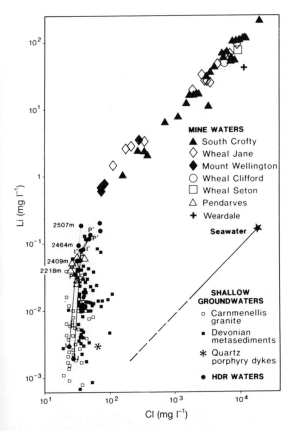

Fig. 16.4. Variation of Li and Cl in shallow groundwaters, minewaters, and HDR circulation waters in the Carnmenellis area, Cornwall. Tie lines for HDR waters link injection waters in open-loop (I) or closed-loop (I′) circulation with corresponding production waters (P and P′ respectively). Sampling depths (2218 m to 2507 m) are given for HDR waters collected downhole in production borehole RH15. Saline water from Weardale is also plotted. (After Smedley *et al.* 1989).

illitic clay secondary phase typical of the alteration of the granite. Thus the concentration of lithium in groundwaters from granite can be used as an indicator of the extent of water–rock reaction and also as an indicator of residence time (Edmunds *et al.* 1986).

The lithium content of groundwaters from shal-

low depths is shown, together with deeper groundwaters from Cornwall, in Fig. 16.4 (Smedley *et al.* 1989), and these are compared with samples from the Devonian 'killas' country rocks. Samples from other areas, including Weardale, are also plotted on the diagram. It is apparent that there are two main trends, one for low salinity and one for high salinity groundwaters. Lithium in rainfall is well below 1 μg l^{-1} and evidence for lithium enrichment resulting from geochemical reactions can be seen even in the most dilute shallow groundwaters giving rise to values in the range 1–80 μg l^{-1}; a similar trend is observed in the country rocks and also in other shallow waters from granites (e.g. Loch Fleet, Strath Halladale). In the saline groundwaters from Cornwall and Weardale very high lithium concentrations are to be found (maximum 227 mg l^{-1} in the Carnmenellis granite) and the Li/Cl ratio reaches a relatively constant value. This seems to be a feature not only of groundwaters in granites, but of other lithologies, each of which has a characteristic aqueous Li/Cl ratio related to an equilibration between the water and rock (Edmunds *et al.* 1986). The distribution of other rare alkali elements, Rb and Cs, in the Carnmenellis groundwaters cannot be accounted for by simple congruent dissolution of the primary minerals of the granite, so that incongruent dissolution or preferential precipitation processes must be invoked (Smedley 1989).

Other occurrences of saline groundwater in granitic rocks

Although the majority of field evidence in the UK for granite water–rock reaction is derived from Cornwall, the Weardale spring also provides supporting evidence for the existence of saline waters circulating at depth within granitic rocks; the compositions of groundwaters from both areas are remarkably similar, reflecting an origin from a similar process. In Weardale the overall composition suggests that some of the salinity may have been derived from the adjacent Carboniferous rocks (Manning and Strutt 1991). Evidence has also been presented (Hamilton-Taylor *et al.* 1988) for a widely distributed source of saline groundwater beneath Cumbria which may be related to deep circulation and reaction in the Lake District granite batholith and/or the surrounding metasediments. Saline groundwater is reported in association with the Shap granite and to be issuing as springs (up to 14 500 mg l^{-1} Cl) around Derwentwater, where it can also be observed as an upward diffusing column in the lake sediments.

The presence of anthropogenic pollution may be clearly recognized in some groundwaters from granites. Many granitic areas form moorland or upland areas. However, in low-lying areas with thin soils, granitic rocks are highly susceptible to pollution from arable farming. This is shown in the Carnmenellis granite, for example, by the higher nitrate, and potassium contents and K/Na ratios of the groundwater which are caused by farming practices. The groundwater samples were chosen for this study because of their low nitrate concentrations but in many cases they approached or exceeded EC limits (10 mg l^{-1} NO$_3$−N).

Evolution of groundwaters in granites

The origin of saline and thermal groundwaters in the Carnmenellis granite in south-west England may be taken as a model for the evolution of salinity and composition by water–rock reaction (Fig. 6.5). With the exception of the Weardale granite, this is the only region in the UK where saline waters unequivocally associated with granite are found. However, the presence of saline groundwaters within crystalline rocks has been described in many other parts of the world—Canada, Finland, Soviet Union—so that the occurrence in Cornwall, supported by Weardale, is not unique. In most places where deep drilling into crystalline rocks has taken place, saline groundwaters have been found and the shallower waters often seem to be genetically related to them. The same applies to hot dry rock (HDR) circulation fluids which also show sub-

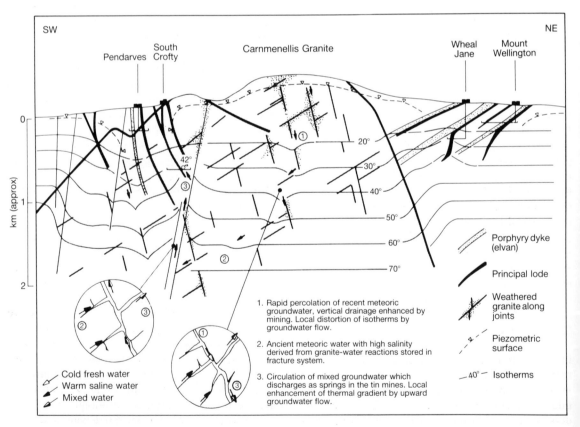

Fig. 16.5. Schematic cross-section through the Carnmenellis granite to illustrate discharge of thermal waters in the tin mines in relation to the depth of the hot dry rock experiment.

dued granite water–rock interaction characteristics produced at temperatures up to 80°C but within very short residence times.

Several suggestions have been made as to the origin(s) of saline water in granitic rocks which have been summarized by Frape and Fritz (1987), Edmunds et al. (1985), Nordstrom et al. (1989a), and Fontes et al. (1989). These include migration of sedimentary formation brines, marine transgressions, residual hydrothermal fluids, breakdown of fluid inclusions, radiolytic decomposition of water, and water–rock interaction (specifically silicate mineral hydrolysis). The Carnmenellis granite can be regarded as a good model for demonstrating salinity development by the latter process, although it is likely that thermally-driven movement of groundwater within the Carnmenellis granite has also introduced extra-granitic fluids from surrounding sedimentary rocks, especially in the earlier, higher temperature hydrothermal phases of water–rock interaction.

In the case of south-west England, it is noteworthy that despite similar thermal output and heat flow throughout the plutons of the batholith, saline fluids only occur (or as yet have only been found) in the Carnmenellis granite. Although there are tectonic considerations which may affect the occurrence of such fluids, the unique feature of the Carnmenellis granite is its higher permeability relative to other plutons, thus facilitating the convective circulation of fluids since the emplacement of the granite right up to the present day (Fehn 1985; Edmunds et al. 1989). It is interesting to note that the other occurrence of saline groundwater associated with granite in the UK, in Weardale (Manning and Strutt 1991), also coincides with an area of abnormally high heat flow which is caused by the granite. Weardale, and the adjacent area of northeast England, has been identified as a prime target for HDR geothermal development where temperatures of 150–200°C should be reached at 4.4–6 km (Downing and Gray 1986; Evans et al. 1988).

A corollary of the Carnmenellis situation is that in the absence of high heat flow, granites of relatively high permeability (if not a contradiction in terms) may be 'flushed' of older saline fluids by dilute meteoric groundwaters to considerable depths. It could be argued that such a case exists at Strath Halladale, Caithness, where dilute groundwaters occur to considerable depths

(~ 300 m). However, where stable tectonic conditions have existed for considerable periods, as in the shield areas of the world, saline waters appear to be the rule rather than the exception. The interrelationship between physical properties, geological history, geothermal history, and the chemical composition of the entrained fluids seems inseparable. The evidence from the UK on the origin of fluids, especially saline fluids in granites, goes some way towards solving a present day granite controversy. However, as in an earlier granite controversy half a century ago (Read 1957), the diversity of world-wide evidence is such that no one single hypothesis will adequately explain all the occurrences of fluids within them.

The search for geothermal energy resources and safe repositories for radioactive wastes has increased knowledge of the occurrence and chemical composition of fluids in granites and crystalline basement rocks enormously. Indeed, 15 years ago, prior to the development of these programmes, such rocks were believed to be essentially 'dry' at depth. It is now known that this is far from reality.

Interest in environmental quality has focused interest on the chemistry of ground and surface waters in upland areas over much of western and northern Britain where granites and other crystalline rocks are important. The natural baseline chemistry of these waters can be closely related to the influence of parent geology. Further studies of shallow groundwater systems on basement rocks are needed to understand their role in buffering surface water quality and in attenuating various kinds of pollution.

The 1980s has been a decade for the establishment of a database concerning the range of composition of groundwaters in these rock types and their geological occurrences. Many diverse hypotheses have been proposed to account for the mechanisms of origin of the chemical compositions of the fluids. Although a number of mechanisms may be necessary to account for all the different types of groundwater composition in granites and basement rocks, it is now time to evaluate the hypotheses more quantitatively. With a continuing greater emphasis upon environmental protection and the search for alternative energy sources, it will be necessary to develop predictive as well as interpretive models to describe the chemical composition and evolution of groundwaters in granites and basement rocks. At the same time, additional field evidence (from both water and rock) is needed from other granitic and basement rocks in the UK and elsewhere to test the hypotheses that have been developed primarily on one particular granite.

Acknowledgements

This chapter is a short review of the extensive studies carried out by many colleagues over the past decade. We wish to thank all those who have contributed to the advances in the understanding of granite–water interaction in the UK and we apologize if there have been any omissions of key results. This paper is published by permission of the Director, British Geological Survey (Natural Environment Research Council).

References

Andrews, J. N., Hussain, N., Batchelor, A. S., and Kwakwa, K. (1987). ^{222}Rn solution by the circulating fluid in a hot dry rock geothermal reservoir. *Appl. Geochem.*, **1**, 647–58.

Black, J. H. (1987). Flow and flow mechanisms in crystalline rock. In *Fluid flow in sedimentary basins and aquifers*, (ed. J. C. Goff and B. P. J. Williams), pp. 185–200, Geol. Soc. Spec. Publ., **34**.

Brady, P. V., and Walther, J. V. (1989). Controls on silicate dissolution rates in neutral and basic pH solutions at 25°C. *Geochim. Cosmochim. Acta*, **53**, 2823–30.

Bromley, A. V. (1989) *Field guide to the Cornubian orefield*. Guidebook for Sixth Int. Symp. in Water–Rock Interaction. Camborne School of Mines.

Cook, J. M., Edmunds, W. M., and Robins, N. S. (1987). *Groundwater contributions to Loch Fleet, Galloway*. Report 87/4. British Geological Survey, Keyworth.

Cook, J. M., Edmunds, W. M., and Robins, N. S. (1991). Groundwater contributions to an acid upland lake (Loch Fleet, Scotland) and the possibilities for amelioration. *J. Hydrol.*, **125**, 111–128.

Couture, R. A., and Seitz, M. G. (1986). Movement of fossil pore fluids in granite basement, Illinois. *Geology*, **14**, 831–4.

Creasey, J., Edwards, A. C., Reid, J. M., Mclead, D. A., and Cresser, M. S. (1986). The use of catchment studies for assessing chemical weathering rates in two contrasting upland areas in northeast Scotland. In *Rates of chemical weathering of rocks and minerals*, (ed. S. M. Colman and D. P. Dethier), pp. 467–502. Academic Press, Orlando.

Downing, R. A., and Gray, D. A. (1986). *Geothermal energy—the potential in the United Kingdom*. HMSO, London.

Edmunds, W. M., and Kinniburgh, D. G. (1986). The susceptibility of UK groundwaters to acidic deposition. *J. Geol. Soc. London*, **143**, 707–20.

Edmunds, W. M., Andrews, J. N., Burgess, W. G., Kay, R. L. F., and Lee, D. J. (1984). The evolution of saline and thermal groundwaters in the Carnmenellis granite. *Min. Mag.*, **48**, 407–24.

Edmunds, W. M., Kay, R. L. F., and McCartney, R. A. (1985). Origin of saline groundwaters in the Carnmenellis Granite: natural processes and reaction during hot dry rock reservoir circulation. *Chem. Geol.*, **49**, 287–301.

Edmunds, W. M., Cook, J. M., and Miles, D. L. (1986). Lithium mobility and cycling in dilute continental waters. *Proc. 6th Int. Symp. Water–Rock Interaction*. Orkustofnun, Reykjavik, pp. 187–91.

Edmunds, W. M., Kay, R. L. F., Miles, D. L., and Cook, J. M. (1987). The origin of saline groundwaters in the Carnmenellis granite, Cornwall (UK): further evidence from minor and trace elements. In *Saline waters and gases in crystalline rocks*, (ed. P. Fritz and S. K. Frape), pp. 127–43. Geol. Assoc. Canada Special Paper No. 33.

Edmunds, W. M., Andrews, J. N., Bromley, A. V., Kay, R. L. F., Milodowski, A., Savage, D., and Thomas, L. J. (1988). *Granite–water interactions in relation to hot dry rock geothermal development*. Investigation of the Geothermal Potential of the UK. British Geological Survey, Keyworth.

Edmunds, W. M., Andrews, J. N., Bromley, A. V., Richards, H. G., Savage, D., and Smedley, P. L. (1989). Application of geochemistry to hot dry rock geothermal development: an overview. In *Geochemistry in relation to hot dry rock geothermal development in Cornwall*, volume 1. Research Report SD/89/2. British Geological Survey, Keyworth.

Evans, C. J., Kimbell, G. S., and Rollin, K. E. (1988). *Hot dry rock potential in urban areas*. Investigation Geothermal Potential UK. British Geological Survey, Keyworth.

Fehn, U. (1985). Post-magmatic convection related to high heat production in granites of southwest England: a theoretical study. In *High heat production granites, hydrothermal circulation and ore genesis*, Inst. Min. Met. B99–112.

Fontes, J.-Ch., Louvat, D., and Michelot, J. L. (1989). Some constraints on geochemistry and environmental isotopes for the study of low fracture flows in crystalline rocks. In *Isotope techniques in the study of the hydrology of fractured and fissured rocks*, pp. 29–67. IAEA Vienna.

Frape, S. K., and Fritz, P. (1987). Geochemical trends for groundwaters from the Canadian Shield. In *Saline water and gases in crystalline rocks*, (ed. P. Fritz and S. K. Frape), pp. 19–38. Geol. Assoc. Canada Spec. Paper 33.

Fuge, R., and Power, G. M. (1969). Chlorine and fluorine in granitic rocks from S. W. England. *Geochim. Cosmochim. Acta*, **33**, 887–93.

Garrels, R. M., and Mackenzie, F. T. (1971). *Evolution of sedimentary rocks*. Norton. New York.

Giggenbach, W. F. (1984). Mass transfer in hydrothermal alteration systems—a conceptual approach. *Geochim. Cosmochim. Acta*, **48**, 2693–712.

Giggenbach, W. F. (1988). Geothermal solute equilibria. Derivation of Na-K-Mg-Ca geoindicators. *Geochim. Cosmochim. Acta*, **52**, 2749–66.

Grigsby, C. O., and Tester, J. W. (1989). Rock–water interactions in the Fenton Hill, New Mexico, hot dry rock geothermal systems. II. Modeling geochemical behaviour. *Geothermics*, **18**, 657–76.

Grigsby, C. O., Tester, J. W., Trujillo, P. E. Jr., and Counce, D. A. (1989). Rock–water interactions in the Fenton Hill, New Mexico, hot dry rock geothermal systems. I. Fluid mixing and chemical geothermometry. *Geothermics*, **18**, 629–56.

Gustard, A., Marshall, D. C. W., and Sutcliffe, M. F. (1987). *Low flow estimation in Scotland*. Report No. 101. Institute of Hydrology, Wallingford.

Hamilton-Taylor, J., Edmunds, W. M., Darling, W. G., and Sutcliffe, D. W. (1988). A diffusive ion flux of non-marine origin in Cumbria lake sediments: implications for element budgets in catchments. *Geochim. Cosmochim. Acta*, **52**, 223–7.

Helgeson, H. C., Murphy, W. M., and Aagaard, P. (1984). Thermodynamic and kinetic constraints on reaction rates among minerals and aqueous solutions. II. Rate constants, effective surface area, and the hydrolysis of feldspar. *Geochim. Cosmochim. Acta*, **48**, 2405–32.

Institute of Hydrology. (1980). *Low flow studies*. Institute of Hydrology, Wallingford.

Kappelmayer, O., and Gerard, A. (1989). The European geothermal project at Soultz-sous-Forêts. In *European Geothermal Update* (ed. K. Louwrier *et al.*), pp. 283–344. Proc. 4th int. seminar on the results of EC geothermal research. Florence 1989. Kluwer, Dordrecht.

Kay, R. L. F., and Bath, A. H. (1982). *Groundwater geochemical studies at the Altnabreac Research Site*. Report ENPU 82–12. Institute of Geological Sciences.

Lasaga, A. C. (1984). Chemical kinetics of water–rock interactions. *J. Geophys. Res.*, **89**, 4009–25.

McCartney, R. A. (1986). Granite–water interactions at 60°–80°C: results from Hot Dry Rock geothermal system field experiments in the Carnmenellis Granite, UK. In *Proc. Fifth Int. Symp. on Water–rock Interaction*, National Energy Authority, Reykjavik, Iceland, 372–5.

McCartney, R. A. (1987). Hot dry rock geothermal systems: geochemical applications of transient field experiments. *Geothermics*, **16**, 419–28.

Manning, D. A. C., and Strutt, D. W. (1991). Metallogenetic significance of a North Pennine Spring water. *Min. Mag.*, **54**, 629–36.

Murphy, W. M., and Helgeson, H. C. (1989). Thermodynamic and kinetic constraints on reaction rates among minerals and aqueous solutions. IV. Retrieval of rate constants and activation parameters for the hydrolysis of pyroxene, wollastonite, olivine, andalusite, quartz, and nepheline. *Amer. J. Sci.*, **289**, 17–101.

Nordstrom, D. K., Olsson, T., Carlsson, L., and Fritz, P. (1989*a*). An introduction to the hydrogeochemical investigations with the International Stripa Project. *Geochim. Cosmochim. Acta*, **53**, 1717–26.

Nordstrom, D. K., Lindblom, S., Donahoe, R. J., and Barton, C. C. (1989*b*). Fluid inclusions in the Stripa granite and their possible influence on the groundwater chemistry. *Geochim. Cosmochim. Acta*, **53**, 1741–55.

Norton, D., and Knapp, R. (1977). Transport phenomena in hydrothermal systems: the nature of porosity, *Amer. J. Sci.*, **277**, 913–36.

Ragnarsdottir, K. V. (1989). Kinetics of dissolution of heulandite at 25°C. In *Proc. Sixth Int. Symp. on Water–rock Interaction*, (ed. D. L. Miles), pp. 567–8. Balkema, Rotterdam.

Read, H. H. (1957). *The Granite Controversy: geological address illustrating the evolution of a disputant*. Murby. London.

Richards, H. G., and Savage, D. (1989). Rate of plagioclase dissolution in the Camborne School of Mines experimental hot dry rock geothermal system, Rosemanowes, Cornwall. In *Proc. Sixth Int. Symp. on Water–rock Interaction*, (ed. D. L. Miles), pp. 577–80. Balkema, Rotterdam.

Richards, H. G., Wilkins, C., Kay, R. L. F., and Savage, D. (1989*a*). Geochemical results from the Rosemanowes HDR system 1986–88. In *Geochemistry in relation to hot dry rock geothermal development in Cornwall*, Volume 2. Research Report SD/89/2. British Geological Survey, Keyworth.

Richards, H. G., Savage, D., and Shepherd, T. J.

(1989*b*). Geochemical prognosis for a commercial depth HDR system in SW England. In *Geochemistry in relation to hot dry rock geothermal development in Cornwall*, Volume 7. Research Report SD/89/2. British Geological Survey, Keyworth.

Savage, D. (1986). Granite-water interactions at 100°C, 50 MPa: an experimental study. *Chem. Geol.*, **54**, 81–95.

Savage, D., Cave, M. R., and Milodowski, A. E. (1985). Interaction of meteoric groundwater with Carnmenellis Granite at 250°C and 50 MPa: an experimental study. In *High heat production granites, hydrothermal circulation and ore genesis*, pp. 315–27. Inst. Min. Met., London.

Savage, D., Cave, M. R., and Milodowski, A. E. (1986*a*). The origin of saline groundwaters in granitic rocks: evidence from hydrothermal experiments. In *Scientific basis for nuclear waste management*, Volume 9 (ed. L. O. Werme), pp. 121–8. Materials Research Society.

Savage, D., Cave, M. R., Haigh, D., and George, I. A. (1986*b*). Dissolution kinetics of laumontite at 150°C, 50 MPa and the relevance to models of mass transfer during diagenesis and low-grade metamorphism. *Proc. Fifth Int. Symp. Water–rock Interaction*, National Energy Authority, Reykjavik, Iceland, 485–8.

Savage, D., Cave, M. R., Milodowski, A. E., and George, I. (1987). Hydrothermal alteration of granite by meteoric fluid: an example from the Carnmenellis Granite, UK. *Contrib. Mineral. Petrol.*, **96**, 391–405.

Savage, D., Bateman, K., Milodowski, A. E., Cave, M. R., Hughes, R. C., Green, K., Reeder, S., and Pearce, J. (1989). Experimental studies of granite–water interaction. In *Geochemistry in relation to hot dry rock geothermal development in Cornwall*, Volume 6. Research Report SD/89/2. British Geological Survey, Keyworth.

Smedley, P. L. (1989). Alkali metal enrichments in groundwaters from the Carnmenellis area, Cornwall. In *Proc. 6th Int. Symp. Water–Rock Interaction*, (ed. D. L. Miles), pp. 647–50. Balkema, Rotterdam.

Smedley, P. L., Bromley, A. V., Edmunds, W. M., Kay, R. L. F., and Shepherd, T. J. (1989). Fluid circulation in the Carnmenellis granite: hydrogeological, hydrogeochemical and palaeofluid evidence. In *Geochemistry in relation to hot dry rock geothermal development*, Volume 4. Research Report SD/89/2. British Geological Survey, Keyworth.

Velbel, M. A. (1986). Influence of surface area, surface characteristics, and solution composition on feldspar weathering rates. In *Geochemical processes at mineral surfaces* (ed. J. A. Davis and K. F. Hayes), ACS Symposium Series 323, Washington D. C., 615–634.

17. Geothermal energy in the United Kingdom

R. A. Downing, R. H. Parker, and D. A. Gray

Introduction

The large increase in the price of oil in the 1970s emphasized the finite nature of conventional energy resources and directed attention towards alternative sources. In the UK, in common with many other countries, this led to a review of the potential for developing geothermal energy, with funding provided by the Department of Energy and the Commission of the European Communities. A primary objective was to estimate the geothermal resources by the mid-1980s. The principal contractor was the British Geological Survey but parts of the programme were sub-contracted to Oxford University, the Open University, Bath University, and the Imperial College of Science and Technology. The overall objective was largely achieved early in 1984 and a summary of the results was published (Downing and Gray 1986) although work continued until 1987 to amplify and clarify some aspects (British Geological Survey 1988). A second major programme was concerned with research into the rock mechanics of the hot dry rock concept. This was contracted to the Camborne School of Mines which carried out an extensive experimental programme at a research site established for the purpose at the Rosemanowes Quarry in Cornwall (Batchelor 1982, 1983; Parker 1989a, 1989b; Camborne School of Mines Geothermal Energy Project 1989a).

The basis for any geothermal exploration programme is an understanding of the areal distribution of heat flow through the upper crust. Consequently this was an important aspect of the resource assessment, extending earlier studies by Richardson and Oxburgh (1979). As the UK is a stable geological area forming part of the continental foreland of Europe, heat flows are relatively low, averaging 55 mW m^{-2}, and geothermal gradients are typically in the range of 25–30°C km^{-1}. The heat flow map currently available (Fig. 17.1) is based on measurements in 210 boreholes that were either specially drilled to measure heat flow or were deep boreholes drilled for hydrocarbon or mineral exploration that provided good temperature measurements. The data from these sources have been supplemented by estimates of heat flow for almost 400 sites where boreholes more than 1000 m deep have been drilled. Measurements of bottom hole temperatures in these boreholes have been interpreted in conjunction with details of both the geology and a data-bank of mean thermal conductivity measurements held by the British Geological Survey (Wheildon and Rollin 1986; British Geological Survey 1988).

The area-weighted mean heat flow in the UK is 54 ± 12 mW m^{-2} (Wheildon and Rollin 1986) and 55 mW m^{-2} can be taken as the mean value for the sedimentary crust. A number of distinct anomalies exist within a variable background. The most obvious anomalies are associated with the granite batholiths of south-west England, northern England, and the Eastern Highlands of Scotland where the cause is heat production from the radioactive decay of uranium, thorium, and potassium in minerals in the granites (Wheildon and Rollin 1986; Lee *et al.* 1987; Gebski *et al.* 1987; British Geological Survey 1988). Elsewhere values are probably only anomalous if they are greater than one standard deviation from the mean, that is over about 70 mW m^{-2}. All the measurements in younger non-metamorphic rocks should be regarded as 'apparent conductive heat flows' influenced to an unquantifiable extent by groundwater flow (Gebski *et al.* 1987; British Geological Survey 1988). Some apparent anomalies in the background field may

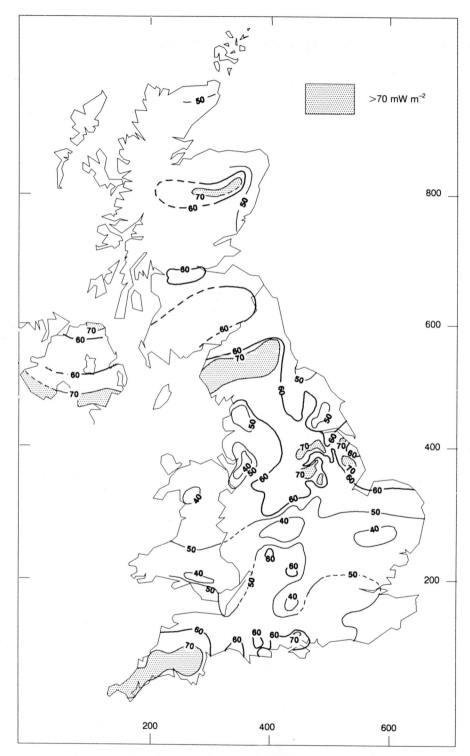

Fig. 17.1. Heat flow map of the United Kingdom (mW m^{-2}).

be due to groundwater rising in regional flow systems. This has been proposed, for example as the explanation in the East Midlands and Lincolnshire, and in the Wessex Basin (Downing *et al.* 1987). As discussed later, some local anomalies are also caused by the flow of groundwater.

With this pattern of heat flow distribution, it is clear that there are two possible sources of geothermal energy in the UK:

1. The heat stored in hot dry rocks, particularly in the granites of Cornwall and northern England, with the heat extracted by stimulating natural fractures at depth and circulating water through them.

2. Low enthalpy resources represented by hot groundwaters in permeable rocks at depths where temperatures of more than 40°C and preferably more then 60°C exist.

Geothermal resources are generally expressed as the amount of heat energy stored in rocks above a specified depth. Most of the variation is due to variations of temperature as the ranges for the specific heat and density of rocks are small. Predictions of deep subsurface temperatures require knowledge of the heat flow field and deep geological structure. By recognizing a limited number of typical geological terrains and using heat flow data and a thermal conductivity model of the upper crust, temperatures at depth can be predicted and the geothermal resources estimated (Gale *et al.* 1984*b*; Gale and Rollin 1986; British Geological Survey 1988, pp. 27–42).

Hot dry rock technology

Background

The exploitation of geothermal energy has required natural groundwater as a heat transfer medium, to bring the energy to the Earth's surface. This thermal energy can be used for electrical power generation, or to provide heating for industry and buildings. Hot dry rock (HDR) technology aims to extract geothermal energy using an artificially introduced heat transfer medium. This is achieved by drilling an 'injection well', creating a

permeable 'reservoir' of rock, and pumping cold water through the reservoir to a 'production well', which is used to abstract the water which has been heated by the reservoir. As geothermal sites where groundwater can be utilized are comparatively rare, the development of HDR technology could open up a very large thermal resource. Clearly, it is important to choose a site which has a rock structure suitable for reservoir creation at a depth which does not incur excessive drilling costs. The reservoir must provide sufficient area for heat transfer between rock and water. It must not have too high a resistance to flow, or too high a water loss, and should not cool down too rapidly and thus prevent the cost of installing the system from being recovered.

One of the most extensive investigations into the development of HDR technology has been carried out in the last decade in Cornwall. In 1977, the Camborne School of Mines (CSM) began work at Rosemanowes Quarry, on the Carnmenellis granite outcrop near Penryn. Since 1980, the CSM Project has been funded mainly by the UK Department of Energy, with a smaller contribution from the Commission of the European Communities. The objectives of the project have been to investigate the engineering requirements for developing HDR reservoirs, and to establish the size and

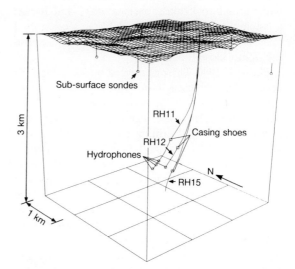

Fig. 17.2. Isometric view of the wells and microseismic sensor positions. Phase 2A and 2B, Rosemanowes:

nature of the HDR resource in south-west England. This has been one of the largest-scale hydrogeological experiments carried out in the United Kingdom. The work has involved staff from a variety of disciplines, including geology, hydrogeology, geochemistry, geophysics, rock mechanics, reservoir engineering, instrumentation technology, well logging engineering, mathematics, and computing. In addition to the work of the Camborne School of Mines, a number of other institutions have contributed to the Project. These include companies such as RTZ Consultants Ltd, Kenting Drilling Services Ltd, and GeoScience Ltd; the British Geological Survey; Imperial College, Sunderland and Sheffield City Polytechnics, Massachusetts Institute of Technology, and the University of Bath.

Summary of the HDR reservoir creation programme

Boreholes 300 m deep were drilled in Phase 1 (1977–1980), and it was demonstrated that it was possible to connect the boreholes by hydraulic stimulation of natural joints in the granite, and to circulate water through these joints (Batchelor 1982). Granite was considered to be a suitable rock for HDR reservoir creation because it had low undisturbed rock mass permeability, and it was possible that it could be stimulated in a controlled manner by pumping water at high pressure into the tightly closed joints.

Phase 2 (1980–88) was carried out at a depth of about 2 km, which is more closely related to conditions which would be required for commercial exploitation of the technology. Phase 2A (1980–83) involved drilling two wells (RH11 and RH12, Fig. 17.2) to a depth of 2.1 km at which the rock temperature was 79°C (Batchelor 1983). A reservoir was created by hydraulic stimulation, but circulation of the system showed that although the stimulated reservoir occupied a large reservoir rock volume (800 million m³), there was a poor connection between the injection and production wells. The reservoir impedance, which is a measure of the resistance of the reservoir to the circulation of the water, was too high (that is 1.8 MPa l^{-1} s^{-1}, measured as the ratio of the pressure drop across the reservoir to the production flow rate, compared with

0.1 MPa l^{-1} s^{-1} which would be more acceptable for a commercial system). The loss of water (70 per cent of the volume injected) was also too high.

Phase 2B (1983–86) included the drilling of a third well (RH15) to a depth of 2.6 km at which the rock temperature was 100°C (Fig. 17.2), and stimulation of the granite joints using a gel of viscosity 50 centipoises (cp) to produce a smaller reservoir, with lower impedance and producing lower water losses (Parker 1989a). This new reservoir was characterized in Phase 2C (1986–88), by carrying out a continuous circulation at different flow rates, and measuring the hydraulic and thermal performance at these flow rates. The results showed that the reservoir was smaller than that required for commercial applications, and that it contained a short circuit between the wells, resulting in premature cooling of the production water (Parker 1989b; Camborne School of Mines 1989a).

Phase 3 began in 1988, with a conceptual design by RTZ Consultants Limited of a 6-km-deep prototype of a commercial system for generating electricity in Cornwall. In addition to participating in the conceptual design study, CSM carried out further research and development, part of which was aimed at manipulating the Rosemanowes reservoir to improve its performance. The Department of Energy announced in 1991 that the next phase of work will concentrate less on research in Cornwall and will involve greater collaboration with a European programme involving France, Germany and the European Commission. The Department's conclusion is that there are still very substantial uncertainties concerning the practicability of HDR in the UK, and it seems unlikely to be competitive in the short to medium term.

The HDR environment at Rosemanowes

The aim of the programme at Rosemanowes has been to develop a technology which has a wide application. High exploration costs will be unacceptable for HDR exploitation, and therefore the technology must be able to adapt to the geological environment in which it is applied. The need to exploit specific localized geological structures to create a reservoir will increase exploration costs, and cause the chance of a sterile operation to be greater. Equally, if such structures could adversely affect the operation of the reservoir by increasing water losses, or providing preferential flow paths, they should if possible be avoided in selecting the site.

Rosemanowes Quarry is in the exposed Carnmenellis granite, in West Cornwall (Fig 17.3). The absence of a sedimentary or metamorphic cover made it possible to ensure that no major geological feature, such as a fault, was visible at its upper surface. The granite has textures ranging from porphyritic nearer the surface to equigranular at a depth of about 2 km. Geophysical data indicate that the base of the granite extends well below a depth of 9 km. *In situ* mechanical properties are:

Uni-axial compressive strength
  (MPa)   : $103 + 32\,z$
Young's modulus (GPa)   : $54 + 4\,z$
Poisson's ratio   : $0.22 - 0.27$
Average density (kg m^{-3}) : 2640

where z is the depth in km.

Two main vertical joint sets (NE–SW, parallel to the trend of the tin/copper lode mineralization, and NW–SE, parallel to tension fractures and strike slip faults known as 'cross courses') have been identified from surface mapping. These joint sets have been identified, with a broad range of strikes, on Borehole Televiewer (BHTV) logs to a depth of 2.6 km, and there are indications of a continuation of joints to at least 3.5 km. These indications are from microseismic data obtained from deeply located events (resulting from shearing caused by penetration of water into joints).

The relationship between the distribution of *in situ* earth stresses and depth has been measured in considerable detail at Rosemanowes, to a depth of 2.5 km. These measurements have been complemented by measurements in local tin mines. Pine and Batchelor (1984) summarized the relationship for *in situ* stresses (in MPa) in the Carnmenellis granite with depth (z in km) as follows:

Maximum horizontal stress, $\sigma_H = 15 + 28\,z$
Minimum horizontal stress, $\sigma_h = 6 + 12\,z$
Vertical stress, $\sigma_v = 26\,z$

Heat flow at the exposed granite surface is about 120 mW m^{-2}, which is just over double the area-weighted mean heat flow in the UK (British Geo-

logical Survey 1984, 1988). The excess heat flow is due mainly to the high radioelement (uranium, potassium, thorium) content of the granite. Three-dimensional heat flow models, which have been based on extensive heat flow measurements and gravity surveys, indicate an almost linear relationship between temperature and depth in the upper 7 km of crust over large portions of the Cornubian granite batholith. With an average surface temperature of 10°C, this results in a relationship for regions close to Rosemanowes:

$$T = 10 + 35\,z$$

where T is the temperature in °C at a depth of z km. The temperature gradients in south-west England are the highest in the United Kingdom, and estimates of the HDR resource in this region have produced figures of at least 1000 TWhr of electricity (Camborne School of Mines 1989*b*). 1000 TWhr is equivalent to the operation of one large (Gigawatt-scale) power station for about 100 years.

In situ hydraulic properties were measured at Rosemanowes at depths up to 2 km, before major hydraulic injections commenced. Low flow rate hydraulic tests at low injection pressures indicated permeabilities between 1 μD and 10 μD at up to 0.7 MPa fluid overpressure. Then permeabilities rose to 60 μD during low pressure injections (3 MPa), and microseismicity resulting from shear slip displacement was first observed during a low flow rate injection (0.5 l s^{-1}), at a wellhead injection pressure of 3.1 MPa (Camborne School of Mines 1984).

Reservoir creation

In order to increase the permeability of the reservoir rock mass, hydraulic stimulation has been carried out on the openhole portion of the boreholes (they are cased with steel tubing for all but the lowest few hundred metres of their length). This process involves injection of water (or viscous gels) at surface injection pressures as high as 15 MPa and flow rates up to 200 l s^{-1}. An important technique used for monitoring this process has been the installation and development of a network of microseismic sondes, both in the deep wells and in boreholes 200 m deep (Fig. 17.2). The

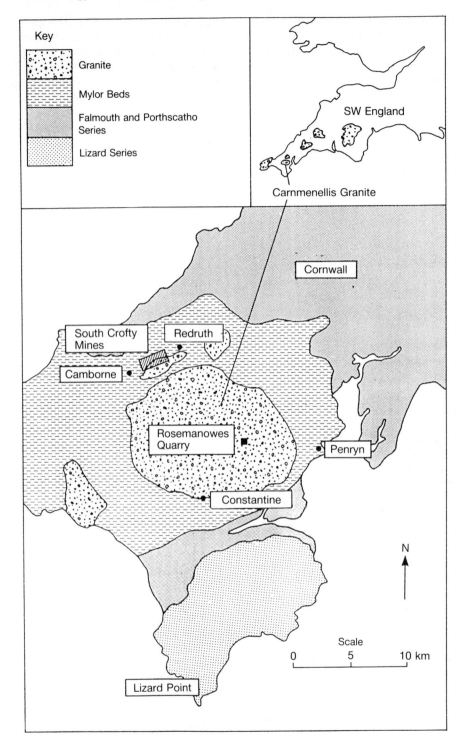

Fig. 17.3. Geological sketch map of the test area.

influence of the water pressure and the anisotropic *in situ* stress field on suitably orientated natural joints causes shear on those joints, and this shear stimulation increases the permeability of the system. Fracturing in unjointed regions of rock is probably confined to small distances from the wellbore. Much effort has been spent on understanding the relationship of the recorded microseismic events to the reservoir development, and it was found that the reservoir grew downwards due to the increasing anisotropy of the *in situ* horizontal stresses with depth (Pine and Batchelor 1984). Figure 17.4 gives a good impression of the extent to which this reservoir growth took place at Rosemanowes. The relationship of microseismic events to hydraulic flow in the reservoir remains the subject of ongoing investigations (Baria and Green 1990).

Experimental work on HDR reservoir stimulation is expensive and time consuming, and numerical modelling of fluid flow through stressed and jointed rock masses is being used to understand and control the physical processes involved. Considerable progress has been made, and the two- and three-dimensional FRIP (Fluid–rock interaction program) family of codes has been developed for use in conjunction with standard oil industry fracturing models (Nicol *et al.* 1990).

Reservoir circulation and characterization

The nature of the HDR reservoir, which has been created by stimulation, has been characterized by causing water to circulate within it over significant periods of time. For example, the present Rosemanowes reservoir was stimulated in 1985, and since then water has been circulated within it under different conditions of injection pressure and injection flow rate (Parker 1989*c*). Figure 17.5 shows the variations in flow rate which have been investigated. From this it will be seen that the maximum injection flow rate (Well RH12) was between 35 and 43 l s^{-1}. At these flow rates, the injection pressure was between 11 and 12 MPa. Hydraulic characteristics such as the decrease in reservoir impedance with injection pressure (and injection flow rate) have been established. Well testing (pressure transient analysis) is well established in the oil industry for reservoir investigations, and has been applied at Rosemanowes. Microseismic activity and accompanying water losses have been found to be dependent on injection pressure. Water losses have varied between 5 and 70 per cent, depending on the state of the reservoir, and in Phase 3A, the use of a downhole pump to lower the pressure in the production well has resulted in water being drawn from storage to give recoveries of over 100 per cent. Water loss appears to have three mechanisms:

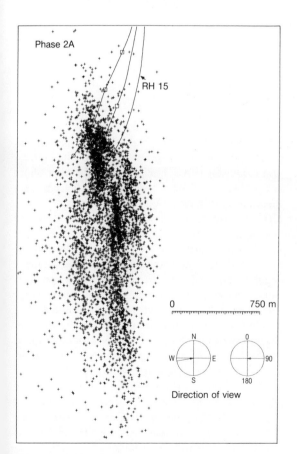

Fig. 17.4. Microseismic events located in Phase 2 (vertical section).

1 steady state diffusive losses when the reservoir is operating in a steady state of circulation over long periods;

2 transient losses into storage, associated with the inflation of microcracks and the dilation of major flowing joints as the reservoir injection pressure increases;

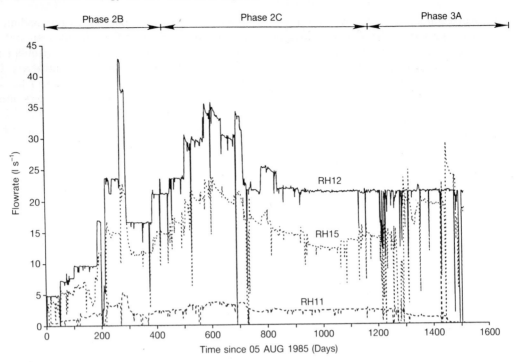

Fig. 17.5. Flowrate since start of circulation in 1985 (RH12: Injection flowrate, RH15 and RH11: production flowrate).

3 reservoir growth associated with additional joint space volume created in the reservoir when joints shear.

Over the last ten years, a total of over 770 000 m³ of water has been lost by these mechanisms.

In addition to proprietary logging of the wells (producing data on mechanical properties of the rock, on the jointing structure, and on the joint mineralization to a depth of 2.6 km), production logs (measuring temperature, flow rate, and pressure downhole) have been run on a regular basis. It is one of the few sites where so much detail has been obtained from borehole logs, particularly logs associated with water flow into and out of boreholes from a jointed rock system.

The thermal performance of the reservoir has been modelled, using simple lumped parameter models and treating the reservoir as a steady state system, together with the available temperature and flow log data. The presence of certain flowpaths with a very small heat transfer area was indicated. These flowpaths collectively carried

10–15 per cent of the total flow and led to the premature cooling of the production water (Nicol and Robinson 1990). The presence of such a 'short circuit' was confirmed by the use of inert tracers injected at different points in the injection well (RH12) and sampled continuously at different points in the production well (RH15) (Richards *et al.* 1990). The use of inert tracers (chiefly sodium fluorescein and the bromide anion) has become a regular diagnostic technique at the CSM project, being aimed at establishing reservoir characteristics such as breakthrough time and volume, and residence time distributions (Kwakwa 1988). A fluorimeter tool has been developed for continuous measurement of fluorescein tracer compositions downhole in the production well.

The chemistry of the water–rock interactions in HDR systems has been the subject of a combined research effort by the HDR Geochemistry Group (consisting of staff from the British Geological Survey, the Natural Environment Research Council Isotope Geology Centre, the University of Bath, and the Camborne School of Mines). The work has involved not only extensive field tests, but also high temperature/high pressure laboratory experiments on water–rock interactions. Progress has been made towards the target of developing a geochemical model of the Rosemanowes reservoir, and extending this to the requirements of a much hotter deep system, in which the increased rates of chemical reactions will probably make geochemical factors much more significant than may appear to be the case in operating the relatively benign Rosemanowes system. The importance of geo-chemical predictions in designing the corrosion-resistance of a deep prototype HDR plant cannot be overemphasized. In addition to predictive modelling, geochemistry (including radon gas dis-solution in the production water) has been used as a diagnostic tool to attempt to monitor changes in reservoir characteristics such as water–rock surface area. Chemical analysis has indicated whether water has been produced from short or long resi-dence time paths in the reservoir. The presence of particulate matter and bacteria in the water pro-duced from the reservoir have been the subject of recent study, in an attempt to explain certain characteristics of the reservoir performance. Par-ticles of joint-filling minerals have been released during certain hydraulic operations, and particles entering the reservoir in injection water have not been recovered in production water. Sulphate-reducing bacteria appear to attack the stimulation gels, with hydrogen sulphide and other reduced constituents indicating residence in a reducing environment. A number of reports on the geo-chemistry have been published (HDR Geochemis-try Group 1990).

Alongside all the experimental work at Rose-manowes, CSM has undertaken a significant instru-ment development programme. This has included the development of a 3-axis clamped tool for locating microseismic events and a seismic source ('sparker') for operation in deep boreholes at temperatures of the order of 200°C.

The way forward for HDR geothermal energy

The results of the Camborne School of Mines project should be considered alongside those of the programmes carried out overseas, with which

CSM has kept in close touch. These include the US programme of the Los Alamos National Laboratory in New Mexico; programmes at Urach and Falkenburg in Germany, and at Le Mayet de Montagne in central France, which have now developed into the joint CEC-French-German project at Soultz-sous-Forêts, north of Strasbourg; several projects in Japan (including the NEDO Hijiori Project and the Tohoku University 'Gamma' Project); and the Chalmers Tekniska Hogskola project at Fjallbacka, north of Gothen-burg in Sweden.

There are still serious technical uncertainties to be overcome if the HDR technology is to be exploited in the United Kingdom or elsewhere. These include the determination of the important *in situ* stress data, temperature and joint character-istics which are to be expected at depths of 6 km in south-west England. Only drilling to these depths will clear up these uncertainties. The ability to stimulate natural joints in granite to produce a HDR reservoir having satisfactory hydraulic and thermal characteristics is the major uncertainty overlying all HDR developments. Only more experience in engineering such stimulations on appropriate sites will present a chance of clearing up this uncertainty. The Department of Energy had decided to pursue this as part of a European collaborative programme.

Low enthalpy resources

Background to low enthalpy prospects

The only surface manifestations of geothermal energy in the UK are the warm springs that issue from Carboniferous rocks at Bath, Bristol, Taffs Well (in the Taff Valley, north of Cardiff), and in Derbyshire, near the margins of the Carboniferous Limestone outcrop; at 46°C, the Bath spring has the highest temperature (Burgess *et al.* 1980; Andrews *et al.* 1982). The temperatures of these springs are due to deep circulation of meteoric water in hydrothermal convection systems that are probably almost entirely in the Carboniferous Limestone. These thermal convection systems indi-cate that geothermal fluids do exist in Upper Palaeozoic rocks. However, in the past these rocks have been buried to considerable depths and sub-

jected to the tectonic forces of the Variscan Oro-
geny. Consequently they are hard and compact
with low matrix permeabilities, generally less than
10 mD and commonly less than 1 mD. Fluid flow
is in fractures which enhance the permeability but
the impossibility of forecasting the position of such
fractures at depth discourages consideration of the
Upper Palaeozoic as a practical source of energy
and attention must be directed to the post-Carboniferous sequence where the rock properties are
more favourable.

The basic requirement for a successful geothermal prospect is a permeable aquifer at a depth
where the thermal gradient supports a temperature
of at least 40°C and preferably more than 60°C.
The higher temperature is suitable for the extraction of heat either directly or by transfer through a

heat exchanger to a secondary water circulation
system. Temperatures in the 40–60°C range
require the use of heat pumps for economic development. When used as the sole supply for a typical
heat load of about 4 MW (thermal), a geothermal
well must produce about 30 l s⁻¹ and very permeable aquifers are required to yield such flow rates
for acceptable drawdowns of aquifer pressure. The
fluids are invariably brines and at inland sites a
second, injection, well is necessary to dispose of
the brine after the heat has been extracted; the
two wells are referred to as a doublet. Reinjection
of the fluid also has the great advantage of maintaining the aquifer pressure. This system of development has been adopted very successfully in the
Paris Basin where over fifty schemes are in use,
providing space heating with water abstracted from
a Jurassic limestone.

In the UK, the only aquifers able to meet the
specifications are the Permo-Triassic sandstones in
the post-Carboniferous basins, or, as they are more
commonly referred to, the Mesozoic basins. The
programme to assess the low enthalpy potential of
the UK was, therefore, focused on these basins,
particularly where the geothermal gradient exceeds
30°C km⁻¹, as in the Wessex Basin and east Lincolnshire. To obtain hot water in these more
favourable areas wells up to 2–2.5 km deep are
required.

Permo-Triassic sandstones as a geothermal reservoir

The investigation of the geothermal potential of
the Permo-Triassic sandstones provided much new
data about the sandstones as reservoir rocks. Four
deep wells were drilled and tested, two near or in
Southampton (Marchwood No.1 and Southampton
No.1), one at Larne in Northern Ireland, and one
at Cleethorpes, near Grimsby, on South Humber-
side (Fig. 17.6 and Table 17.1). In addition individual specific programmes, including drill-stem
tests and reservoir appraisals from geophysical logs
and core samples, were made in many oil and
mineral exploration boreholes. The studies were
concerned with assessing the regional distribution
and nature of the sandstones and their aquifer
properties (Downing and Gray 1986; British Geological Survey 1988).

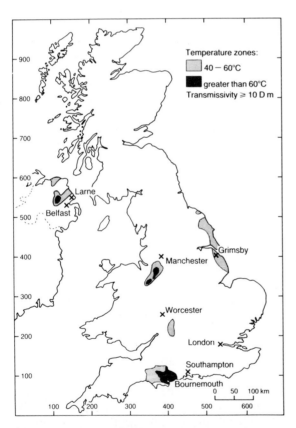

Fig. 17.6. Potential low enthalpy geothermal
fields in the UK. Defined by a temperature of
more than 40°C and a transmissivity of more
than 10 Dm.

Wessex Basin

The Wessex Basin was recognized at an early stage as a favourable area for more detailed study. The Sherwood Sandstone Group, the only geothermal aquifer in the basin, attains thicknesses of more than 250 m north-west of Weymouth. It is divided into two formations—the Budleigh Salterton Pebble Beds and the overlying Otter Sandstone Formation (Holloway *et al.* 1989). The principal aquifers are in the Otter Sandstone which is a fine- to medium-grained sandstone that displays cyclic sedimentation, commonly conglomeratic at the base and grading upwards into sandstones; mudstone and siltstone interbeds are common. The sandstones are believed to be fluvial channel sands deposited in braided streams. Interpretation of gamma and sonic logs indicates that the formation as a whole represents a major fining-upward cycle. However, near the depocentre, west of Dorchester, the lower part consists mainly of anhydrite and anhydritic sandy shale, possibly representing an origin in a sabkha environment (Holloway *et al.* 1989).

The cyclic sequence is characterized by discrete thin bands of very permeable uncemented sandstones which are the main water-bearing horizons. The porosity of these sandstone bands is about 20–25 per cent and the average permeability up to about 500 mD, although individual maximum values of core samples is about 5 D. Most of the water abstracted from the Marchwood Well entered the well from three sandstone bands each about 2 m thick, and 90 per cent of the yield of the

Table 17.1   Principal details of low enthalpy geothermal wells in the UK

	Marchwood	Southampton	Larne		Cleethorpes	
Status of well	Exploration	Development	Exploration		Exploration	
Basin	Wessex	Wessex	Northern Ireland		Lincolnshire	
Total depth (m)	2617	1827	2880		2100	
Target lithology	Triassic sandstones	Triassic sandstones	Triassic sandstones	Permian sandstones	Triassic sandstones	Basal Permian Sands
Depth of reservoir (m)	1666–1725	1729–1796	968–1616	1823–2264	1100–1497	1865–1891
Thickness of reservoir (m)	59	67	648	441	397	26
Thickness of pay zone (m)	6	16	286	Virtually nil	250	8.5
Completion details	Well screen	Open-hole	Cased off	Perforated casing	Cased off	Open-hole
Transmissivity (Dm)	4	3.3	7	< 1	60	< 2
Reservoir temperature (°C)	73.6	76	40	65	53	64
Well head temperature (°C)	71.6	73	—	—	c.50	—
Bottom hole temperature (°C)	84.6	76.2	—	91.2	—	68
Flow rate of test (l s^{-1})	30	20	—	—	20	—
Pressure reduction after 30 days (MN m^{-2})	3.7	3.1	—	—	c.0.60*	—
Salinity of water (g l^{-1})	103	125	200	—	35–80	c.220

* After 15 hours.

Southampton Well came from two bands with a total thickness of about 11 m (Downing *et al.* 1984). Test pumping the Southampton Well, using the Marchwood Well as an observation well, indicated the transmissivity of the aquifer is about 5 Dm (8.5 m^2 d^{-1}) and the coefficient of storage 3.8×10^{-5} (Allen *et al.* 1983). Both wells yielded about 30 l s^{-1} for pressure reductions of 3.7 MN m^{-2} (equivalent to a drawdown of 360 m) over 33 days and 4.2 MN m^{-2} (equivalent to about 400 m) over 5 days at Marchwood and Southampton respectively.

The two geothermal wells at Southampton, together with more recent hydrocarbon exploration boreholes, have shown that Southampton is near the eastern limit of the Sherwood Sandstone Group. The transmissivity increases towards the west and interpretation of geophysical logs and core samples suggests that the transmissivity in the deeper part of the basin, near and to the north-west of Bournemouth, is probably about 20 Dm, in an area where the temperature exceeds 60°C (Allen and Holloway 1984).

The permeability of the sandstones is determined to a large extent by the grain-size, the extent of sorting, and the extent of cementation. In terms of both the proportion of fine-grained sediments and the grain-size of the sandstones, the northern part of the Wessex Basin has less favourable reservoir characteristics. The degree of cementation is the most important factor controlling reservoir quality in the southern part of the basin. Diagenetic processes tended to reduce the primary intergranular porosity by early compaction and pressure solution, grain overgrowths, and cementation. Anhydrite cement preserved the form of the pore space during compaction and prevented precipitation of calcite in the pore spaces. Later dissolution of earlier cements, particularly anhydrite, as well as the framework grains, created secondary porosity which has had a major influence on the aquifer properties particularly of the thin, discrete, permeable bands referred to earlier (Knox *et al.* 1984; Milodowski *et al.* 1986; Strong and Milodowski 1987; British Geological Survey 1988).

Apart from the extreme western part of the basin, near the outcrop, groundwater in the Sherwood Sandstone is a brine with maximum salinities of 250 g l^{-1} north-west of Bournemouth, where the sandstone is at its maximum depth, suggesting at least an element of density settlement. At the present time the very saline brine may be virtually stationary with any flow following preferred flow routes around the periphery of the saline area where salinities are less than 200 g l^{-1}. The current relatively high heat flow in the southern part of the Wessex Basin could be due to the transfer of heat by groundwater on the rising limb of a regional flow system (Downing *et al.* 1987).

The brines in the Wessex Basin, both in the Sherwood Sandstone and the overlying Mesozoic formations, differ from those in other parts of the UK; the isotopic chemistry indicates a different evolution and the Br/Cl ratio implies that they originated from evaporites that did not have a marine origin (Edmunds 1986). This suggests that a dominant component of all the waters may have originated as a Triassic groundwater. The explanation may be that during the long period of subsidence in the Jurassic, and subsequently the Cretaceous, compaction forced fluids from the Triassic upwards by piston displacement into the overlying younger rocks, displacing part of their pore-waters as it did so. This possibly imprinted the signature of pore-waters derived from the Triassic on waters now found in Jurassic rocks and possibly also in the Cretaceous, although in the latter case it tends to be masked by a more recent influx of meteoric water.

The solutes in the brines within the Sherwood Sandstone are likely to be basically Triassic in age but they may have been diluted by two influxes of meteoric water during the late Cimmerian uplift and in Tertiary to Recent times—influxes that possibly created the secondary porosity in the sandstones referred to earlier.

East Yorkshire and Lincolnshire Basin

Two geothermal aquifers occur in the East Yorkshire and Lincolnshire Basin—the Basal Permian Sands and Breccia and the Sherwood Sandstone Group. The Permian aquifer attains thicknesses of 20–60 m in the coastal zone of east Lincolnshire where the depth exceeds 1800 m and the temperature 60°C. (Gale *et al*, 1983). The sequence in this area was believed to consist of aeolian sands. Its potential as an aquifer was investigated by the Cleethorpes Well. However, this well proved a sequence 26 m thick consisting of fluviatile deposits. Only 8.5 m represented permeable sands and sandstones with a mean porosity of 16 per cent and values generally less than 20 per cent. A drill-stem test indicated a transmissivity of less than 2 Dm—too low for economic development. The salinity of the water was 220 g l^{-1}. As the Basal Permian Sands sequence encountered at Cleethorpes is considered to be atypical of east Lincolnshire, the well did not indicate the potential of the aquifer in the region, although it did demonstrate that the prospects for development in and around Grimsby—one of the main heat loads—are poor.

The Sherwood Sandstone Group increases in thickness from about 100 m, near Nottingham, to over 400 m in east Yorkshire. The temperature of the mid-point of the aquifer exceeds 40°C in the coastal zone of Yorkshire and Lincolnshire. In the Cleethorpes Well, the Sandstone proved to be mainly medium-grained, almost 400 m thick, generally friable, and not well-cemented; porosity values are mainly between 23–30 per cent. A preliminary production test, carried out by gas-lifting with nitrogen (Downing *et al.* 1985), yielded 20 l s^{-1} for a drawdown of only 5 m; the temperature of the water was 53°C. A transmissivity of 60 Dm was calculated from the data provided by the test implying that the well would probably yield 30 l s^{-1} for a drawdown of no more than 70 m if developed over 20 years without the pressure being maintained by reinjection.

As the sandstone is not well cemented the rock has retained its primary porosity to a considerable extent. The influx of recent meteoric waters, extending as far downgradient as Cleethorpes, has dissolved some of the earlier cements and part of the grain framework, thereby enhancing the porosity. Fibrous illite significantly reduces the permeability of the sandstone by forming a bridging mesh across pores (Milodowski *et al.* 1987). In air-dried core samples these fibres collapse against the pore walls and this is believed to be the main reason for the difference between the transmissivity assessment of 220 Dm based on core analysis, and the 60 Dm calculated from the pumping test.

The salinity of the water in the Sherwood Sandstone increases with depth from about 35 to 80 g l^{-1}. Stable isotopes of hydrogen and oxygen indicate it is of meteoric origin but the Br/Cl ratio implies the salinity may be derived from evaporites. It is of very different composition from that in the Basal Permian Sands. The Permian water has a much higher concentration of 220 g l^{-1}. Relative to chloride, sodium is depleted but the converse applies to calcium, magnesium, and strontium. The Br/Cl ratio is enriched relative to sea water, again a contrast with the water in the younger Sherwood Sandstone. The data are too limited to identify the history of the Permian brine definitely but ultrafiltration of a formation water from the underlying Coal Measures seems to be a strong possibility (Downing *et al.* 1985). The marked difference between the compositions of the water within the Sherwood Sandstone and Basal Permian Sands indicates there is no connection between the two formations, certainly in the vicinity of Cleethorpes. The presence of thick evaporites in the intervening Permian sequence lends credence to this view. But the high heat flow over east Lincolnshire may be linked to groundwater rising in the Carboniferous (Downing *et al.* 1987); if so, there may be a connection between the two formations, possibly by a limited number of well-spaced fractures.

Northern Ireland

The Sherwood Sandstone Group attains thicknesses in excess of 500 m in Northern Ireland but the producing thickness is much less and probably restricted to only some 200–250 m near the top of the formation (Bennett 1983). The geothermal gradient is about 30°C km^{-1} and over appreciable areas the temperature of the aquifer exceeds 40°C, although 60°C is attained only in a limited area to the north of Lough Neagh. The distribution and thickness of the Lower Permian sandstones are not well known.

The prospects for geothermal energy in Northern Ireland are coloured by the poor results from the Larne No. 2 Well. Tests of both the Sherwood Sandstone and the Lower Permian sandstones gave unexpectedly low transmissivities (Downing *et al.* 1982) relative to those estimated from logs and/or tests in neighbouring deep boreholes at Ballymacilroy and Newmill. Interpretation of drill-stem tests, in conjunction with geophysical logs, of the Sherwood Sandstone in the Ballymacilroy borehole suggested a tranmissivity of 15 Dm for about 200 m of the formation. A 45 m thick interval in the Permian gave a transmissivity of 3 Dm. At Larne the transmissivity of the Sherwood Sandstone was no more than 7 Dm (lower than the assumed areal average of 10–20 Dm) and the porosity of the Lower Permian sandstones (which were almost 450 m thick) was only 12 per cent due to extensive diagenetic cement. Values should be greater than 20 per cent to ensure a successful yield from a geothermal well. It is possible that the permeability of the Permo-Triassic sandstones decreases towards the centre of the basin.

Worcester Basin

The Permo-Triassic sandstones are over 1000 m thick over much of the Worcester Basin and attain maximum values of more than 2000 m. As the mean permeability is considered to be 100 mD, the transmissivity can be expected to be of the order of 100 Dm. The insulating Mercia Mudstone is relatively thin, which influences the temperature of the water in the sandstones, and the gradient in the thermally conductive sandstones is only 18°C km^{-1}. Hence, despite their great thickness, temperatures of 60°C are unlikely to be attained anywhere in the basin and are generally in the range 30–45°C. Heat pumps would be required if the resource were to be developed.

Chemical data about groundwaters in the deeper parts of the Worcester Basin are limited to a few analyses from a single borehole. They suggest that the formation waters are less mineralized than those in other Permo-Triassic basins. The salinity ranges from about 6000 mg l^{-1} in the Triassic Wildmoor Sandstone to 22 000 mg l^{-1} in the Permian Bridgnorth Sandstone. These relatively low values probably reflect one or more periods of flushing by meteoric water (Smith and Burgess 1984).

Cheshire Basin

In the Cheshire Basin the Permo-Triassic sandstones attain thicknesses of over 2500 m and occur at depths of up to 3000 m but the geothermal gradient is only about 20°C km^{-1}. Nevertheless, the temperatures of water in the deepest parts of the basin exceed 60°C over some 300 km^2 and in places possibly attain 80°C; 40°C is exceeded in an

area of about 750 km^2. There is no direct information about the aquifer properties but interpretation of geophysical logs suggests that the total water-bearing thickness of the sandstones is about 500 m and the transmissivity should exceed 10 Dm. Conditions thus appear favourable for obtaining hot waters from the basin although, because of the low thermal gradient, deep wells would be necessary and, given the high salinity of the fluids and the inland location, reinjection wells would be required to dispose of the spent brine, (Gale *et al.* 1984*a*).

Low enthalpy resources

The 'Geothermal Resources' of the principal Permo-Triassic aquifers were assessed from:

$$[\theta\varrho_w c_w + (1 - \theta)\varrho_m c_m]V\,(T_m - T_g)\ \text{joules}$$

where θ = porosity

ϱ_w and ϱ_m are densities of water and matrix respectively (g m^{-3})

c_w and c_m are specific heats of water and matrix respectively (J g^{-1} °C^{-1})

V = volume of aquifer (m^3)

T_m = mean reservoir temperature (°C)

T_g = mean annual ground temperature (assumed to be 10°C)

Calculations based on this formula give the total heat stored in the Permo-Triassic sandstones at temperatures of more than 40°C, as 220 × 10^{18} joules (Gale *et al.* 1984*b*). The proportion more likely to be available for development is represented by the 'identified resources', calculated by multiplying the geothermal resources by a recovery factor defined (Lavigne 1978) as:

$$[(T_m - T_r)\,/\,(T_m - T_g)]\ F$$

where T_r is the reject temperature of the heat extraction system and F is a factor related to the hydraulic properties of the aquifer and the method of extracting the hot water.

Most of the heat in an aquifer is stored in the rock rather than the water and if development is by means of two wells, one for abstraction and one for reinjection, the injected fluid takes up heat

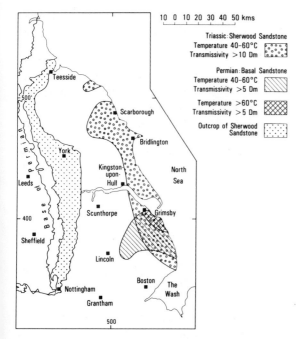

Fig. 17.7. Low enthalpy geothermal resources of the Permo-Triassic sandstones in the East Yorkshire and Lincolnshire Basin.

from the rocks and sweeps it towards the abstraction well thereby increasing the total yield. In this situation, a value for *F* of 0.33 has been accepted in the Paris Basin but as there is no practical experience of the operation of a doublet system in the UK, a more conservative figure of 0.25 was assumed to assess the resources and 0.1 was adopted for development by a single well when the brine is discharged to the sea or an estuary after the heat has been extracted. The identified resources in the Permo-Triassic sandstones are given in Table 17.2, assuming development by doublets. At more than 60°C they amount to about 5×10^{18} joules and in the 40–60°C range to almost 50×10^{18} joules, assuming the use of heat pumps which would allow rejection of the brine at 10°C. Areas of greatest geothermal potential have been referred to as 'potential geothermal fields' (Fig. 17.6); the criteria are temperatures of more 40°C and transmissivities of at least 10 Dm, although with present day prices for alternative fuels a transmissivity of more than 20 Dm would be necessary for economic development.

It would be impractical to develop all the identified resources as this would require many wells distributed over an entire aquifer and realistically the amount of energy that could be developed is smaller than that given in Table 17.2; the scale of development would, of course, also depend upon a match between the geographical distribution of heat loads, and available geothermal resources.

Over 50 per cent of the identified resources are in the Sherwood Sandstone Group in east Yorkshire and Lincolnshire (Figure 17.7) which represents the major readily accessible low enthalpy heat store in the UK, equivalent to almost 1000 million tonnes of coal. The high yield obtained from the Sherwood Sandstone in the Cleethorpes Well emphasized the potential of this aquifer at shallow, more accessible depths where the permeability is likely to be higher. Although the water temperatures are lower, a greater recovery of energy, because of higher flow rates, is a compensating factor. However, economic studies indicate that group heating schemes using hot water from intermediate depths are not yet attractive given the rather unusual features of such schemes and the risk perceived by potential developers in the capital intensive, inital stages (Energy Technology Support Unit 1986).

The Wessex Basin is the most favourable area for water resources at more than 60°C (Fig 17.8). The only operational geothermal scheme in the UK is in this basin at Southampton where the Southampton No.1 Well (Table 17.1) has been

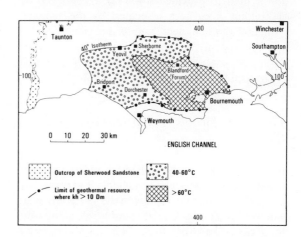

Fig. 17.8. Low enthalpy geothermal resources of the Sherwood Sandstone in the Wessex Basin.

Table 17.2 Identified resources of the Permo-Triassic sandstones at temperatures of over 40°C

	Temperature (°C)			
	40–60*		60†	
	10^{18} joules	Mtce	10^{18} joules	Mtce
East Yorkshire and Lincolnshire	26.2	974	0.2	7
Wessex	2.8	105	1.8	69
Worcester	3.0	112	—	—
Cheshire	8.9	331	1.5	56
Northern Ireland	6.7	249	1.3	48
Totals	47.6	1771	4.8	180

Estimates assume development with doublets.
* Use of heat pumps assumed and hence a reject temperature of 10°C.
† Assumes heat pumps would not be used and hence rejection at 30°C.
Mtce is million tonnes of coal equivalent.
A scale is given to the figures by bearing in mind that the annual use of electrical energy in the UK is about 10^{18} joules.

developed for the City Council by a commercial company. Although the well is capable of yielding 20–30 l s⁻¹ in the short-term, analysis of the test pumping data indicated the presence of relatively impermeable flow barriers near the well—possibly faults or facies changes—that would cause a rapid decline of the aquifer pressure in the future (Allen *et al*. 1983). Consequently, it is being developed at a maximum rate of 12 l s⁻¹ which should ensure a life for the scheme of some 20 years. The brine is at a temperature of 73°C and the heat is transferred by a heat exchanger to a 7 km hot water distribution system to provide 26 000 GJ a⁻¹. 122 000 m² of building floor area is heated by the scheme. Ultimately, when heat pumps are incorporated into the system, the energy yield should double. After the heat has been extracted the brine is discharged to waste in the adjacent estuary.

The future for low enthalpy resources in the UK

In the ten years since the geothermal programme began, the energy scene has changed appreciably. Oil prices were rising rapidly in the 1970s and were expected to exceed US$100 per barrel, with parallel rises in other fossil fuels. This has not materialized and oil prices now (May 1991) stand at about $20 per barrel while gas prices are significantly lower in real terms than in 1982. However, conventional energy resources are finite and in the longer term prices will rise in real terms. The future for low enthalpy geothermal energy, as a supplement to local requirements for energy, depends very much on the price of alternative fossil fuels but also on a more positive attitude to the use of this form of energy. It is still seen as a high risk venture requiring a high initial capital outlay for the deep wells and associated engineering works. The fact that 30 per cent of the total energy used in the UK is for water and space heating emphasizes the appropriate role low enthalpy resources could play in meeting at least part of this demand if economic factors become favourable in the future.

The programme of investigation in the 1980s identified the extent and size of the resources and demonstrated that the validity of the development technique is not in question—the Permo-Triassic sandstones can be developed at the temperatures and flow rates necessary, where and when economic conditions are appropriate.

Most of the potential resources are in Triassic sandstones at relatively shallow depths where, although the temperature is lower, the permeability and hence the specific capacity of wells, is greater, thereby allowing a greater recovery of heat. However, heat pumps would be necessary to exploit these resources. The development of geothermal energy in Southampton for space heating is providing valuable practical experience of the operation of a scheme based on a deep well.

Conclusions

This review of work in the UK on hydrogeologically based alternative sources of energy has indicated the dependence of the future development on the availability and cost of conventional sources. Low enthalpy geothermal resources are available, but at a price which is currently too high for exploitation. For hot dry rock to become a true resource, further technical development of the technology is necessary. Should that be successful, the resource in south-west England could be very large, and areas such as the Tyne/Tees region, where the buried Weardale granite looks promising, could also become suitable for exploitation.

Acknowledgements

The section on low enthalpy resources reviews part of a programme to assess the geothermal resources in the UK carried out by the British Geological Survey between 1977 and 1988; it was funded by the Department of Energy and the Commission of the European Communities. The author (R. H. Parker) of the section on hot dry rock geothermal energy is grateful to the UK Department of Energy, for their continued support for the Camborne School of Mines project. The work described was carried out under contract as part of the Department of Energy's Renewable Energy Research and Development programme, managed by the Energy Technology Support Unit (ETSU). The views and judgements expressed are those of the author and do not necessarily reflect those of ETSU or the Department of Energy.

References

Allen, D. J., Barker, J. A., and Downing, R. A. (1983). The production test and resource assessment of the Southampton (Western Esplanade) Well. *Invest. Geotherm. Potent. UK*. Institute of Geological Sciences, Keyworth.

Allen, D. J. and Holloway, S. (1984). The Wessex Basin. *Invest. Geotherm. Potent. UK*. British Geological Survey, Keyworth

Andrews, J. N., Burgess, W. G., Edmunds, W. M., Kay, R. L. F., and Lee, D. J. (1982). The thermal springs of Bath. *Nature* (London), **298**, 361–3.

Baria, R. and Green, A. S. P. (1990). Microseismics: a key to understanding reservoir growth. In *Camborne School of Mines international conference on hot dry rock geothermal energy*, Camborne, Cornwall, June 1989, (ed. R. Baria) pp. 363–77. Robertson Scientific Publications, London.

Batchelor, A. S. (1982). The creation of hot dry rock systems by combined explosive and hydraulic fracturing. *Int. conf. on geothermal energy*, Florence, May 1982, pp. 321–342. BHRA Fluid Engineering, Bedford.

Batchelor, A. S. (1983). Hot dry rock reservoir stimulation in the UK: an extended summary. In *European Geothermal Update* (ed. A. S. Strub and P. Ungemach), pp. 681–71. Reidel, Dordrecht.

Bennett, J. R. P. (1983). The sedimentary basins in Northern Ireland. *Invest. Geotherm Potent UK*. Institute of Geological Sciences, Keyworth.

British Geological Survey. (1984). Catalogue of Geothermal data for the land area of the United Kingdom, 2nd edn. (ed. A. J. Burley, W. M. Edmunds, and I. N. Gale). *Invest. Geotherm. Potent. UK*. British Geological Survey, Keyworth.

British Geological Survey. (1988) Geothermal Energy in the United Kingdom: review of the British Geological Survey's programme 1984–1987. *Invest. Geotherm. Potent. UK*. British Geological Survey, Keyworth.

Burgess, W. G., Edmunds, W. M., Andrews, J. N., Kay, R. L. F., and Lee, D. J. (1980). The hydrogeology and hydrochemistry of the thermal waters in the Bath-Bristol Basin. *Invest. Geotherm. Potent. UK*. Institute of Geological Sciences, Keyworth.

Camborne School of Mines Geothermal Energy Project. (1984). *Low flow rate hydraulic tests*. Phase 2A report. (Report no. 2–52, unpublished).

Camborne School of Mines Geothermal Energy Project. (1989a). *2C–5 Final Report Phase 2C (1986–1988)*. ETSU G 137F. Department of Energy, London.

Camborne School of Mines Geothermal Energy Project. (1989b). *2C–7 Resource evaluation*. Volume 3: *Gravity and thermal modelling*. ETSU G 137–P14C. Department of Energy, London.

Downing, R. A. and Gray, D. A. (1986). *Geothermal energy—the potential in the United Kingdom*. HMSO, London.

Downing, R. A., Burgess, W. G., Smith, I. F., Allen,

D. J., Price, M., and Edmunds, W. M. (1982). Geothermal aspects of the Larne No 2 Borehole. *Invest. Geotherm. Potent. UK.* Institute of Geological Sciences, Keyworth.

Downing, R. A., Allen, D. J., Barker, J. A., Burgess, W. G., Gray, D. A., Price, M., and Smith, I. F. (1984). Geothermal exploration at Southampton in the UK—a case study of a low enthalpy resource. *Energy, Exploration and Exploitation*, **2**, 327–42.

Downing, R. A., Allen, D. J., Bird, M. J., Gale, I. N., Kay, R. L. F., and Smith, I. F. (1985). Cleethorpes No. 1 Well—a preliminary assessment of the resource. *Invest. Geotherm. Potent. UK.* British Geological Survey, Keyworth.

Downing, R. A., Edmunds, W. M., and Gale, I. N. (1987). Regional groundwater flow in sedimentary basins in the United Kingdom. In *Fluid flow in sedimentary basins and aquifers* (ed. J. C. Goff and B. P. J. Williams), pp. 105–25, Special Pub. 34. Geological Society London.

Edmunds, W. M. (1986). Geochemistry of geothermal waters in the U.K. In *Geothermal Energy—the potential in the United Kingdom* (ed. R. A. Downing and D. A. Gray), pp. 111–23. HMSO, London.

Energy Technology Support Unit (ETSU). (1986). *Geothermal aquifers*. Department of Energy R and D programme 1976–1986. ETSU for Department of Energy, London.

Gale, I. N. and Rollin, K. E. (1986). Assessment of the geothermal resources. In *Geothermal energy—the potential in the United Kingdom* (ed. R. A. Downing and D. A. Gray), pp. 132–47. HMSO, London.

Gale, I. N., Smith, I. F., and Downing, R. A. (1983). The post-Carboniferous rocks of the East Yorkshire and Lincolnshire Basin. *Invest. Geotherm. Potent. UK.* Institute of Geological Sciences, Keyworth.

Gale, I. N., Evans, C. J., Evans, R. B., Smith, I. F., Houghton, M. G., and Burgess, W. G. (1984a). The Permo-Triassic aquifers of the Cheshire and West Lancashire basins. *Invest. Geotherm. Potent. UK.* British Geological Survey, Keyworth.

Gale, I. N., Rollin, K. E., Downing, R. A., Allen, D. J., and Burgess, W. G. (1984b). An assessment of the geothermal resources of the United Kingdom. *Invest. Geotherm. Potent. UK.* British Geological Survey, Keyworth.

Gebski, J. S., Wheildon, J., and Thomas-Betts, A. (1987). Investigations of the UK heat-flow field 1984–1987. *Invest. Geotherm. Potent. UK.* British Geological Survey, Keyworth.

HDR Geochemistry Group. (1990). *Joint report by*

Camborne School of Mines, British Geological Survey and University of Bath to UK Department of Energy and Commission of the European Communities*. Research report no SD/89/2, Vols 1-7, British Geological Survey, Keyworth.

Holloway, S., Milodowski, A. E., Strong, G. E., and Warrington, G. (1989). The Sherwood Sandstone Group (Triassic) of the Wessex Basin, Southern England. *Proc. Geol. Assoc.*, **100**, 383–94.

Knox, R. W. O'B., Burgess, W. G., Wilson, K. S., and Bath, A. H. (1984). Diagenetic influences on reservoir properties of the Sherwood Sandstone (Triassic) in the Marchwood Geothermal Borehole, Southampton, England. *Clay Minerals*, **19**, 441–56.

Kwakwa, K. A. (1988). Tracer measurements during long-term circulation of the Rosemanowes HDR geothermal system. In *13th annual workshop on geothermal reservoir engineering*, pp. 245–52. Stanford University, California.

Lavigne, J. (1978). Les ressources géothermiques françaises possibilités de mise en valeur. *Ann. Mines*, April 1978, pp. 1–16.

Lee, M. K., Brown, G. C., Webb, P. C., Wheildon, J., and Rollin, K. E. (1987). Heatflow, heat production and thermotectonic setting in mainland, UK. *J. Geol. Soc. London*, **144**, 35–42.

Milodowski, A. E., Strong, G. E., Wilson, K. S., Allen, D. J., Holloway, S., and Bath, A. H. (1986). Diagenetic influences on the aquifer properties of the Sherwood Sandstone in the Wessex Basin. *Invest. Geotherm. Potent. UK*, British Geological Survey, Keyworth.

Milodowski, A. E., Strong, G. E., Wilson, K. S., Holloway, S., and Bath, A. H. (1987). Diagenetic influences on the aquifer properties of the Permo-Triassic sandstones in the East Yorkshire and Lincolnshire Basin. *Invest. Geotherm. Potent. UK.* British Geological Survey, Keyworth.

Nicol, D. A. C. and Robinson, B. A. (1990). Modelling the heat extraction from the Rosemanowes HDR reservoir. *Geothermics*, **19**, 247–57.

Nicol, D. A. C., Randall, M. M., and Hicks, T. W. (1990). Stimulation modelling. In *Camborne School of Mines international conference on hot dry rock geothermal energy*, Camborne, Cornwall, June 1989. (ed. R. Baria), pp. 289–98 Robertson Scientific Publications, London.

Parker, R. H. (1989a). *Hot dry rock geothermal energy. Phase 2B Final Report of the Camborne School of Mines Project*. Volumes 1 and 2. Pergamon Press, Oxford.

Parker, R. H. (1989b). The Camborne School of Mines hot dry rock geothermal energy project. *Scientific Drilling* (1989). Vol 1. pp.34–41

Parker, R. H. (1989*c*). Characterisation of the Rosemanowes HDR geothermal reservoir using an extended circulation programme. In *European geothermal update, Proc. 4th int. sem. on results of EC geothermal energy research and demonstration*, Florence, 27–30 April 1989. (Ed. K. Louwrier, E. Staroste, J. D. Garnish, and V. Karkoulias), pp. 141–153. Kluwer Academic Publishers, Dordrecht.

Pine, R. J., and Batchelor, A. S. (1984) Downward migration of shearing in jointed rock during hydraulic injections. *International Journal of Rock Mechanics, Mining Sciences and Geomechanics Abstracts.* **21**, (5), 249–63.

Richards, H. G., Kwakwa, K. A., and Oskui, G. P. (1990). Flowpath characterisation of the Rosemanowes HDR reservoir using inert tracers. In *Camborne School of Mines international conference on hot dry rock geothermal energy*, Camborne, Cornwall, June 1989. (ed. R. Baria), pp. 557–74 Robertson Scientific Publications, London).

Richardson, S. W. and Oxburgh, E. R. (1979). The heatflow in mainland UK. *Nature*, (London), **282**, 565–7.

Smith, I. F. and Burgess, W. G. (1984). The Permo–Triassic rocks of the Worcester Basin. *Invest. Geotherm. Potent. UK*. British Geological Survey, Keyworth.

Strong, G. E. and Milodowski, A. E. (1987). Aspects of the diagenesis of the Sherwood Sandstone of the Wessex Basin and their influence on reservoir characteristics. In *Diagenesis of sedimentary sequences* (ed. J. D. Marshall), pp 325–37, Spec. Pub. 36, Geol. Soc. London.

Wheildon, J. and Rollin, K. E. (1986). Heatflow. In *Geothermal energy—the potential in the United Kingdom* (ed. R. A. Downing and D. A. Gray), pp. 8–20. HMSO, London.

18. Hydrogeological assessments for underground mines

J. W. Lloyd, R. I. Jeffery, and N. H. Neill

Introduction

In recent years, a large proportion of capital investment has been concentrated in maintaining and extending coal production within the concealed coalfields of Yorkshire and the East Midlands. The development of these reserves since 1975 has necessitated the construction of shafts and drifts up to 8 m in diameter to depths in excess of 1000 m and shallower drifts for mine access. All of these shafts and drifts have passed through water-bearing strata. The range and nature of permeability, aquifer type, depth, and thickness of the water-bearing strata have varied greatly, from highly permeable low strength, uncemented Permo-Triassic sandstones, to more discretely fractured high strength limestones and Upper Coal Measures sandstones. The depth of the water-bearing strata has varied from surface to 750 m as has the potential for groundwater ingress to the excavations.

The economics for the development of a deep mine, or for an extension to an existing mine, are such that the shaft sinking or drift construction cycle is very firmly on the critical path for the overall development and hence early return on very substantial total capital investment. For example, a deep shaft to a depth of 1000 m can cost up to £20 million to construct. Although high, this figure may only represent less than 10 per cent of the total capital investment which can extend into many hundreds of millions of pounds. As a result, it is essential that hydrogeological and geotechnical conditions for construction are identified accurately at an early stage in a mine project to enable the most efficient, safe, and cost effective methods of construction and ground treatment to be designed and implemented to offset groundwater ingress.

In addition to the constructional phase the safe working of a mine can depend upon the potential groundwater inflow situation, in that mining operations may have to be tailored to negate inflow, or in the worst situation mining may not take place. As with the construction of mine access structures, therefore, the hydrogeological controls relevant to mining operations require assessment.

In the following discussion, the research carried out to determine reliable methods of down-hole *in situ* testing, data analysis, and groundwater flow assessment is reviewed. The research has been developed over the last fifteen years to quantify the risk of groundwater inflow to coal mining situations in typical Permo-Triassic and Coal Measures strata in the UK. The implications and application of the research are also discussed in relation to the design and implementation of ground treatment strategies that will allow the safe expedient passage of excavations through water-bearing strata.

Development of *in situ* testing techniques

In the early 1970s aquifer packer testing techniques for deep site investigations utilized methods derived from the oil industry because equivalent standard geotechnical equipment was insufficiently robust. Since that time effective testing methods have been developed and have formed an important part of the overall research, as the tests provide the base data for inflow assessment.

Early testing using geotechnical style equipment with pump-packer assemblies was developed by Chalmers *et al.* (1979) and used successfully to depths of 200 m. The equipment incorporated a downhole pump and single or straddle packer assembly and was eventually modified for successful use to depths of 400 m (Daw 1984). The main testing emphasis, however, has been on the use of robust oilfield based downhole drill-stem test (DST) assemblies.

In drill-stem tests the drill-pipe is run dry and the test is carried out by opening and shutting the shut-in valve to allow and stop, alternately, the entry of groundwater from the test section into the drill-pipe. Groundwater heads are measured in the test section and flows calculated by applying the inflow heads to drill-pipe volume.

Initially, these down-hole assemblies were of the conventional oil industry type incorporating a downhole choke to restrict and linearize inflow characteristics, a down-hole shut-in valve, and compressional solid rubber packers that are expanded against the wall of the borehole by the application of vertical weight imposed onto a bottom hole assembly. Such equipment is not resettable and has to be raised to the surface after every test. Instrumentation within such systems is simple, using down-hole clockwork Amerada gauges and a surface bubble hose to quantify flow rate. In strata of moderate to high strength, exhibiting low to intermediate permeability, compressional DST procedures were found to work reasonably well and reliable hydrological parameters were acquired. Unfortunately, severe limitations with the system were experienced when testing weakly cemented strata particularly where irregular borehole geometry had developed. The selection of test horizons therefore was controlled strictly by the configuration and strength of the borehole walls rather than the aquifer characteristics of the strata.

In the early 1980s downhole tests conducted in Warwickshire for the first time used braided steel reinforced inflatable rubber packers that were inflated by means of a rotating down-hole pump system (Lloyd and Jeffery 1983). These inflation packers provide significantly greater flexibility for hydrogeological testing than the compression packers. The membranes lack the rigidity of the compression packers and are more readily able to conform to the eccentricities of borehole geometry and thus provide an efficient seal to a test interval. Furthermore, because the expansion of the inflatable packer system is not related to string weight it is also possible in cases of 'dry' or low permeability tests to reset the test assembly without pulling the

equipment back to surface. The incorporation of transducers to record data rather than Amerada gauges has greatly enhanced testing performance. A diagram of the assembly is given in Fig. 18.1.

The inflation packer system was first used to investigate the low strength, highly permeable Permo–Triassic strata at a depth of 333–400 m at Asfordby where permeabilities were found to be 10^{-10}–10^{-5} m s^{-1} and strengths of 4–40 MPa. Excellent data for low to intermediate permeability layers were obtained in sandstones and siltstones exhibiting irregular borehole geometry. However, problems were encountered when testing the highly permeable Sherwood Sandstone for which early flow rates through the down-hole assembly were found to be very high, often in excess of 3 l s^{-1}.

Because of the complicated configuration within the assembly, back pressures due to flow constriction under high flows were found to occur in the test section below the hydraulic shut-in tool so that incorrect flows were initially calculated. To alleviate the problem a conductor wireline gauge (CWL) containing a transducer was introduced into the drill-string at a position immediately above the hydraulic shut-in tool after the packers were set. With the use of the transducer, heads in the drill string were measured at intervals of less than 10 s and flow rates accurately calculated, significantly enhancing the flexibility of the testing system. Using the CWL, static groundwater head is determined by allowing inflow to the drill-pipe until heads in the pipe and the formation equilibrate.

During the Asfordby site investigation, a series of injection tests was also carried out to assess both hydraulic conductivity and the grouting capability of these sandstones. The results proved to be very poor, with the values of hydraulic conductivity generally an order of magnitude lower than that derived from DST testing.

The most important research conclusion from the deep aquifer testing at Asfordby was the need to simplify the test system. Of primary importance was the need to remove the choking effect and, secondly, the need to develop a flexible system more readily applicable to the testing of materials with a very wide range of hydraulic conductivity. To achieve these criteria, the downhole assembly was totally redesigned. The down-hole testing equipment currently in use is illustrated in Fig. 18.1. The system utilizes up to three braided steel

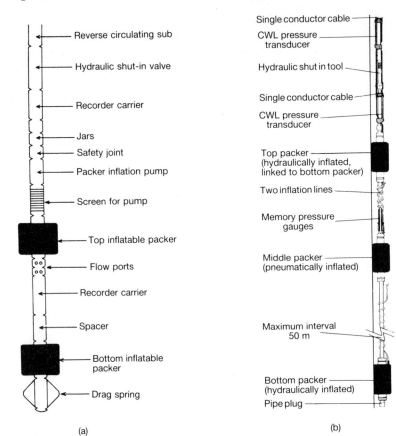

Fig. 18.1. (a) Initial DST inflation string (b) Current DST inflation string.

reinforced rubber inflatable packers, that can offer two different packer straddles per test assembly from 3–50 m. All of the packers are inflated externally using either water or nitrogen. The system offers no restrictions to flow and has an equivalent flow area of 0.0026 m². Instrumentation is via two quartz transducers that are strapped to the outside of the test assembly and are tapped into the assembly above and below the valve via two flow ports. The sampling rates of the gauges are less than 3 per second and all down-hole transient pressure data are accessed and stored directly onto a surface microcomputer. The downhole assembly is lowered into the borehole on a 73 mm diameter testing string that allows fluid to be evacuated from the test assembly by swabbing. The use of down-hole swabbing allows the test assembly to be fully resettable and permits the cleaning of test sections where mud invasion occurs. Further, by swabbing, clean samples can be obtained for chemical analysis. A typical test procedure for strata exhibiting significant hydraulic conductivity is shown in Fig. 18.2(a).

The equipment currently in use is capable of measuring hydraulic conductivity in the range 10^{-4}–10^{-11} m s^{-1}. The added flexibility of the third packer has enabled up to 20 individual tests to be conducted at one site within a 48 hour period without pulling the test assembly to the surface. This flexibility coupled with the use of heavy duty PQ wireline drilling to protect a borehole wall, has allowed a considerable increase in the number of tests that can be conducted even in the most weakly cemented sandstones, that, prior to the

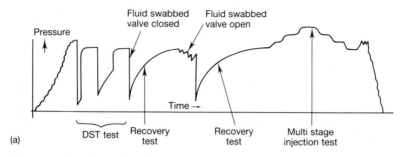

(a)

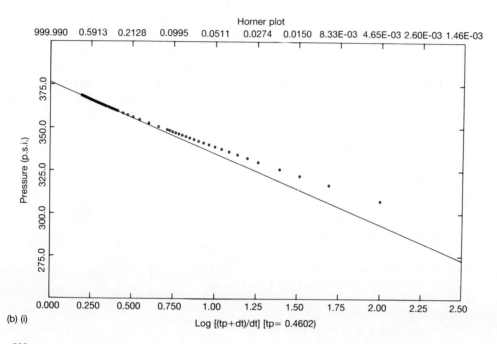

(b) (i)

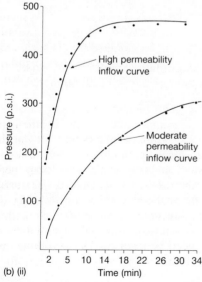

(b) (ii)

Fig. 18.2. (a) Test schedule shown in terms of pressure and time

Fig. 18.2. (b) (i) Horner plot results (ii) Radial flow plot results.

development of the test assembly, could not be tested independently.

Hydraulic conductivity determination

The DST data are interpreted using traditional oil-field techniques described by Horner (1951) and Earlougher (1977). Such methods are similar to the curve fitting and straight line approximation analyses used in most groundwater investigations to evaluate hydraulic conductivity. In addition to hydraulic conductivity the interpretation techniques provide an estimate of formation pressure or static groundwater head.

As with many analytical techniques they lack the flexibility of numerical solutions and so radial models have been developed to simulate some DST data (Lloyd and Jeffery 1983). To date the emphasis has been placed on the inflow tests because measurement techniques provide an accurate record of both flow and pressure data, which is obviously not the case in terms of flow during shut-in. The model developed has been based upon the well-known radial flow model of Rushton and Redshaw (1979) with modifications to account for variable flow.

The application of the model has proved successful for non-restricted hydraulic valve tests where flows have been small to moderate. Where tests have produced large inflows, which because of the depth of the tests are associated with rapid rises in water levels in the drill-pipe and formational pressure changes, the model simulations have not proved satisfactory. The reason is attributed to the omission of a turbid flow factor for flow in the drill-pipe, which is being rectified. Examples of test data and interpretations are given in Fig. 18.2(b).

The DST hydraulic conductivity determinations are clearly bulk values for test sections. Because of costs and other constraints they are limited in number although every attempt is made to make them representative. The data are supplemented by horizontal and vertical hydraulic conductivity determinations of core material in the laboratory. The two data sets are then reviewed together with lithological, fracture, and geophysical data to determine a horizontal and vertical layered hydraulic conductivity distribution for inflow

assessment (Jeffery and Daw 1989). The distribution determined is inevitably subjective.

Shaft and drift inflow assessments

Groundwater modelling techniques have been adopted for inflow assessments in Britain because of the layered hydraulic conductivity conditions encountered in the British coal-fields. Initially resistance analogue modelling was used and proved successful in a study in the Coventry coal-field (Lloyd *et al.* 1983). The modelling demonstrated the importance of ground layering on inflow volumes as shown in Fig. 18.3. Subsequently finite difference models have been developed to provide more flexibility in analysis (Edwards 1985, 1988; Lloyd and Edwards 1988; Jeffery *et al.* 1989).

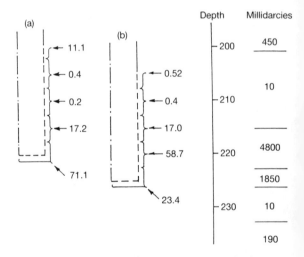

Fig. 18.3. Examples of inflows to shaft (per cent); (a) 13800 m³ day⁻¹ (b) 12700 m³ day⁻¹.

For the calculation of inflows to a vertical shaft three-dimensional groundwater flow may be approximated as being radial and axisymmetric about the shaft axis. Using a radial (*r-z*) model, flow may be analysed both in the radial (*r*) and vertical (*z*) directions. Because the advancement of shaft construction is slow, only steady-state flow conditions need be invoked. The governing equation is therefore:

$$\frac{\partial}{\partial r}(k_r \frac{\partial h}{\partial r}) + k_r \frac{1}{r} \cdot \frac{\partial h}{\partial r} + \frac{\partial}{\partial z}(k_z \frac{\partial h}{\partial z}) = 0 \qquad (1)$$

where: $h(r,z)$ is the groundwater head distribution
kr, kz are hydraulic conductivities in the radial and vertical directions applied to each layer defined in the model.

The flow domain is shown on Fig. 18.4. The lined section of the shaft is considered impermeable with flow only to the open section where, because any inflows are rapidly removed by pumping, the pressure conditions may be set at atmospheric with ($h = z$).

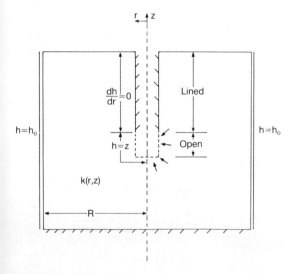

Fig. 18.4. Flow domain used for radial flow assessment to a shaft.

The lower and lateral radial boundaries depend upon the local hydrogeology. Normally a very low permeability layer can be identified below the shaft base to which a no-flow horizontal boundary condition ($\partial h/\partial z = 0$) can be applied. A fixed head condition is set at the radial boundary (R) where the head value is equivalent to the level of the static groundwater head prior to shaft construction ($h = h_0$). In the absence of hydrogeological data to define R the effect of the boundary imposition may be examined and minimized by sensitivity analysis.

As noted below, the model developed has been successfully tested against a number of shaft inflow studies, notably for Wearmouth and Selby. Under

most conditions it has been found suitable but has proved unstable on the rare occasions where locally unconfined conditions have developed immediately adjacent to a shaft. To handle such situations a saturated-unsaturated groundwater flow model has been developed.

The assessment of drift inflows poses more difficulties than shaft inflows because no simple geometrical relationship exists between a drift and the hydraulic conductivity layering. For comprehensive assessments three-dimensional flow modelling is undoubtedly preferred, but such modelling is complicated. It requires the combination of heterogeneous hydrogeology and an inclined cylindrical inflow opening (Fig. 18.5(a)) and for most sites, the limited data available do not warrant the use of such models.

Edwards (1988) has developed the requisite three-dimensional model but considers that the excessive computational effort required, combined with the data constraints, make the application of such a model impracticable. As a result he has

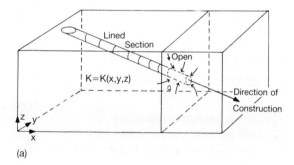

(a)

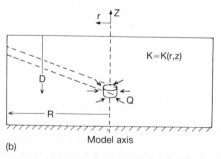

(b)

Fig. 18.5. (a) Three dimensional representation for drift inflows (b) Radial (r-z) approximation for drift inflows (after Edwards, 1988).

invoked a radial flow approximation for drift inflows.

Comparisons between the three-dimensional and the radial models for theoretical drift scenarios show that the latter model provides reasonable inflow estimates but fails to represent flows immediately adjacent to the drift's open section adequately. In Fig. 18.5(b) the geometrical approximation adopted for the radial drift model is shown. In the computation of inflow, the same boundary criteria are applied as in the shaft radial flow model and the influence of the lined section of the drift is ignored.

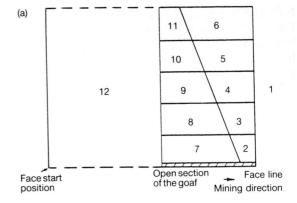

Inflows to longwall mines

While the application of inflow assessments to shafts and drifts relates to the constructional phase of operations, the assessments for longwall mines relate to the mining phase with respect to width of panel workings, mine safety, and overall viability (Fawcett *et al.* 1984; Singh *et al.* 1986). The analyses of inflows are more difficult than for shafts, or even drifts, for the following reasons:

(1) multiple panels and roadways exist;

(2) mining rates are such that non-steady flows must be considered;

(3) the longwall mining method, with hanging-wall collapse and strata reconsolidation, causes hydraulic conductivity to vary with time;

(4) subsidence above the longwall face results in fracture formation and bedding-plane separation;

(5) the time dependent nature of (3) and (4) means that quantitative hydrogeological data relating to these parameters are virtually non-existent.

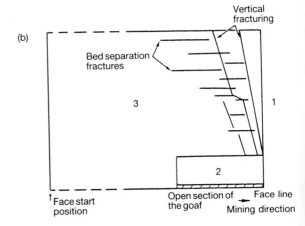

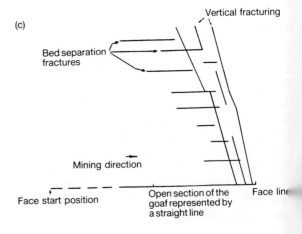

Fig. 18.6. Vertical sections showing strata fracturing above a longwall coal face represented by (a) equivalent permeable media, (b) equivalent permeable media and discrete fractures, and (c) discrete fractures (Goaf is a mining term for the space left after coal has been mined; usually the roof of the seam collapses into this space).

Neill (1987, 1988*a*, *b*) has extensively researched the concepts involved in longwall mine inflows and has represented them by using two-dimensional flow models set vertically in the direction of mine advancement (Fig. 18.6(a)).

For the initial modelling a conceptual ground geometry, as shown in Fig. 18.6(a), has been determined based upon British Coal's subsidence theory. Each numbered zone is assigned an equivalent hydraulic conductivity. The flow mechanisms are such that fractures are generated from the mine face upwards into the overlying strata. The extent of the fracture length and the hydraulic conductivity of the fracture, combined with the presence or otherwise of water-bearing strata above the mine, will determine the inflow to the mine under the appertaining groundwater head conditions. Where significant inflows are possible, the distribution of the hydraulic conductivity generated results in rapid initial flows and partial desaturation of certain areas. The overall fracture induced flow mechanism is further complicated by bedding plane separation, which is an important feature of subsidence. The separations provide discrete time-dependent groundwater storage

zones that can fill or empty in response to intersection by fractures.

To account for the mechanisms described above Neill (1987, 1988*a*, *b*) has applied the following partially saturated flow equation to the geometrical distribution shown in Fig. 18.6(a) and developed a non-steady state finite difference equivalent hydraulic conductivity model in which the hydraulic conductivity and storage terms are saturation dependent:

$$\frac{\delta}{\delta y}\left(k_y(\phi)\,\frac{\delta h}{\delta y}\right) + \frac{\delta}{\delta z}\left(k_z(\phi)\,\frac{\delta h}{\delta z}\right)$$
$$= C(\phi)\,\frac{\delta h}{\delta t} - q(q,z,t) \quad (2)$$

where: k_y and k_z are horizontal and vertical hydraulic conductivity
C is specific moisture capacity
ϕ is pressure head
q is inflow (or outflow) at node (y,z)
h is $\phi + z$ and is groundwater head
t is time

As shown in Fig. 18.7 the application of the

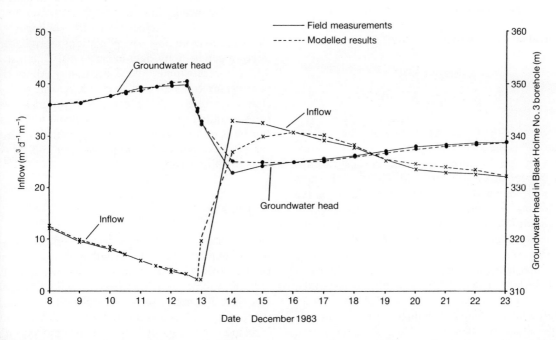

Fig. 18.7. Results of the inflow modelling for A1 face, Wistow Colliery. Continuous lines: field measurements; dashed lines: modelled results. The model is a 2-D vertical model at right angles to the advancing face.

equivalent hydraulic conductivity model proved successful in simulating inflows at Wistow Colliery (Neill 1990). In carrying out the calibration, the initial inflows following an inrush were found to be the most difficult to represent. It was postulated that the equivalent hydraulic conductivity concept was inadequate to represent flows conveyed by rapidly developing fractures that might transfer large amounts of groundwater very quickly to the working face along discrete conduits. To examine the significance of fractures three scenarios were considered:

(1) a dual porosity medium;

(2) discrete fractures and impermeable matrix;

(3) discrete fractures and permeable matrix.

Method (3) in which the flow in the fractures and in the permeable matrix are considered as separate regimes within the same continuum has been adopted as the most appropriate. Each regime is discretized into separate volume elements and the flow equations solved for each such volume element. Flow in the permeable matrix is described by the normal transient fully saturated fluid flow equations for intergranular flow with flow in the individual fractures described by:

$$\frac{\delta}{\delta x^*}\left(K^* \frac{\delta h}{\delta x^*}\right) = S_{sf} \frac{\delta h}{\delta t} - q(x^*,t) \qquad (3)$$

where x^* is the local coordinate along the fracture line, K^* is the hydraulic conductivity, in the direction of the fracture, S_{sf} is the specific storage coefficient of the fracture and $q(x^*,t)$ is the flow rate of sources or sinks per unit volume of the medium. Unsaturated flow has not been incorporated into the model which has been based on the code developed by Huyakorn *et al.* (1987).

The style of geometry adopted for the discrete fracture-permeable matrix modelling is shown in Fig. 18.6(b). The inability to simulate unsaturated flow necessitated the modification of the code to represent the filling and drainage of bedding-plane separations. In theory, a bedding-plane separation may be represented as a fracture but it is not possible to give the fracture a storage value with

this model. Therefore, in the modelling approach adopted, the important storage has been approximated as being in the porous matrix adjacent to the fracture.

The mechanism of fracturing that occurs above a longwall face is not clear. The stress-strain patterns induced by longwall mining may be estimated (NCB 1975) although the form in which the strain is manifest as fractures is unpredictable. It may be distributed uniformly with several small fractures, or more probably, concentrated at a few points where larger fractures are formed. A variety of geological factors coupled with mining practice play their part. As any inflow to the mine will be in part dependent upon the nature of the fracturing, theoretical fracture generation has been examined.

A two dimensional, random line network generation code developed by Rouleau (1988) has been adapted to the longwall situation to obtain stochastic representation of strata fractures. Typical results of such an output are given in Fig. 18.6(c). Data are input as the means and standard deviations of the fracture parameters.

The computer programs developed have shown that inflows to longwall coal faces can be simulated successfully by adapting existing groundwater flow models and considering hydraulic conductivity changes due to mining. The limitations of available fracture conductivity data mean that it is impossible to forecast mine inflows accurately, although worst cases can be calculated. Therefore, a stochastic fracture network generation program has been modified to provide a probabilistic tool that can be used to perform sensitivity analyses. This can be developed further to provide the capability of probabilistic assessment of inflow.

Application

The application of the shaft or drift inflow assessments coupled with rock strength criteria is shown on the British Coal decision diagram in Table 18.1.

If the inflow model suggests an inflow rate of less than 30 l min^{-1}, then excavation can proceed without special precautions. Inflow rates of this level are handled easily; any water is removed with the excavated material. If inflow rates are estimated to be 60 l min^{-1}, arrangements have to be

Table 18.1 Decision chart for ground treatment in shafts and drifts

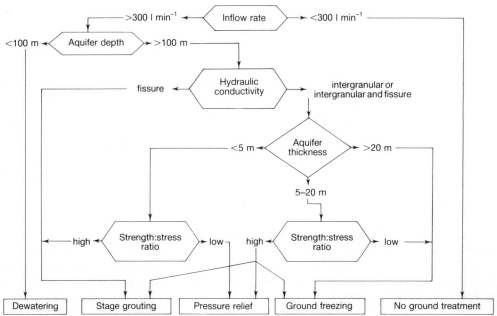

specified to pump water from a sump out of the excavation. In the British Coal specification, inflow rates of up to 300 l min⁻¹ can be tolerated using temporary pump systems operating from within a sump. Inflow rates generally above this figure require an alternative sinking strategy, although rates greater than 300 l min⁻¹ can be tolerated locally, for 10–20 m of open section if the aquifer thickness is thin and not vertically extensive.

When evaluating the most appropriate excavation strategy for estimated inflows greater than 300 l min⁻¹ the hydrological parameters must be assessed in conjunction with data on the nature of the hydraulic conductivity and the geomechanical strength characteristics of the associated strata. The evaluation is made in terms of the risk assessment strategy (Table 18.1) which is broken down into a series of levels:

Level 1 Inflow rate

Level 2 Aquifer depth

Level 3 Nature of hydraulic conductivity

Level 4 Aquifer thickness

Level 5 Strength/stress ratio

Level 6 Method of ground treatment

The overall depth of the proposed excavation and the disposition of the aquifers within this depth range are the first of the major factors to be considered once an inflow assessment has been completed (Table 18.1). For example while freezing may be considered the obvious choice if significant inflows are estimated, the same method may not represent the most cost-effective approach if a high inflow is estimated only near the base of a very deep excavation.

In instances where grouting is considered appropriate, the nature of the hydraulic conductivity is critical. The presence, size, and frequency of intergranular zones will determine the choice of cement and/or chemical grout. Where chemical grouting is essential then it is necessary to define the size of openings further to decide whether a particulate chemical grout will be used or whether a highly clarified material is necessary. Temperature and the chemistry of the groundwaters are also major influences in the formulation of grout design. Another aspect is whether the aquifer material requires additional strengthening to enable safe excavation. Most chemical grouts will only provide minimal additional strength and if the aquifer is under high hydrostatic pressure benefits can be obtained by combining a groundwater depressurizing system with the grout injection.

In the case of ground freezing the nature of the hydraulic conductivity is not so important; neither is the actual magnitude of potential inflow above 300 l min^{-1}. Critical factors are the quality and temperature of the groundwater and the thermal properties of the rock or soil within the aquifer. The mobility of the groundwater can be a restriction on the use of ground freezing where steep groundwater head gradients exist causing high natural flow rates as found in karstic systems.

Deep well dewatering may offer an alternative approach in some instances and here the main parameters of importance are the nature, permeability, and depths of the aquifers and the groundwater chemistry. The latter is obviously important from the point of view of disposal of large volumes of groundwater. Generally, however, pressure relief in combination with other ground treatment, such as grouting, will be more appropriate. A client's specifications for pumping limits from within the shaft can also dictate the ground treatment methods. Whilst ground freezing can be considered as a total exclusion method, grouting and dewatering or depressurizing well systems will always leave some residual inflow to the shaft.

Conclusions

During the past ten years, advances in *in situ* testing have allowed the assessment of inflow characteristics to be made with an increasing amount of confidence. Qualitative assessments as to the nature of flow into an excavation can be made by correlating data derived from flexible systems of *in situ* tests with that from borehole logs and laboratory data. Interpretations are sufficiently advanced to determine hydraulic conductivity of test sections in downhole testing; however, the distribution of the hydraulic conductivity over the length of a test section is still subjective.

The methodology for the assessment of inflows into shafts has advanced considerably over the past ten years, the main constraint being the subjectiv-

ity of some of the base data. Inflow assessment techniques for drifts are also good within the context of the data available. In the longwall mine studies, the modelling techniques may be said to have overtaken the current understanding of the hydrogeology.

As with groundwater pollution and other facets of hydrogeology, the apparent ability to represent ground conditions mathematically has not been matched with adequate field data. There is a need for research into methods for the remote measurement of aquifer characteristics, particularly time variant changes in such parameters for mining purposes.

Acknowledgements

The authors would like to thank British Coal for permission to publish the case history data and interpretations. Research undertaken by Dr N. H. Neill, was supported by the Scientific and Engineering Research Council. The groundwater flow models were developed at the School of Earth Sciences, University of Birmingham and the help of Dr D. N. Lerner is acknowledged.

References

Chalmers, A., Daw, G. P., and Scott, R. A. (1979). A modified form of aquifer depletion-recovery test for assessing potential water makes into deep excavations. *Proc. 4th Congress*, **2**, pp. 67–72. ISRM, Montreaux.

Daw, G. P. (1984). Application of aquifer testing to deep shaft investigations. *Q. J. Eng. Geol.*, **17**, 367–79.

Earlougher, R. C. (1977). *Advances in well test analysis*. Soc. Pet. Eng. Monograph 5, Henry L. Doherty Series, New York.

Edwards, M. G. (1985). Two-dimensional numerical modelling of inflows to a drift through multilayered permeability ground. *Proc. Int. Mine Water Cong.*, Vol. 1, 455–66. Int. Mine Water Assoc.

Edwards, M. G. (1988). Mathematical modelling of groundwater inflows to shafts and drifts under construction. PhD thesis, University of Birmingham.

Fawcett, R. J., Hibberd, S., and Singh, R. N. (1984). An appraisal of mathematical models to predict water inflows into underground coal workings. *Int.*

J. Mine Water, **3**, 33–54.

Horner, D. R. (1951). Pressure build-up in wells. *Proc. Third World Petroleum Cong.*, Section II, pp. 503–23. E. J. Brill, Leiden, Holland.

Huyakorn, P. J., White, H. O., and Wadsworth, T. D. (1987). *TRAFRAP-WT: A two dimensional finite element code for simulating fluid flow and transport of radionuclides in fractured porous media with water table boundary conditions*. Hydrogeologic Inc., Alberta, Canada.

Jeffery, R. I. and Daw, G. P. 1989. Hydrogeological investigations and assessments for shaft sinking. In *Shaft mining*, 231–40. Inst. Min. Met.

Jeffery, R. I., Lloyd, J. W., and Edwards, M. G. (1989). Assessment of shaft inflow characteristics for deep aquifers associated with coal mining in the United Kingdom. In *Shaft mining*, pp. 241–7. Inst. Min. Met.

Lloyd, J. W. and Edwards, M. G. (1988). Estimation of groundwater inflow to an underground mining operation. *Int. J. Mine Water*, **7**, 25–47.

Lloyd, J. W. and Jeffery, R. I. (1983). Deep aquifer testing methods and data interpretation. *Z. Dt. Geol. Ges.*, **134**, 871–84.

Lloyd, J. W., Rushton, K. R., and Jones, P. A. (1983). An assessment of groundwater inflows into a proposed shaft and drift in the Warwickshire Coalfield using packer test data in an electrical analogue model. *Int. J. Mine Water*, **2**, 1–18.

National Coal Board (1975). *Subsidence engineer's handbook* (2nd edn), NCB Mining Dept., London.

Neill, N. H. (1987). *Hydrogeological studies of groundwater inflow problems related to longwall mining—steady state modelling* (Volume 1). Rept Dept. of Geol. Sci., Univ. of Birmingham.

Neill, N. H. (1988a). *Hydrogeological studies of groundwater inflow problems related to longwall mining—steady state modelling* (Volume 2). Rept Dept. of Geol. Sci., Univ. of Birmingham.

Neill, N. H. (1988b). *Hydrogeological studies of groundwater inflow problems related to longwall mining—time variant modelling*. Rept Dept. of Geol. Sci., Univ. of Birmingham.

Neill, N. H. (1990). The computer modelling studies of the prediction of groundwater inflow to longwall coal mines. PhD thesis, University of Birmingham.

Rouleau, A. (1988). *A numerical simulator for flow and transport in stochastic discrete fractured networks*. National Hydraulic Research Institute, Saskatoon, Canada.

Rushton, K. R. and Redshaw, S. C. (1979). *Seepage and groundwater flow*. Wiley, Chichester.

Singh, R. N., Hibberd, S., and Fawcett, R. J. (1986). Studies in the prediction of water inflows to longwall mine workings. *Int. J. Mine Water*, **5**, 29–44.

19. The international scene—the involvement of British hydrogeologists

R. W. Simpson

Introduction

The rapid growth in population in the developing world in the 1980s has significantly increased the demand for water, and has created the need for water resource development of both surface water and groundwater. In attempting to meet this demand, a significant impetus has been given by the United Nations International Water Supply and Sanitation Decade (1981–1990). This focused the attention of governments and donors on the problems involved in providing potable water supplies and adequate sanitation facilities for the world's population by 1990. As a consequence, the need to identify, evaluate, and exploit ground-water for water supply has increased and hydrogeologists have been engaged on a range of projects throughout the world. Although some projects are undertaken with the aim of examining or protecting the environment, de-watering mines and engineering structures, or developing geother-mal energy, almost all overseas hydrogeological programmes are concerned with the provision of water supplies.

The study, design, construction, and operation of groundwater schemes requires a variety of skills drawn from both the scientific and engineering professions; this chapter deals solely with the work of hydrogeologists overseas.

British hydrogeologists in the developing world

Background

There has been a long tradition of British hydro-geologists working in the developing world upon aid-funded projects for domestic and agricultural water supply. This tradition follows that of their colleagues in the civil engineering profession who have designed and supervized the construction of many of the world's major water supply schemes. Having English as their mother tongue and the world-wide trade connections of the UK, British hydrogeologists have been well placed to work anywhere in the world. There have been many opportunities to participate in multi-lateral and bi-lateral aid programmes, because of Britain's donor status, and also in the many projects financed by British charitable agencies.

Type of work

The majority of the assignments undertaken over-seas by British hydrogeologists has been with inter-national civil engineering consulting firms whose headquarters are in Britain. A few assignments, however, have been with firms based in other European countries. Most of the work is associated with urban water supply, irrigated agriculture, and rural development. The majority of projects are carried out in the arid zone because groundwater is the only available water source. Next in import-ance are those areas in the tropics which experi-ence pronounced dry seasons when groundwater is needed to supplement irrigation water supplies or to provide a safer and more reliable supply to domestic consumers. Whatever the climatic environment, a more secure source of supply can provide a better quality of life by improving health and sanitation and by enabling industry and agri-culture to develop.

Training

A further reason why British hydrogeologists are involved in so many projects is the high standard of hydrogeological training available in Britain for both British and foreign nationals. Over the last twenty-five years a large number of students have undertaken graduate and post-graduate courses in hydrogeology, the best known being the MSc courses at the Universities of Birmingham and London. Half the students attending these two courses are from overseas and, having acquired new skills, put them into practice on returning to their own countries. Overseas hydrogeologists obtain training whilst working alongside British and other expatriate staff on large-scale groundwater projects. Most of the aid-funded projects contain a large training programme involving fieldwork, office experience, and occasionally training in another country. The British expatriate staff acquire new skills on these projects through using expensive and sophisticated methods of analysis that are often required by such projects.

Type and location of hydrogeological projects

There is a paucity of published information regarding the type, size, or location of projects undertaken overseas by British hydrogeologists for the majority of the project reports is unpublished and available only to the client for whom the work was undertaken. Lists of dam and water supply projects are compiled by The Association of Consulting Engineers and The International Commission on Large Dams but unfortunately there is no organization that catalogues hydrogeological schemes. Shaw (1989) includes three overseas groundwater projects in a broad compilation of case studies in engineering hydrology but the principal source of generally available information is the small number of papers published in journals and conference proceedings.

To obtain a broader picture, the author contacted the principal British organizations that employ hydrogeologists overseas to obtain brief details of the projects undertaken by them during the 1980s. From the responses received, 205 projects were undertaken in 66 countries by the organizations sampled and it is estimated that this accounts for approximately 80 per cent of all British overseas hydrogeological activities.

Categorizing the projects identified by the survey is difficult because of the variety of environments and conditions under which they were undertaken and the purposes for which they were carried out. Classification by location and type of work seemed most appropriate, first by continent and then by country. This indicated that 35 per cent had been undertaken in the Middle East, 28 per cent in Africa, and 27 per cent in Asia. The rest of the world only accounted for 10 per cent of the total.

The survey showed that within the Middle East, work had been undertaken in 13 countries, with 31 per cent of the projects in Oman, 17 per cent in Jordan, and 13 per cent in Yemen People's Democratic Republic (PDR). In Africa, projects have been carried out in 24 countries with 16 per cent in Nigeria, 9 per cent in Egypt, and 7 per cent in each of Kenya, Gambia, and Zimbabwe. Asian projects were in 13 countries; 23 per cent in Malaysia, 18 per cent in Bangladesh, and 13 per cent in Indonesia. Thus the countries in which hydrogeologists have been most active in terms of projects completed are Oman (11 per cent), Malaysia and Jordan both with 6 per cent, Bangladesh (5 per cent), and Nigeria and Yemen PDR each with 4 per cent.

It is not possible within the scope of this chapter to describe all the projects analysed in the survey. Only those that have made a significant impact during the 1980s are referred to, for example by employing innovative techniques, or advancing hydrogeological knowledge in relation to hydrological processes or a geological environment, or successfully developing groundwater resources, or improving the standard of living or the environment of a particular area.

The projects have been broadly classified according to location and environment and hence in terms of geographical and geological setting. The chapter is sub-divided into major regions each of which has, in general, an individual set of hydrogeological characteristics. They are Africa, the Middle East, Asia, the Americas and the Caribbean, and the Pacific.

Africa

Introduction

Large areas of the developing world lack a cover of sedimentary rocks (Fig. 19.1) and hence suffer from unfavourable groundwater conditions. In Africa, this problem is accentuated because of the extensive distribution of crystalline basement rocks (Fig 19.2). Apart from the north and a few isolated areas where sedimentary aquifers occur, many of the hydrogeological studies in Africa are concerned with the basement rocks and their weathered mantle.

Groundwater resources in these basement rocks are limited and well yields are low. Consequently, researchers and practitioners have had to examine alternative methods of optimizing groundwater development potential by investigating improved means of locating groundwater and of increasing well yields. This has involved the use of remote sensing, aerial photography, and geophysical surveys in exploration, and on the well design side it

has included experimentation in the design and construction of large diameter wells and collector wells, the use of thermoplastic well casing, and the introduction of improved hand-pumped wells.

Exploiting groundwater from the weathered and hard rock aquifers of the African basement

Throughout Africa, the outstanding problem facing hydrogeologists has been to locate and develop sufficient groundwater supplies not only in the arid Sahelian region but also in the rest of the continent where basement rocks predominate. The search for water in these rocks and their weathered derivatives, has centred around the need to locate the maximum thickness of superficial weathered deposits, to attain the greatest depth of water bearing material, and to determine the alignment and depth of fracture systems (Fig. 19.3). The quest for water in such locations has been reported by the British Geological Survey (1989*a*), Clark

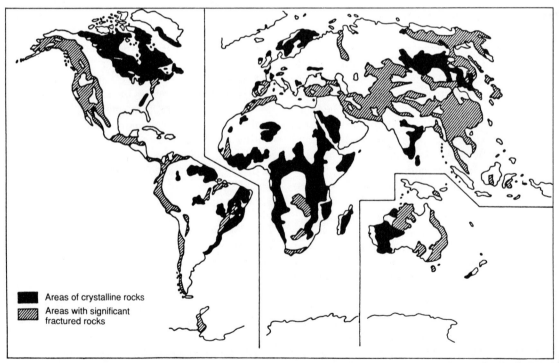

Fig. 19.1. World-wide distribution of basement rocks (reproduced from British Geological Survey 1988).

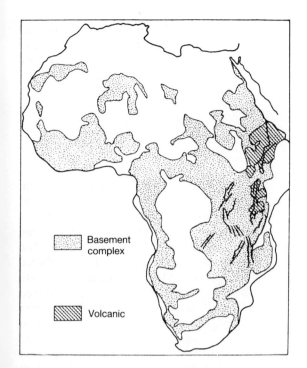

Fig. 19.2. The distribution of the crystalline basement complex and major rift features (solid lines) in Africa (after Jones 1985).

(1985), Hazell *et al.* (1988), and Jones (1985). The use of magnetic, gravity, and seismic surveys in Zimbabwe has greatly assisted exploratory drilling programmes in identifying the presence of favourable aquifers. Valuable research was undertaken by the British Geological Survey over a five-year period, including the organization of seminars and workshops to disseminate the research findings. Whilst this research was in progress, consulting firms were implementing rural water supply schemes throughout the basement areas using many of the exploration techniques that had evolved from the recommendations of the research. Schemes that incorporated remote sensing, geophysical surveys, and exploration drilling included projects in Nigeria, Malawi, Zambia, and Kenya.

Many of the traditional water sources for dispersed rural communities are hand-dug wells whose yield is limited by the low productivity of the aquifer penetrated, the small depth of water below the water-table, the ingress of silt and sand into the well, and the lack of efficient water lifting equipment. Dug wells have a few advantages compared with boreholes such as their simple relatively cheap construction, the socio-economic benefits of self-help construction and their suitability for

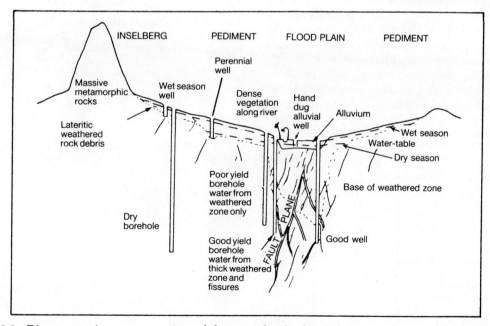

Fig. 19.3. Diagrammatic representation of the complex hydrogeology of basement rocks (after Clark 1985).

abstracting water from shallow low-permeability aquifers. Nevertheless, they generally have lower yields and are more susceptible to drought and pollution. Improved designs are being introduced to reduce pollution risk and to incorporate hand-pumps (Fig. 19.4). Hand-dug wells, however, cannot fully exploit the basement aquifers and consequently alternative designs are being examined which provide better hydraulic characteristics. A survey of microbiological data collected in Malawi between 1984 and 1987 from over 300 boreholes and 60 unprotected dug-wells, indicated that boreholes were significantly less polluted than dug-wells. About 60 per cent of boreholes could supply water complying with the World Health Organization guideline value of zero faecal coliform (FC) in a 100 ml volume of water, compared with less than 10 per cent of dug-wells (Lewis and Chilton 1989). Using a more realistic guideline for untreated water of 25 FC per 100 ml, then around 97 per cent of boreholes met this criterion whilst only 15 per cent of unprotected wells could achieve the same standard (Fig. 19.5).

To obtain higher individual well yields from these aquifers of low transmissivity, the British Geological Survey has undertaken long-term investigations to establish criteria for the site selection, design, and construction of collector well systems (British Geological Survey 1988). These wells incorporate horizontal lateral pipes which radiate from near the base of a central caisson or from the type of large diameter hand-dug wells that have been made in South Asia. They have been successfully constructed in Zimbabwe, Malawi, and Sri

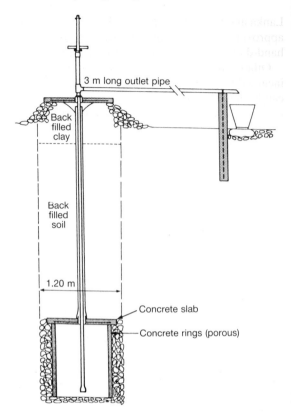

Fig. 19.4. Protected dug well in Malawi (reproduced from British Geological Survey 1988).

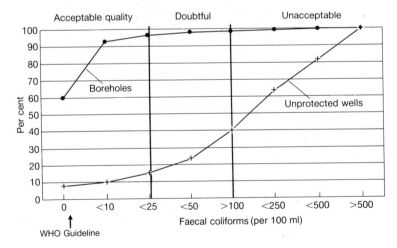

Fig. 19.5. Comparison of bacteriological quality of boreholes and unprotected wells (reproduced from Lewis and Chilton 1989).

Lanka and tests have indicated that their yields are approximately three times higher than those from hand-dug wells in a similar environment.

Other improvements in well design have included the selection of well casing material that can be manufactured locally to replace expensive imported steel. In Malawi and Zambia thick walled PVC pipe is now being maufactured and, by using this cheaper well casing, it has proved possible to install low-cost boreholes deeper into the weathered basement rocks than hand-dug wells. This has the benefit of obtaining sufficient aquifer penetration to provide acceptable well yields. These higher yielding boreholes not only provide reliable domestic water supplies but their yields are occasionally sufficient to allow small-scale irrigated agriculture. An additional benefit of using non-ferrous well casing and screen is a reduction in the iron content of the well water. Studies in Malawi by Lewis and Chilton (1989) indicate that many cases of unsatisfactory iron concentrations are not due to naturally-occurring iron minerals, but rather to contamination of the water by materials which were used in the construction of the borehole (Fig. 19.6).

Although well water contamination may result from contact with some well casing materials and from surface pollution, one of the recurrent problems encountered during the development of basement aquifers is the poor water quality caused by the high mineral content derived from trace ele-

ments in volcanic rocks. The principal elements that frequently occur in concentrations too high for potable water are fluoride, manganese, and iron. Of particular concern are the high fluoride concentrations that occur in both west and east Africa at levels that are far too high for domestic consumption. In Nairobi, more than half the 1500 boreholes that used to provide domestic and industrial supplies to the city until major surface water sources were constructed, had fluoride concentrations in excess of the World Health Organization's guideline level of 1.5 mg l^{-1} for drinking water (Howard Humphreys and Partners 1986).

Rural and urban water supply

British hydrogeologists have been engaged on rural water supply schemes across Africa (Chilton and Smith-Carington 1984; Chilton *et al*. 1982; Grey *et al*. 1985) but much of the work is unpublished. Undertaken principally by consulting engineering firms on behalf of public and private sector clients, the schemes have ranged from water source systems for individual establishments, such as farms, clinics, or schools, to reticulation schemes for agricultural estates, villages, or small towns including, in some cases, procurement and installation of equipment. Many of the largest rural water supply schemes have been undertaken in West Africa and have extended over areas as large as states and

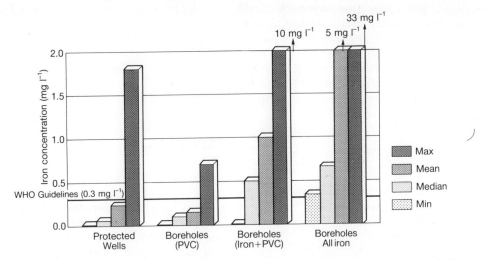

Fig. 19.6. Effect of source construction on iron content of groundwater (after Lewis and Chilton 1989).

even countries. From 1981 to 1987 British contractors engaged upon turnkey rural water supply projects employed several British consultants to undertake groundwater source identification and development, particularly in Nigeria and Cameroon, involving several hundred villages and towns. The contractors included Biwater and Paterson Candy and the consultants Sir M. MacDonald and Partners (1982a, 1986), Watson Hawksley (1982, 1986), and Hydrotechnica (1987).

Not many groundwater sources in Africa are able to meet the demands of urban centres and therefore the majority of large towns and cities derive their supplies from surface water sources. Some major towns, however, may rely in part on wells and many developments have benefited from the involvement of British hydrogeologists particularly in East and West Africa.

Irrigation water supplies

As in the case of municipal supplies, insufficient groundwater generally exists in aquifers in Africa to provide irrigation supplies. The exceptions are in North Africa where extensive and productive aquifers occur along the northern slopes of the Atlas mountains in Morocco, Tunisia, and Algeria. Further east, the thick Nubian sandstones underlying Libya (Wright *et al*. 1982), Egypt, and Sudan are being progressively tapped for large scale agricultural and mining projects. In Libya, the world's largest current civil engineering project, the Great Man-Made River Project, is designed to abstract groundwater from large wellfields in the Sahara and transmit 2 Mm^3 day^{-1} to the Mediterranean coast for irrigation and domestic supplies. British hydrogeologists are involved in the design of the wellfields and also the subsurface drainage aspects of the irrigated areas.

Mine water supplies

In Central and Southern Africa, extensive and detailed groundwater investigations have been undertaken for the water supply of mining projects including diamond and silver mines in Botswana and copper mines in Zambia. This work has, in several cases, involved the design and supervision

of the drilling of large numbers of wells, as well as the detailed analysis of the groundwater resource potential using computer simulation models.

Geothermal energy

Since 1985, the British Geological Survey has been associated with the Kenya Rift Valley Geothermal Project. Studies, extending from the Tanzanian border to northern Kenya, have been concerned with obtaining a better understanding of the nature and movement of the thermal waters in the rift and of the cool ambient waters that recharge the thermal fields. On a regional scale the Rift Valley exhibits the hydrogeological features to be expected of a valley-interfluve system, with lateral groundwater flows from the main rift escarpments to discharge areas on the floor of the rift; groundwater also flows axially away from the rift floor culmination at Lake Naivasha. This pattern is modified by major rift faults which are barriers to lateral flow, leading to longer deeper flow paths, and by faults in the rift floor which tend to align flow paths within the rift along its axis.

Chemical analysis of geothermal waters from the volcanic centres of Eburru, Olkaria, Longonot, and Suswa can be interpreted in terms of a mixing series between waters derived from the rift escarpments and water from Lake Naivasha; there is no evidence of a unique deep thermal water. The geothermal areas at Eburru, Olkaria-Domes, Longonot, Suswa, and Magadi appear to be individual geothermal fields with local heat sources giving rise to separate convection cells in which the geothermal waters originate as mixtures of ambient groundwater and lake or, in the case of Magadi, river water (Allen *et al*. 1989).

Middle East

Introduction

Apart from the large riverine basins of the Nile, Euphrates, and Tigris, perennial supplies of water in the Middle East are very limited. A small number of reliable springs occur in the mountainous areas of the Levant and Arabian Gulf coast but elsewhere water is extremely scarce. Throughout the Middle East, and the Arabian peninsula in particular, the climate is arid and an acute water shortage has imposed a limit on industrial and agricultural development in many parts of the region.

Traditional sources of water, other than springs, have been the hand-dug wells and subsurface galleries (aflajs or qanats) which tap the alluvial fan or river (wadi) deposits. Such aquifers contain renewable resources and have received much attention recently in order to provide improved supplies for both urban and agricultural users. On a broader scale, investigations began in the late 1970s and early 1980s to explore the groundwater potential of the extensive sedimentary basins of the region. Although the fossil waters which they contain represent a non-renewable resource, these sources are now being exploited on a large scale, particularly in the Arabian peninsula.

Alluvial aquifers

There are many locations in the Middle East where alluvial basins have provided water for many years to urban and agricultural areas. These basins can be categorized as: riverine (i.e. those recharged by a major river—such as those alongside the Nile and Euphrates rivers); intermontane (those fed by springs and small seasonal wadis in mountainous areas such as those which surround Sana'a and Damascus); or coastal plains (such as the Batinah and Salalah Plains in Oman and the Tihama Plain of Yemen and Saudi Arabia along the Red Sea, all of which are recharged by run-off from the foothills of neighbouring mountains). Significant recent hydrogeological studies and investigations have taken place in the wadi systems and coastal plains of the United Arab Emirates, Oman, Saudi Arabia, the Yemen Arab Republic (AR), and the People's Democratic Republic of Yemen. In Oman, the Ministry of Agriculture has selected several coastal plain areas where groundwater investigations are being undertaken to assist in the assessment of agricultural potential. Similarly, in Saudi Arabia and Yemen AR, the coastal plain to the west and the wadi deposits to the east of the mountains bordering the Red Sea are being examined to determine the groundwater potential for

agricultural use. Many of the cities and towns around the coastline of Southern Arabia obtain their public water supplies from alluvial groundwater schemes designed by British hydrogeologists. In the United Arab Emirates, Sir William Halcrow and Partners have designed the water supplies to Al Ain and Sharjah; in Oman, Sir M. MacDonald and Partners have designed supplies for Muscat, Nizwa, Sur, Sohar, and Buraimi, and in Yemen PDR, John Taylor and Sons designed the water supply of Al Mukalla and, with Sir M. MacDonald and Partners, the facilities for Greater Aden (Davey 1989). Geophysical and drilling techniques have been used to determine the distribution of the alluvial deposits, and water balance studies and pumping tests have been undertaken to estimate well yields. A similar approach has been followed for the intermontane plains and coastal plains but recharge estimates in these environments have proved more difficult than in wadis because of the problems associated with determining the amounts of infiltration to the aquifers and defining their thickness, extent and hydraulic characteristics. Nevertheless, groundwater supplies to urban and small-scale irrigation areas have been provided by the construction of wellfields, recharge dams, or artificial recharge schemes.

Estimation of recharge in arid areas

The assessment of groundwater potential is very much dependent upon a reliable estimate of recharge. In arid areas recharge events are rare and the amounts are difficult to measure; consequently an accurate estimation of the replenishment of the groundwater resource is not usually possible. Some of the more detailed water resource studies in Saudi Arabia and Oman have included the monitoring and evaluation of many of the parameters governing the hydrological processes in order to obtain water balances of individual wadis or basins. Much of this analysis has relied upon obtaining a correlation between rainfall, runoff events, and changes in groundwater levels to estimate recharge rates and hence potential abstraction rates. Because the shortage of data often precludes detailed analysis, much research into the hydrological processes has been necessary.

Rainfall studies in the mountains of Oman and Saudi Arabia have been undertaken by Imperial College (Wheater and Bell 1983; Wheater *et al.* 1989) to obtain rainfall depth and duration values. Infiltration studies using naturally occurring radio-active and chemical tracers have been undertaken at selected sites in the arid zone by the British Geological Survey (Chapter 5 and Edmunds *et al.* 1988). The results of this research has provided guidelines for many others working on hydrogeo-logical projects in this type of environment.

Fossil groundwater basins

Traditional groundwater sources have provided water for both domestic and small scale agricul-tural supplies but recently several expensive groundwater schemes have developed deep aqui-fers by exploiting modern survey and drilling tech-niques. The majority of these aquifers, however, have a limited life because they are not now being replenished; their resources were derived during previous pluvial periods.

These deep and extensive aquifers predomi-nantly comprise sandstone and limestone facies and form large basins extending across North Africa and the Arabian peninsula (Fig. 19.7). Although the initial discovery and investigation of groundwater in these basins took place in the 1960s and 1970s, principally during oil exploration pro-

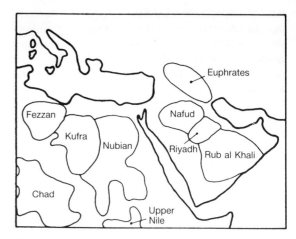

Fig. 19.7. Fossil groundwater basins in north-east Africa and the Middle East (after Simpson 1985).

grammes, it was in the late 1970s and early 1980s that detailed studies and investigations commenced to assess the groundwater potential.

The Tertiary Umm Er Radhuma (UER) lime-stone formation (Fig. 19.8), which extends from northern Saudi Arabia to Oman in the south, was investigated in Saudi Arabia between 1977 and 1980 (Groundwater Development Consultants 1980; Bakiewicz *et al.* 1982). This programme involved a large team of British hydrogeologists which undertook extensive, investigations, includ-

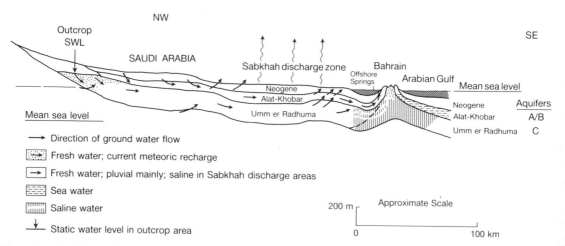

Fig. 19.8. Hydrogeological cross-section of the post-Cretaceous aquifer system in eastern Saudi Arabia and Bahrain (reproduced from Walton 1987)

ing the application of deep, well-drilling technology and the construction of computer models to design wellfields and simulate fossil groundwater gradients. Rapid development of deep groundwater took place in Saudi Arabia not only in this formation but also in the older Wasia, Minjur, Saq, and Wajid aquifers that outcrop in the eastern half of the kingdom. The majority of the exploitation has been for public water supplies and agriculture, including wheat farms which are irrigated by centre pivot irrigation systems based on individual boreholes. A large number of farms have been created and the 1980s has seen the over-development of several of these major aquifer systems. Near Riyadh, boreholes 400 m deep have been drilled in the Wasia to provide water for the city. Sophisticated water treatment is necessary to reduce the carbon dioxide and mineral content, and lower the temperature before it can enter supply (Sir M. MacDonald and Partners 1982*b*).

The water quality of all these aquifers in and near the outcrop is generally good although down gradient, in the confined zones, the salinity increases eventually to non-potable levels. In the UER total dissolved solids are generally less than 1500 mg l^{-1} but along the Arabian Gulf coast they increase to more than 5000 mg l^{-1}, and the aquifer ceases to provide water suitable for consumption. Poor quality water also occurs in the Wasia Sandstone, below the UER (Fig. 19.9).

On the Arabian Gulf coast, Qatar obtains public water supplies from the upper UER and the overlying Rus formation (Pike 1983, 1985) and Bahrain (Fig. 19.8) from the Rus and overlying Khobar (Wright *et al*. 1983; Walton 1987).

Further south, in Oman, the UER and adjacent formations have been exploited for the supply of water to farms and isolated military establishments from boreholes ranging in depth from 300–800 m (Gamble and Biggin 1987).

To the north in Jordan, wells 500 m deep (Fig. 19.10) were drilled in the Cambrian Disi-Saq sandstone in the early 1980s for the water supply of Aqaba (Simpson 1982) and throughout the decade more and more wells have been drilled in the aquifer on both sides of the Jordan/Saudi Arabian border for agricultural supplies. The groundwater resources in this aquifer are not being replenished and unfortunately some over-development has now taken place similar to that in other aquifers further south and east.

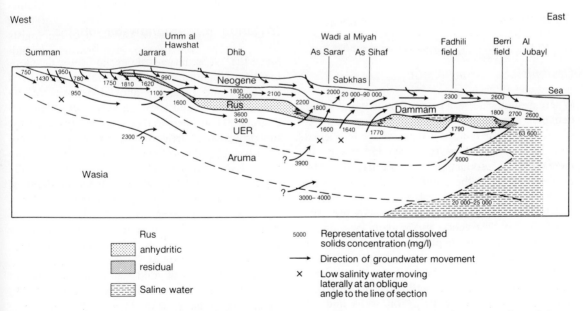

Fig. 19.9. Hydrogeological section of the aquifer system in eastern Saudi Arabia (reproduced from Bakiewicz *et al*. 1982).

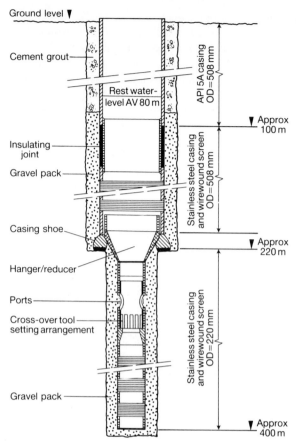

Ground level

Cement grout

Rest water-level AV 80 m

API 5A casing OD=508 mm

▼ Approx 100 m

Insulating joint

Gravel pack

Stainless steel casing and wirewound screen OD=508 mm

Casing shoe

▼ Approx 220 m

Hanger/reducer

Ports

Cross-over tool setting arrangement

Stainless steel casing and wirewound screen OD=220 mm

Gravel pack

▼ Approx 400 m

Fig. 19.10. Design of production borehole at Qa Disi in Jordan (after Simpson 1982).

Groundwater management

Where groundwater is a scarce resource and water demands are high the competition for water is severe. Often the conflict is between the municipal and the agricultural users as the latter seek to increase crop production through irrigation by encroaching upon the aquifers traditionally used for domestic supply. The lack of a water resources management and licensing system usually prevents any control of abstraction and in many locations in the region serious depletion of groundwater resources is taking place. One example is in the capital of the Yemen Arab Republic, Sana'a, where existing municipal wellfields are suffering from interference from irrigation wells and no legislation exists to control unlicensed abstraction (Charalambous 1981).

Asia

Introduction

Large-scale groundwater development in Asia has predominantly centred upon the alluvial basins of the continent's large rivers, such as the Indus, Ganges, and Brahmaputra.

In these and some smaller basins the development of groundwater has taken place for the purpose of providing irrigation water in both the public and private sectors. Traditionally, shallow wells have supplied domestic water and the deeper wells have been sunk for agriculture. Recently another type of well has had to be introduced, the scavenger well, in order to remove fresh water overlying saline groundwater that has built up beneath poorly managed surface water irrigation areas.

Other aquifers that are used in the region are the basement rocks that cover much of South Asia and the volcanic and carbonate rocks which predominate in South-East Asia and the islands in the Eastern Pacific. In most cases the aquifer potential is relatively low and groundwater resources are only used for village and small-scale irrigation supplies.

Irrigation from groundwater sources

Since the 1960s, large-scale groundwater development for irrigation purposes has taken place from the thick and extensive alluvial deposits of the Indus, Ganges, and Brahmaputra rivers. Approximately 50 000 deep wells have been installed for the provision of irrigation water supply in these river basins and also for water-table control in the lower Indus. In general terms, the wells, locally referred to as 'tubewells', are drilled to a depth of up to 100 m and installed with a 152 or 203 mm diameter screen/casing string, a 305 or 356 mm diameter pump housing, and a gravel pack. Their yields are approximately 2 ft^3 s^{-1} (57 l s^{-1}) giving rise to the description of '2 cusec tubewells'. Originally constructed of mild steel, the more recent wells are made of glass reinforced plastic (GRP) except occasionally for the lower casing which may still be mild steel. Continuing improvements to the design of the wells are taking place either to reduce the present rapid deterioration of the well yield (Bakeiwicz *et al.* 1985) or to minimize the cost of installation and operation (British Geological Survey 1989*b*). British consultants are engaged in the planning and implementation of these large-scale tubewell irrigation projects as well as the design and supervision of the construction of the tubewells. Many of the regional studies have involved large-scale groundwater models to simulate the recharge and flow mechanisms of the alluvial aquifer systems and incorporate subroutines to represent the cropping patterns in order to obtain a balance between the cropwater demands and the aquifer resources (Sir M. MacDonald and Partners 1982*c*; Howard Humphreys and Partners 1984). In Pakistan where large areas are underlain by saline water (Fig. 19.11), it has been necessary to install wells in irrigated areas to provide water-table control and to also skim fresh water from above the saline water that has built up under many of the irrigated areas (Milne *et al.* 1983; British Geological Survey 1989*c*).

Countries in Asia other than Pakistan, India, and Bangladesh are not blessed with extensive alluvial aquifers although river basins in Thailand (Howard Humphreys and Partners 1989*a*) and Indonesia (Groundwater Development Consultants 1989) have been developed for intensive irrigation with individual wellfields consisting of more than 100 wells. Other sedimentary aquifers, particularly in Indonesia, have been used to provide supplementary irrigation from wells yielding up to $60 \ l \ s^{-1}$ which penetrate thick sequences of sands and gravels and where rainfall amounts are high. Elsewhere in the region groundwater is being exploited for rural and urban water supply.

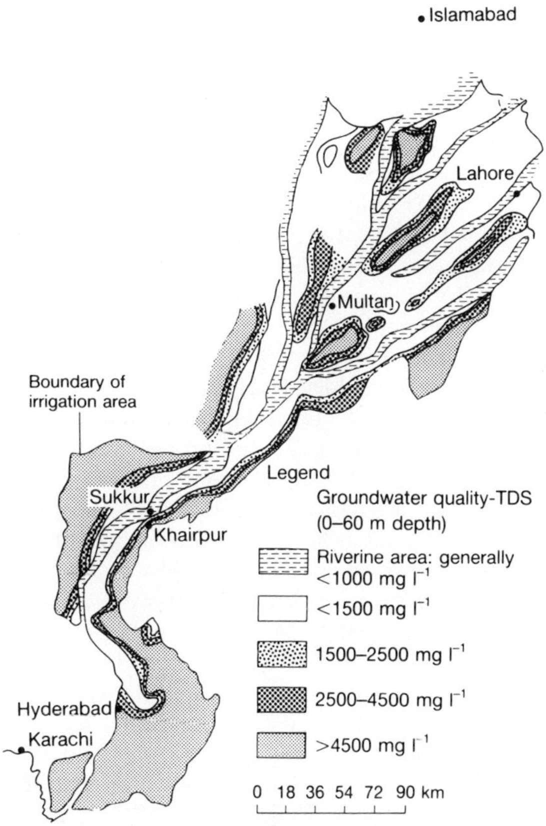

Rural water supply

Large, regional, rural development projects, involving water supply components, have been undertaken in several countries in Asia. In Brunei, an extensive groundwater and surface water study identified and developed sources for rural water supply (Watson Hawksley 1988) and other large scale studies have been carried out in Malaysia by Sir M. MacDonald and Partners (1990) and Acer (1989), and in Indonesia by Watson Hawksley (1981). The rural water supply schemes are usually

Fig. 19.11. Groundwater quality (salinity) of the Indus plains (reproduced from Bakiewicz *et al.* 1985).

part of an integrated rural development project involving upgrading of many of the infrastructural components. In the case of the Malaysian project a turnkey service was offered including both design and construction of water supply facilities. The groundwater sources for these rural schemes, include alluvial deposits throughout the region, weathered basement rocks in the Indian sub-continent and volcanic-sedimentary sequences in South-East and East Asia around the Pacific rim.

The groundwater potential of the basement rocks is low and techniques for developing these poor aquifers are similar to those being used in Africa. The British Geological Survey (1989*a*) has undertaken research in India and Sri Lanka to improve the methodology for the exploration and development of groundwater and to make a better assessment of its potential. Geophysical and drilling investigations have led to the construction of collector wells similar to those developed for the basement areas of Africa. Development of groundwater from the volcanic-sedimentary sequences of Indonesia (Fig 19.12) has many constraints as described by Lloyd *et al.* (1985). Aquifer delineation in these inhomogeneous sequences is difficult but well yields can be sufficiently high to provide water for both urban and irrigation supplies (Binnie and Partners 1984*a*). Nearby, another complex hydrogeological regime is the karst limestone area of Greater Yogyakarta where acute domestic water shortages occur because neither surface water nor shallow water resources exist. Only deep groundwater is available for public water supply but drilling successful wells is difficult in an area where groundwater flow takes place through a deep sinkhole and multi-layered cavern system along irregular solution channels. Detailed speleological surveys were necessary to identify the flow paths and to assist in assessing where collapse structures could be used for groundwater-fed storage ponds and where well drilling could take place effectively (Sir M. MacDonald and Partners 1982*d*).

Americas and the Caribbean

Introduction

During the 1980s, British hydrogeologists have not been so active in the Americas and the Caribbean as in previous decades. Work has continued, however, in Central America on public water supply schemes and small-scale irrigation projects, and in the Caribbean, on domestic water supply and groundwater protection investigations. Other activities have included the preparation of hydrogeological maps of several Caribbean islands as a contribution to a hydrogeological atlas of the Caribbean sponsored by UNESCO; institutional support to water utilities in groundwater management; and collaborative studies to assess the impact of agriculture and sewage disposal on groundwater quality.

Public water supply

For a number of years, British hydrogeological assistance has been provided, predominantly through British aid programmes, to help in developing groundwater sources for the water supply of

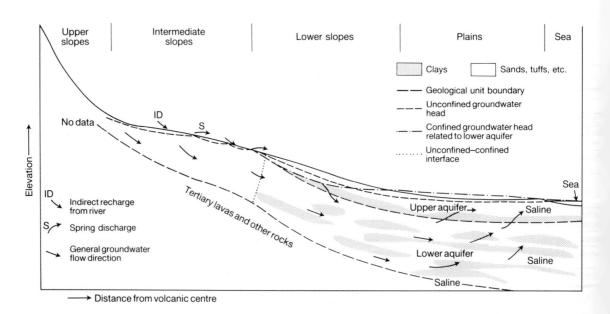

Fig. 19.12. Possible groundwater flow in a volcanic-sedimentary sequence (after Lloyd *et al.* 1985).

the capital cities and many of the provincial towns of Costa Rica, Honduras, and El Salvador. Although annual rainfall in the Cordillera Central Mountains of Central America can exceed 7500 mm a^{-1}, there are several reasons why surface water sources cannot be used economically for public water supply in many of the Central American countries.

The region is volcanically active and the poorly consolidated volcanic deposits are easily eroded and transported down the rivers. This not only causes the rivers to have high sediment loads, with a need for expensive water treatment, but the rivers are deeply incised with steep gradients which precludes the possibility of constructing large impounding reservoirs economically. Development of groundwater sources is therefore more feasible than surface water sources. The capital cities of San Jose, Costa Rica; Tegucigalpa, Honduras; and San Salvador, El Salvador, with the assistance of British hydrogeologists, are developing wellfields which obtain reliable public water supplies from sequences of lavas and ignimbrites (Foster *et al.* 1985). Locating and developing these sources has required extensive groundwater investigations involving the design of special drilling techniques suitable for drilling through and coring hard volcanic rocks, stratigraphic correlation of complex lava flow sequences, identification and delineation of aquifers, hydrogeochemical studies,

simulation of aquifer behaviour, and the assessment and estimation of water balance parameters. The programmes were carried out by Howard Humphreys and Wallace Evans and Partners, in the late 70s and early 80s, and by Sir William Halcrow and Partners Ltd in 1985 for San Jose, San Salvador, and Tegucigalpa respectively. The success of these projects has helped to provide these rapidly growing urban developments with a safe and secure supply of water.

Another interesting groundwater environment that has been examined by British hydrogeologists is the fresh water lens found on several of the coral limestone islands in the Caribbean. Many of the islands have little topographical relief and the thin fresh water lens rests upon and is surrounded by saline water. Consequently, abstraction of groundwater has to be by means of shallow wells, galleries, or ditches so that water-table drawdown is minimized and the risk of up-coning saline water is reduced. The design of suitable abstraction schemes has developed over the years and the principal hydrogeological activity of the 1980s has been to derive operating rules for the management of the lenses by estimating recharge from rainfall, often using land-use models as shown in Fig. 19.13 (Thomson and Foster 1986), and then by determining the thickness and extent of the fresh water lens on the basis of hydrogeochemical characteristics. Recently these hydrochemical studies have

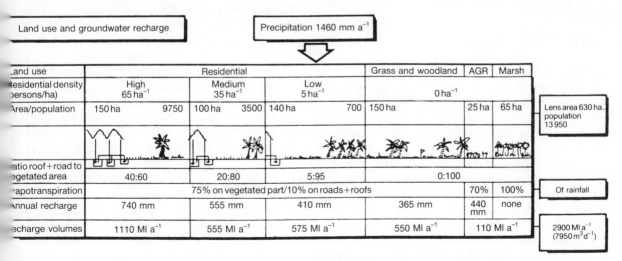

Fig. 19.13. Bermuda central lens: groundwater recharge mechanisms and quality (after Thomson and Foster 1986).

included detailed bacteriological and chemical sampling because of the increasing level of pollution of groundwater from agricultural and waste disposal activities.

Groundwater pollution

Aquifers on small limestone islands are at risk from solid waste disposal, seepage from unsewered sanitation and stormwater systems, and contamination from agricultural practices. All these risks have been identified on both Bermuda and the Bahamas (Thomson and Foster 1986; Howard Humphreys and Partners 1989b). Further detailed investigations are required in both countries to evolve a suitable groundwater management strategy which will protect the limited groundwater resources from the present serious pollution hazards. Working with local government agencies, British hydrogeologists are helping to provide an understanding of the hydrogeochemical processes that take place at the fresh/saline water interface, under both urbanized and intensively developed agricultural areas. Collaborative research programmes in Mexico and Barbados, as well as Sri Lanka, are currently being undertaken by the British Geological Survey to assess groundwater pollution caused by low cost sanitation and the intensive use of fertilizers and pesticides in irrigated agriculture.

Geothermal energy

The British Geological Survey first became involved in high temperature geothermal resources in the 1970s when, in co-operation with Merz and McLellan, it surveyed a geothermal prospect on St Lucia, in the Caribbean, in a volcanic island arc setting. Seven exploration wells were drilled and several were tested proving the existence of a vapour-dominated reservoir in a fractured hydrothermally altered dacite. Commercial development of the site is now under consideration. Further exploration programmes were conducted in Montserrat and Panama.

A detailed study of the alteration mineralogy of rocks in the geothermal field at Miravelles, in Costa Rica, was undertaken as it was considered

to be representative of the Central American volcanic province, where hydrothermal activity occurs in a volcanic zone associated with lithospheric plate subduction. The hydrothermal activity is linked to recent volcanism in an early Quaternary caldera. The salinity of the hydrothermal fluid is 2000–3000 mg l^{-1} of chloride and the reservoir temperature is 220–270°C. Mineral assemblages in the volcanic reservoir rock, that has been altered by the fluids, reflect the evolution of the hydrothermal conditions, and the nature of these assemblages allowed construction of a more refined model of the geothermal system which should be valuable during development of the field by further drilling (Milodowski *et al.* 1989).

In Peru, in the southern Andes and Altiplano, preliminary studies have concluded that thermal waters in Palaeozoic basement rocks have a temperature of only 140°C and do not represent an exploitable resource. However, it is possible that the large andesitic strato-volcanoes and dacite domes of the high Andes do host active hydrothermal systems.

Pacific

General

There is little current activity involving British hydrogeologists in the Pacific region. Institutional support by staff from the British Geological Survey and the Institute of Hydrology has been provided for a number of years to strengthen government departments in Fiji, the Solomon Islands, and the Gilbert and Ellis islands in their assessment and development of groundwater resources. British consultants have undertaken water resources and water supply studies in the Gilbert Islands (Lloyd *et al.* 1980) and in the small volcanic islands forming the Cook Islands (Binnie and Partners, 1984b). Groundwater sources of the Cook Islands are limited to small gallery systems located in the coastal deposits (Edworthy 1985).

Two hydrothermal systems were investigated by BGS in Vanuatu, in the south Pacific, where the heat source is a Quaternary basaltic volcanic centre. In one of the systems, sea water is circulating in the basaltic reservoir at 180°C. The second system is in a central graben where groundwater in a karstic limestone is mixing with a connate brine and a deeper thermal component, that may attain temperatures of 200–215°C at depth (Bath *et al.* 1986).

Europe

In spite of an increase in the number of groundwater projects in continental Europe, the involvement by British hydrogeologists is still small. Almost all projects are undertaken by in-country nationals and, unlike the developing world, the use of expatriate skills is very limited. The majority of British assignments has been in the tourism and transport sectors and has been concerned with groundwater resource feasibility studies on behalf of local consultants.

Conclusion

During the 1980s, British hydrogeologists have been active in many countries around the world, undertaking projects ranging from groundwater exploration and development through to its protection, management, and control. Projects have ranged from individual wells or boreholes, providing water supplies for village livestock, to the development of major aquifer systems, including the design and construction of wellfields incorporating several hundred wells.

Most of the activity has been equally divided between Africa, Asia, and the Middle East. The latter region was the most active in the early 1980s but by the end of the decade economic changes had reduced the number of projects. Asia continues to be an area of increasing activity which will probably continue to grow in the 1990s. However, the region that will probably see the greatest change with respect to the involvement of British hydrogeologists will be Europe. Although it appears that no hydrogeological projects were undertaken by British nationals in the 1980s in Eastern Europe, and few in Western Europe, there are now signs that groundwater specialists from Britain are being invited to participate in European projects especially in Eastern Europe. It will be interesting to see how the pattern of assignments for British hydrogeologists in the international

sector develops in the 1990s; it may well be that horizons will be set on the east.

References

Acer (1989). *Groundwater advisory mission, Malaysia rural water supply project.* Overseas Development Administration, London

Allen, D. J., Darling, W. G., and Burgess, W. G. (1980). *Geothermics and hydrogeology of the southern part of the Kenya Rift Valley with emphasis on the Magadi-Nakura area.* Report SD/89/1, British Geological Survey, Keyworth.

Bakiewicz, W., Milne, D. M., and Noori, M. (1982). Hydrogeology of the Umm Er Radhuma aquifer,Saudi Arabia, with reference to fossil gradients. *Q. J. Eng. Geol.*, **15**, 105–26.

Bakiewicz, W., Milne, D. M., and Pattle, A. D. (1985). Development of public tubewell designs in Pakistan. *Q. J. Eng. Geol.*, **18**, 63–77.

Bath, A. H., Burgess, W. G., and Carney, J. N. (1986). The chemistry and hydrology of thermal springs on Efete, Vanuatu, SW Pacific. *Geothermics*, **15**, 277–94.

Binnie and Partners (1984a). *Central Java water resources study.* P2AT Department of Public Works, Java.

Binnie and Partners (1984b) *Water resources and water supply of Rarotonga.* Govt. of the Cook Islands.

British Geological Survey (1988). *The Collector Well Project 1983–1988, final report.* No. SD/88/1. Keyworth.

British Geological Survey (1989a). *The basement aquifer research project 1984–1989, final report.* No. WD/89/15. Keyworth.

British Geological Survey (1989b). *The pilot study into optimum well design, IDA 4000 Tubewell II Project, Bangladesh*, Volume 6, Summary of the Programme and Results. Report WD89/14, Keyworth.

British Geological Survey (1989c) *Pilot studies in scavenger well use.* Left Bank Outfall Drain Project. Keyworth.

Charalambous, A. N. (1981) Problems of groundwater development in the Sana'a Basin, Yemen Arab Republic. *Improvement of methods of long term variations in groundwater resources and regimes due to human activities*, pp. 265–74. Pub. No. 136, Int. Assoc. Hydrol. Sciences.

Chilton, P. J. and Smith-Carington, A. K. (1984). Characteristics of the weathered basement aquifer in Malawi in relation to rural water supplies. In

Challenges in African hydrology and water resources, pp. 57–72. Publication No. 144, Int. Assoc. Hydrol. Sciences.

Chilton, P. J., Grey, D. R. C., and Smith-Carington A. K. (1982). *Manual for integrated projects for rural groundwater supplies*. UNDP/Malawi Government.

Clark, L. (1985). Groundwater abstraction from Basement Complex areas of Africa, *Q. J. Eng. Geol.*, **18**, 25–34.

Davey, J. C. (1989) Wellfield development for urban supplies in PDR Yemen. *J. Instn Water Env. Mgmt*, **3**, 413–22.

Edmunds, W. M., Darling W. G., and Kinniburgh, D. G. (1988). Solute profile techniques for recharge estimation in semi-arid and arid terrain. In *Estimation of natural groundwater recharge* (ed. I. Simmers), pp. 139–57. D. Reidel, Dordrecht.

Edworthy, K. J. (1985). Groundwater development for oceanic island communities. In *Hydrogeology in the Service of Man*, 65–75. Memoirs 18th Congress, Int. Assoc. Hydrogeol, Cambridge.

Foster, S. S. D., Ellis, A. T., Losilla-Penon, M., and Rodriguez-Estrada H. V. (1985). Role of volcanic tuffs in groundwater regime of Valle Central, Costa Rica, *Ground Water*, **23**, 795–801.

Gamble, D. H. F., and Biggin C. M. (1987). Groundwater supply and advanced water treatment in Oman, *J. Instn Water Engrs Sci.*, **41**, 55–74.

Grey, D. R. C., Chilton, P. J., Smith-Carington, A. K., and Wright, E. P. (1985). The expanding role of the hydrogeologist in the provision of village water supplies: an African perspective. *Q. J. Eng. Geol.*, **18**, 13–24.

Groundwater Development Consultants (1980). *Umm er Radhuma study*. Govt of Saudi Arabia.

Groundwater Development Consultants (1989). *Madura Groundwater Irrigation Project*, Interim Report. Ministry of Public Works, Indonesia.

Sir William Halcrow and Partners Ltd (1985). *Amateca Groundwater Project*, Water Authority of Tegucigalpa, Honduras.

Hazell, J. R. T., Cratchley, C. R., and Preston, A. M. (1988). The location of aquifers in crystalline rocks and alluvium in Northern Nigeria using combined electromagnetic and resistivity techniques. *Q. J. Eng. Geol.*, **21**, 159–75.

Howard Humphreys and Partners. (1984). *Groundwater evaluation and monitoring, South West Bangladesh regional rural development project*. Bangladesh Agricultural Development Corporation.

Howard Humphreys and Partners (1986). *Third Nairobi water supply project, regional studies*. Hydro-geology Report. Nairobi City Commission, Kenya.

Howard Humphreys and Partners (1989*a*) *Sukhothai groundwater development project, final report*. Royal Irrigation Department, Thailand and Overseas Development Administration, London.

Howard Humphreys and Partners (1989*b*). *New Providence sewerage master plan study*. Water and Sewerage Corporation of Bahamas.

Hydrotechnica (1987). *Rural water supplies programme*. Kwara State, Nigeria, Biwater Nigeria Ltd.

Jones, M. J. (1985). The weathered zone aquifers of the basement complex areas of Africa. *Q. J. Eng. Geol.*, **18**, 35–46.

Lewis, W. J. and Chilton, P. J. (1989). The impact of plastic materials on iron levels in village groundwater supplies in Malawi. *J. Inst Water Env. Mgmt*, **3**, 82–8.

Lloyd, J. W., Miles, J. C., Chessman, G. R., and Bugg, S. F. (1980). A groundwater resources study of a Pacific Ocean atoll–Tarawa, Gilbert Islands. *Water Resources Bull.* **16**, 646–53.

Lloyd, J. W., Pim, R. H., Watkins, M. D., and Suwara, A. (1985). The problems of groundwater assessment in the volcanic-sedimentary environment of Central Java. *Q. J. Eng. Geol.*, **18**, 47–61.

Sir M. MacDonald and Partners (1982*a*) *Kwara State rural water supplies project*. Biwater Treatment Ltd.

Sir M. MacDonald and Partners (1982*b*) *Riyadh additional water supply: Wasia wellfield*. Ministry of Agriculture and Water, Riyadh.

Sir M. MacDonald and Partners (1982*c*). *N.W. Bangladesh groundwater modelling study*, Bangladesh Agricultural Development Corporation.

Sir M. MacDonald and Partners (1982*d*). *Gunung Sewu cave survey*. Ministry of Public Works, Indonesia.

Sir M. MacDonald and Partners (1986). *Benue State rural water supplies*, Paterson Candy International Ltd.

Sir M. MacDonald and Partners (1990). *Malaysia rural water supplies*. Biwater Shellabear International Ltd.

Milne, D. M., Stoner R. F., and Bakiewicz, W. (1983). Fresh groundwater skimming. In *Groundwater and Man*, Memoirs 16th Congress, Int. Assoc. Hydrogeol. Sydney, Australia.

Milodowski, A. E., Savage, D., Bath, A. H., Fortey, N. J., Nancarrow, P. H. A., and Shepherd, T. J. (1989). *Hydrothermal mineralogy in geothermal assessment: studies of Miravelles Field, Costa Rica and experimental simulations of hydrothermal alteration*. Report WE/89/63, British Geological Survey, Keyworth.

Pike, J. G. (1983). The planning of water resources development in an Arabian Gulf state: Qatar, and case studies, *J. Instn Water Engrs Sci.*, **37**, 75–88.

Pike, J. G. (1985) Groundwater resources and development in the central region of the Arabian Gulf. In *Hydrogeology in the Service of Man*, pp. 46–55. Memoirs 18th Congress Int. Assoc. Hydrogeol, Cambridge.

Shaw, E. K. (1989). *Engineering hydrology techniques in practice*. Ellis Horwood, Chichester.

Simpson, R. W. (1982). Aqaba draws on desert water, *Middle East Water and Sewerage Journal*, **6**, 85–8.

Simpson, R. W. (1985). Arab water resources—an assessment. In *Water resources; technology and application*. Arab-British Chamber of Commerce Economic Report No. 6. London.

Thomson, J. A. M. and Foster, S. S. D. (1986). Effect of urbanisation on groundwater of limestone islands: an analysis of the Bermuda case, *J. Instn Water Engrs Sci.*, **40**, 527–40.

Walton, N. R. G. (1987) The water supplies of Bahrain Island—21 years on. *J. Inst Water Env. Mgmt*, **1**, 231–8.

Watson Hawksley Ltd (1981). *Java water supply sector study*. Asian Development Bank.

Watson Hawksley Ltd (1982). *Eastern Nigeria basement study*. Ministry of Water Resources. Nigeria.

Watson Hawksley Ltd (1986). *Cameroon rural water supply*. Biwater International.

Watson Hawksley Ltd (1988). *Brunei rural water supply and irrigation project*. Ministry of Development. Brunei.

Wheater, H. S. and Bell, N. C. (1983). Northern Oman flood study, *Proc. Instn Civ. Engrs*, **75**, 453–73.

Wheater, H. S., Larentis, P., and Hamilton, G. S. (1989). Design rainfall characteristics for southwest Saudi Arabia, *Proc. Instn Civ. Engrs*, **87**, 517–38.

Wright, E. P., Benfield, A. C., Edmunds, W. M., Kitching, R. (1982). Hydrogeology of the Kufra and Sirte basins, eastern Libya. *Q. J. Eng. Geol.*, **15**, 83–103.

Wright, E. P., Izatt, D., and Lori, I. (1983). *Hydrogeology and groundwater in Bahrain*. Paper No. 10. Arab Water Technology Conference, Dubai.

Index